Cambridge Studies in Social Anthropology

General Editor: Jack Goody

T0246332

23

Australian Kin Classification

For other titles in this series turn to page 568

Australian
Kin Classification

HAROLD W. SCHEFFLER

Professor of Anthropology, Yale University

CAMBRIDGE UNIVERSITY PRESS

Cambridge

London New York Melbourne

CAMBRIDGE UNIVERSITY PRESS
Cambridge, New York, Melbourne, Madrid, Cape Town, Singapore, São Paulo

Cambridge University Press
The Edinburgh Building, Cambridge CB2 8RU, UK

Published in the United States of America by Cambridge University Press, New York

www.cambridge.org
Information on this title: www.cambridge.org/9780521219068

First published 1978
This digitally printed version 2007

A catalogue record for this publication is available from the British Library

Library of Congress Cataloguing in Publication data

Scheffler, Harold W.
Australian kin classification.

(Cambridge studies in social anthropology; no. 23)
Bibliography: p.
Includes index.
1.   Australian aborigines – Social life and customs.
2.   Kinship – Australia.   3.   Kinship – Terminology.
I.   Title.
GN666.S33   301.42'1'0994   77–78391

ISBN 978-0-521-21906-8 hardback
ISBN 978-0-521-04052-5 paperback

CONTENTS

TABLES

PREFACE

This book is an attempt to demonstrate that the categories by which the aboriginal peoples of Australia
order their social lives are predominantly kin categories - a moot point in social anthropology - and,
beyond this, to reveal the structures and the relations
among the structures of Australian systems of kin
classification. The project is, in a way, a continua-
tion of one initiated by A. R. Radcliffe-Brown in 1910
and to which he devoted much of his professional attention for many years thereafter.

Radcliffe-Brown aspired to demonstrate, as he said
in 1951, that all Australian "kinship and marriage
systems" are "varieties of one general type." Although
his language was taxonomic, his vision was more in
keeping with the contemporary structuralist program
which, in the area of kinship studies, may be described
as an attempt to isolate a set of elementary structures
of which, in varying combinations, all kinship systems
are constructed, and to order the empirical diversity

among these systems by showing how any one may be derived from any other by certain rules of transformation or permutation. I will argue that Radcliffe-Brown did not accomplish what he set out to do, largely because of certain misconceptions about the structural principles he supposed all Australian kinship systems have in common, and because he (like so many other anthropologists before and after) fell into the methodological trap of confounding structural semantic and sociological accounts of systems of kin classification. By avoiding this methodological error, I hope to provide a superior account of the structures and the relations among the structures of Australian systems of kin classification, and thereby not only to set the stage for more adequate sociological accounts of them, but also to improve our understanding of the structures of Australian societies.

It is not, I think, merely an accident that Radcliffe-Brown chose to devote so much attention to Australian societies. When he began his project Australia was generally regarded as the refuge of the most archaic forms of human society, and many scholars argued forcefully, if not compellingly, that these forms are not based in concepts of genealogical connection and, therefore, have nothing in common with "kinship" as we and many other peoples know it. The comparability of these forms to social structures elsewhere in the world was therefore very much a matter

of speculation and dispute. For a number of good
reasons, Radcliffe-Brown supposed that systems of
ordering social relationships by reference to genealogical relationships among individuals, that is to say
"kinship systems," occur in all human societies and are
especially prominent in the institutions of so-called
primitive or tribal societies. It was therefore a
major challenge to him to show that although the kinship institutions of Australian societies may be somewhat unusual in comparative perspective, they are
nevertheless kinship institutions, and, moreover, they
exhibit structural continuities with the kinship institutions of other peoples.

Although Radcliffe-Brown's interpretation of the
Australian data is rejected by many anthropologists,
so are the competing interpretations. The very nature
of Australian social institutions and of their comparability to the institutions of other peoples is still
very much in dispute. For anyone who believes, as I
do, that kinship systems are a constant feature of
human societies, the challenge of the Australian data
still remains.

It might be thought that one of the reasons why
the major theoretical issues have not been resolved is
lack of adequate data, and perhaps additional research
in the field is required. Without wishing to minimize
the potential value of additional field research, I
would argue that, to the contrary, perhaps the principal

difficulty has been the use by anthropologists, either
explicitly or implicitly, of inadequate and misleading
lexical-semantic theory - in the light of which certain
critical linguistic facts (although perhaps noticed and
reported) have not been seen as especially relevant and
have therefore not been taken sufficiently into account
in theoretical arguments. In the light of more adequate
lexical-semantic theory, the available data for at least
some Australian languages are, I believe, reasonably
adequate to permit us to reject some of the competing
interpretations and to establish beyond a reasonable
doubt that concepts of kinship and systems of kin classification are central and universal features of Australian social structure. These data, moreover, provide
a reasonably secure foundation on which to construct
models of the structures and the relations among the
structures of Australian systems of kin classification.
I do not suppose that, in all their aspects, the models
presented herein will withstand critical evaluation and
will not need to be revised in the light of further
ethnographic and linguistic research. I will be pleased
if ethnographers and linguists engaged in research in
Australia are stimulated by my efforts to improve upon
them by improving the quantity and quality of the relevant data and to suggest modifications, where necessary,
to the models presented herein.

The lexical-semantic theoretical concepts and
methods of structural semantic analysis that inform

this study have been explicated elsewhere, both in earlier studies by the same author (see especially Scheffler and Lounsbury 1971), and in numerous works on lexical semantics by other authors, some of which are noted in the References. However, the most immediately relevant concepts and methods are explained as they are introduced in the text and bibliographic references are provided to sources where they are again explained and their applications in other cases are illustrated.

For the benefit of those who advocate truth in packaging, I may add that the title of this study has been chosen with care. I have not attempted to provide anything like a definitive account of all the meanings of any Australian kinship term or set of terms, or of the relations among all those meanings. What I have attempted is to provide formally adequate accounts of the kin-class significata of certain Australian language expressions and of the relations among those significata. I am aware that many of these expressions have yet other kinds of meanings and are usable in ways other than to designate kin classes or members of kin classes. There are, however, at least two reasons why these other meanings need not be taken into account in this study, either in the specifications of the definitions of the kin classes or in the process of specifying those definitions. One of the reasons is that they are different kinds of meanings, whose specifications

are not genealogical. The other reason is that the
other kinds of meanings are structurally dependent on
the kin-class significata. These structurally dependent meanings are specifiable only in terms of or by
reference to the structurally central meanings. Therefore, adequate comprehension of the structurally dependent meanings must wait upon adequate comprehension
of the structurally central meanings. But more of this
in Chapter 1.

NOTATION

The various orthographic and notational conventions
employed in the text and tables are as follows. Australian language expressions are represented ortho-
graphically as they are in the sources from which the
data have been taken, although it is probable that in
many instances these representations are not phonolog-
ically, much less phonemically, accurate. This has the
obvious advantage of making it easier for the reader to
refer back to the original sources, should he or she
wish to do so. Moreover, the inaccuracies present few
problems, because the present focus is on meaning
rather than on form, and because the few problems presented by apparently inaccurate orthographic representations are readily resolvable. English glosses for
Australian language expressions appear in single quotes,
for example, 'sister' for an Australian expression
whose structurally primary genealogical designatum is

the same as that of English sister. Where, as often
happens, there is no appropriate simple English expression to use as a gloss, or where using a simple English
expression might prove confusing, more complex English
expressions are introduced, also in single quotes.
Many Australian kinship terms designate two or more
related kin classes, one included in the other. Where
the reference is to the more inclusive class (sometimes
described as the widened or expanded class), the Australian expression or its English gloss is represented
in upper case letters, for example, SISTER for the full
genealogical range of an Australian expression glossed
as 'sister'.

Kintypes, or various kinds of genealogical relationships, are represented by single or juxtaposed
upper case initials, often those of English kinship
terms. Thus,

F = father	S = son	B = brother	H = husband
M = mother	D = daughter	Z = sister	W = wife
P = parent	C = child	Sb = sibling	Sp = spouse

Therefore, MB is to be read as "mother's brother," MZ
as "mother's sister," and MBS as "mother's brother's
son."

The letters (m) and (w) are used for man and woman
respectively, to designate either ego or one of his or
her kinsmen or kinswomen, and in each case specify the
sex of the party designated by the symbol. In the text
and in the various tables listing the kintype denotata

of Australian kinship terms, these symbols are sometimes prefixed to kintype notations. In these contexts
they serve only to indicate the sex of ego who so
employs the terms in question. Where a kintype notation occurs in the text or tables without such a prefix,
it may be understood that the sex of ego is immaterial,
unless the sex of ego is otherwise indicated, for example, "a man's MBD," "a woman's FZS." Thus, in the text
and in the tables, these symbols may be read as follows:

mMBS = male ego's mother's brother's son

wMBS = female ego's mother's brother's son

MBS = any person's (male or female ego's)
mother's brother's son.

In contrast, in the equivalence rule formulas (m) and
(w) are used more broadly to designate either the sex
of ego or that of his or her kinsman or linking kinsman
to whom the rule applies. Therefore, unless interpretation of the symbols is otherwise restricted (as noted
below), the (m) may be read (in the equivalence rule
formulas) as "male ego, or any male relative of ego
or of any relative of ego," and the (w) may be read as
"female ego, or any female relative of ego or of any
relative of ego." But in these formulas it is often
necessary to distinguish specifically nonterminal
positions in the genealogical chains (i.e., a linking
kinsman). A single dot (.) and a sequence of three
dots (...) are used for these purposes. A single dot
preceding a male or female sign means this sign must

be interpreted as ego and precludes any other interpretation. A sequence of three dots before a male or female sign means this sign must <u>not</u> be interpreted as ego, but must be interpreted as a linking kinsman of the appropriate sex but of unspecified genealogical position. A single dot following a kintype abbreviation means that the given abbreviation must be understood as referring to the designated relative only. Three dots following a kintype abbreviation mean that the given abbreviation cannot be interpreted as referring to the designated relative, but must be interpreted as a linking relative. Contrast the following when they occur in the equivalence rule formulas:

MB = anyone's mother's brother

mMB = any man's mother's brother

.mMB = male ego's mother's brother

.mMB. = male ego's mother's brother as a
       designated relative

MB... = anyone's mother's brother as a link to
       some more distant relative

mMB... = any man's mother's brother as a link to
       some more distant relative

...mMB = a male linking relative's mother's
       brother

.mMB... = male ego's mother's brother as a link
       to some more distant relative.

In componential definitions of kin classes, the single dot (.) represents the relationship of class intersection and may be read as "and"; (v) represents the relationship of class union and may be read as "or"; ($\circ^7$) represents male, and ($\circ$) represents female. The tilde (~) represents negation and may be read as "not." In the equivalence rule formulas, the equal sign (=) serves to link the two parts of a rule which are the simple reciprocal corollaries of one another; the sign may be read as "and conversely."

The locations of the various linguistic or "tribal" groups mentioned in the text are shown on Map 1.

ACKNOWLEDGMENTS

The research on which this study is based was supported by grants from the National Science Foundation (GS 28091) and the Australian Institute of Aboriginal Studies, and by a Research Fellowship in Social Anthropology, Research School of Pacific Studies, The Australian National University, in 1971-72.

Numerous colleagues have assisted in many ways. I am especially grateful to Emeritus Professor A. P. Elkin, University of Sydney, who generously permitted me to consult and to cite from his unpublished field reports. Professor A. L. Epstein, then Head of the Department of Anthropology, The Australian National University, arranged the Research Fellowship. Louise Hercus, John von Sturmer, and Nicolas Peterson made it possible for me to meet and work with many excellent aboriginal informants in South Australia, Queensland,

and the Northern Territory. Members of the staffs of
several libraries and museums were generous with their
time, knowledge, and facilities. Portions of this
study were presented in seminars at Monash University,
the University of Sydney, and The Australian National
University, where I received many useful comments.
Jeremy Beckett, Les Hiatt, Annette Hamilton, Dianne
Barwick, Michael Allen, and W. E. H. Stanner were all
helpful in many ways, as were Barbara Sayers, Mrs. D. F.
Thomson, and Miss Judith Wiseman.

In the United States I received assistance from
Lauriston Sharp, Mervyn Meggitt, Kenneth Hale, Nancy
Munn, Warren Shapiro, Bruce Rigsby, and Floyd Lounsbury,
and Michael Silverstein. Portions of this study were
presented in seminars at Yale University, the Graduate
Center of the City University of New York, the University of Arizona, Arizona State University, and the
University of New Mexico, where again I received many
useful comments.

I am especially indebted to Susan Bean, Jimmy
Holston, and the two anonymous professional reviewers
for the Cambridge University Press, all of whom read
the manuscript carefully and sympathetically and made
many useful comments on how it might be improved.

Elizabeth Kyburg and Robin Wrzosek did their usual
excellent work in typing the difficult manuscript.

My Australian informants included Peret Arkwookerum
(Wik Ngantjera), Mick McLean (Wongkunguru, Southern

Aranda), his wife Kathleen McLean (Arabana), Arthur
McLean (Arabana), Leslie Russel and Johnny Reis (Wongkunguru), Ben Murray (Dieri), Angus and Eileen McKenzie
and May Wilton (Adjamathanha), Sammy Johnson and Tommy
Low (Walbiri) and Bob Holroyd (Thayorre). It was a
great pleasure to work with all these people and to
learn from them.

H.W.S.

QUEENSLAND

NEW SOUTH WALES

Murawari

Wongaibon

Wilyakali

VICTORIA

Kurnai

Ompela

Mongkan

Yir
Yoront

Marula

Yantruwanta

Wongkonguru Yaurawaka

Arabana   Dieri

Adjamathana Piladapa

Antakarinya

Yaralde

Murngin

NORTHERN
TERRITORY

Walbiri

Aranda

SOUTH AUSTRALIA

Pitjantjara

Jaŋkundjara

Mari'ngar
Murinbata

Forrest
River

Wunambal

Ngarinyin

Nyulnyul

WESTERN

AUSTRALIA

Karadjeri

Kariera

Mardudhunera

500 km

0

Chapter 1

PRELIMINARY CONSIDERATIONS

The nature of Australian aboriginal social systems has
been debated for more than a century and widely diver-
gent interpretations have been offered both for particular cases and for Australian societies in general. The
controversy centers on certain systems of classification
by means of which the people order their social lives.
Three interpretations have predominated. One body of
opinion has it that in Australia social relations are
governed, by and large if not wholly, by considerations
of kinship. Proponents of this interpretation, most
notably A. R. Radcliffe-Brown, maintain that Australian
social categories are, virtually exhaustively, kinship
categories - meaning by "kinship" egocentric relations
of genealogical connection.[1] Another interpretation is
that in Australia social relations are governed, again
by and large if not wholly, by considerations of membership of groups such as lineages, clans, moieties, and
sections or subsections - the latter two being known
also as "marriage classes." Proponents of this inter-

pretation argue that those Australian expressions sometimes described as "kinship terms" are more appropriately described as "relationship terms." The relationships
they signify are social and are between groups and individuals as members of groups.[2] A third interpretation
attempts to mediate between these two extremes by arguing that, although some Australian societies may be
ordered wholly by kinship, most of them are ordered by
both kinship and group relations. Where both are present they may be so well integrated that persons acting
in kinship capacities are also fulfilling their rights
and duties as members of related social groups.[3]

The central thesis of this study is that the first
interpretation is correct; the other two are fundamentally mistaken. The groups they take as structurally
fundamental, or as structurally independent of systems
of kin classification, are instead structurally dependent on kin classes and superclasses. This, it seems to
me, was Radcliffe-Brown's interpretation, although he
has been understood in other ways (cf. Dumont 1966).
It must be acknowledged, however, that Radcliffe-Brown
did not fully demonstrate the validity of his claims
(cf. Meggitt 1972). He was unable to do so because of
certain inadequacies in his conception of the structures
and the relations among the structures of Australian
systems of kin classification (see Chapter 2).

The first step in evaluating Radcliffe-Brown's interpretation of Australian society must be to discover

the structures and the relations among the structures
of Australian systems of kin classification. This is
the subject of Chapters 3 through 10. Although one aim
of this study is to clarify our understanding of the
nature of Australian society, in these eight chapters
I say very little about social relationships, except
insofar as they have to do with marriage, and even then
only insofar as marital relationships or potential or
prospective marital relationships are demonstrably relevant to designation by a kinship term. The reason is
that these chapters are concerned with the structures
and the relations among the structures of systems of kin
classification, and not with any relationships that may
exist between these semantic structures and features of
social structure. For reasons given elsewhere (see
Scheffler 1972a, 1972b, 1973; Scheffler and Lounsbury
1971) and in Chapter 2 below, I believe it is essential
to keep separate social-structural analysis and the
structural-semantic analysis of systems of kin classification - and this is the procedure followed here.
Therefore, the reader who is interested in social structure rather than in the formal semantic analysis of systems of kin classification may find the next ten chapters not to his or her taste. But if the central thesis
of this study is correct, detailed understanding of the
structures and the relations among the structures of
Australian systems of kin classification is prerequisite
to understanding much of anything else about Australian

social structure.

The argument of Chapter 12 is that Australian moiety, section, and subsection systems are structurally derived from the superclasses of certain systems of kin classification. Functionally, these systems have very little to do with marriage but "serve as mechanisms for summarizing [and extending] kinship and the social behavior of which it is the determining factor" (Elkin 1933b: 90; see also Radcliffe-Brown 1930-31: 440). The social relationships they "summarize" are predominantly those of men in relation to the world of the Dreaming.[4] The argument of Chapter 14 is that the so-called lineages, clans, or patrilines of certain societies are simply small aggregates of men related to particular Dreamings (totems) and to one another through patrifiliation (see also Fortes 1969: 115 ff.). The external kinship relations of these sets of patrifilial kin link the sets and their Dreamings to one another in a system of complementary social relationships. Here, as Fortes (1969: 120) put it, we have "an incontestable limiting case of a social structure wholly comprised within a framework of kinship institutions."

The first step must be to demonstrate that the systems of classification analyzed in Chapters 3 through 10 are systems of kin classification, that is, systems wherein individuals are classified by reference to features of the genealogical connections assumed to exist between them and a specified individual (ego or the pro-

positus). Of course, to have such a system, a culture
must posit relations of genealogical connection; that
is, it must posit the existence of human genitors and
genetrices.

Australia, however, is the fabled land whose peoples are "ignorant of the relationship between sexual
intercourse and conception" and therefore "of the facts
of physical paternity" - or so it is said (cf. Montagu
1974 [1937]). If this were true Australian cultures
could have no concepts similar to the Western concepts
genitor and father; they could have no concepts of
paternal genealogical connection, although they might
have concepts of maternal genealogical connection. And
lacking concepts of paternal genealogical connection,
their languages could have no words that could be
glossed as 'father', 'father's sister', and so on, although again they might have words that could be glossed
as 'mother', 'mother's brother', and so on. Even so,
Spencer and Gillen (1899: 58) insisted that Australian
languages "have no words equivalent to our English words
father, mother, brother, etc."; that is, Australian
languages do not include kinship terms. This claim is
not difficult to refute.

AUSTRALIAN THEORIES OF PROCREATION
Received anthropological opinion (until recently)
has been that the typical Australian theory of procreation is that a woman conceives (becomes pregnant) when

she is entered by a "spirit-child" associated with a totemic site or spirit center. Although it is acknowledged that in some areas a relationship between sexual intercourse and pregnancy is "recognized," in these instances the alleged function of sexual intercourse is only somehow to prepare the woman for the entry of the spirit-child. Thus, it is alleged (cf. Montagu 1974) the "standard" Australian theory of procreation does not allow for the existence of physical fathers or genitors, and in the view of many anthropologists we may speak of "fathers" in Australian society only in a "social" sense (meaning mothers' husbands but not genitors).[5]

There is, however, considerable evidence to the contrary. In recent years it has become apparent that many anthropologists have seriously misunderstood and misrepresented theories about how human beings come to instantiate totemic beings as theories about human reproduction. Too often, also, a preoccupation with what Australians "do not know" has been allowed to obscure investigation, reporting, and discussion of what they "do know," that is, of the "facts" and relations among the "facts" that they themselves posit. The epistemological bias that turns such "facts" into magic and superstition (cf. Montagu 1974: 164; Spiro 1968: 247) has been exposed by Schneider (1965), Leach (1967), Horton (1967), Barnes (1973), and others, and need not be discussed here.[6]

If we allow for the moment that some reports of an exclusively spiritual theory of conception may be correct - and do not succumb to the temptation to argue that this is the orthodox, standard, and only genuinely Australian theory of conception (cf. Montagu 1974: 163) - it may be said that three sorts of theories of human reproduction are reported in the ethnographic literature. First there is the theory noted above. The second theory is that repeated sexual intercourse is essential to the process of conception; semen mixing with blood in the uterus gradually forms a fetus that blocks the flow of menstrual blood and interrupts the menses (see, for example, Thomson 1936). One variation on this theory is that the substance of the fetus is derived solely from the genitor's semen; the genetrix contributes no substance but only nourishment to the growing fetus (Howitt 1904). The third theory builds on the second and holds that, once the fetus has been formed, it is entered by a "spirit-child." According to some peoples this entry occurs intrauterine at the time of the quickening (during the fourth month of pregnancy), but others say it occurs at or shortly after the time of birth (Warner 1958 [1937]; Meggitt 1962, 1972; Peterson 1969, 1972; Hiatt 1965; Goodale 1971; T. Strehlow 1964, 1965, 1971a).

Some ethnographic accounts make it clear that theories of the third kind are really two theories, a strictly physical-sexual account of human reproduction,

and an account of how each individual acquires an inal-
ienable identity within a totemic cosmological system.
We might, following Leach (1967), describe this latter
theory as a kind of "religious dogma," but we must recognize also that both theories are equally social dogmas
or doctrines,[7] and that the two theories are complementary rather than opposed. They are not mutually contradictory doctrines about the same thing. One is a doctrine about human reproduction, the other a doctrine
about "immortal souls." Therefore, this latter theory
or doctrine is not at all comparable to the virgin birth
dogma of Christianity (cf. Leach 1967; Spiro 1968: 249).
It is, however, quite comparable to the Christian doctrine that all men have souls that are given to them at
conception by God - although in the Australian theory
the "soul" or totemic identity is not acquired at conception but some months later, at the time of the quickening or at the time of birth.

It seems probable that most if not all reports of
theories of the first kind (purely spiritual conception)
are nothing more than incomplete accounts of theories
of the third kind. We can be quite certain that this
is true in many instances, for example, the case of the
so-called Tully River Blacks or Dyirbal and the case of
the Aranda.

Dixon (1968) has shown that Roth's (1903) account
of the Dyirbal theory is at best incomplete and that the
Dyirbal are not now, and were not in Roth's time, "ig-

norant" of physical paternity – except of course insofar
as they do not posit as matters of fact things that are
knowable only with the aid of microscopes and through
controlled, clinical experimentation. The Dyirbal language features the verb bulmbinyu 'to be the male pro-
genitor of', and according to Dixon it "has clear reference to the particular act of copulation that induced
a conception." Presumably Dixon finds no reasons to
suppose that this expression was coined to signify a
new concept introduced into the culture after Roth made
his cursory observations. Dixon accounts for the incom-
pleteness of Roth's account by suggesting that there are
"two levels of belief concerning human conception," a
"basic level" and a "mystical level," and he says "the
mystical level of belief may well be the only level
normally explicitly acknowledged, the basic level being
more implicit."[8] It seems clear, however, that what we
have here are not "two levels of belief" about the same
thing, each accounting for the same fact or facts (conception per se), but instead two complementary theories,
one about conception per se and the other about the
instantiation of totemic beings in individual human
beings.

The case of the Aranda is similar. Spencer and
Gillen's (1927) report that the Aranda are ignorant of
physical paternity was denied by C. Strehlow (1913),
Pink (1936), and Roheim (1938). Montagu (1974: Chapter
3) attempted to explain away their objections. But T.

Strehlow has recently (1964, 1971a) provided a full,
reliable, and philosophically sophisticated account of
the Aranda theory of human reproduction that confirms
and extends C. Strehlow's, Pink's, and Roheim's accounts. The Aranda say that each person has two "lives"
or "souls," one animal and mortal, the other spiritual
and immortal. The first is created through sexual intercourse; the second is part of the "life" of one of
the immortal supernatural ancestors (one of the totemic
beings) and it enters the already formed fetus. This
second soul "decided the personality of the child after
birth; the totem of each individual and his personal
links with the world of Eternity were determined by the
soul that took up residence in him." Also, the individual acquired "the physical characteristics and the
whole personality" of this totemic ancestor, so that
"all Aranda men, women, and children were believed to
have been completely recreated in the images of those
totemic ancestors who had become reincarnated in them"
(T. Strehlow 1964: 730). Strehlow further states:
"The embryo or fetus that had been begotten by its
father was regarded in exactly the same way as the
young plant that had burst forth from the seed cast by
the wind upon a sacred site. It was only after the
embryo or the young plant had come into being that a
second 'spirit' or 'soul' was able to enter into these
mortal forms. This second 'spirit' or 'soul' was in
both cases the all-important one; for the second 'spir-

it' or 'soul' alone was immortal, and it consequently gave to what otherwise would have been a creature or a thing of evanescent worthlessness those glorious attributes which alone determined its true worth and status in the eternal scheme of things" (1971a: 596).

The Aranda entertain no doubts that each person has a physical father or genitor, no less than a physical mother or genetrix, and they assign considerable social significance to these identities (see T. Strehlow 1947: 120, 133; 1964: 730, note 10; 1971a: 597-8; Pink 1936). For other social and especially ritual purposes, however, it is one's "conception totem" that is relevant, and it is in relation to these purposes that "physical paternity is normally dismissed as being virtually of no practical importance" (Strehlow 1971a: 596). Thus, it seems, Aranda men do not deny physical paternity, but only its relevance to certain aspects of one's being; and according to Strehlow (1971a: 597), even "these denials used to be made in their most sweeping form only in the presence of the women, the children, and the younger men, and any inquisitive white outsiders" (emphasis added).

The Aranda, then, are no more or less "ignorant" of "the facts of physical paternity" (or of physical procreation in general) than are most other peoples, and their theory of human reproduction is not very different from the theories of many other peoples (cf. Scheffler 1973: 748-51). Of course, the Aranda theory

does not posit precisely the same facts as are posited
in the clinical-biological theory of human reproduction,
but this is only to be expected and does not prohibit
us from saying that the Aranda have some "knowledge"
(ethnobiological knowledge) of "facts" of physical
paternity and maternity. This theory, however, is
complemented by another, and together they account for
more than mere animal reproduction. They go well beyond
this to provide also an account of at least some of the
significant social divisions and similarities among men
and to relate these divisions and similarities to cosmological conceptions and categories. This latter part
of the Aranda theory and, no doubt, the theories of many
other Australian peoples has too often been mistaken
for a mystical or spiritual theory of human procreation,
and thus a great deal of anthropological ingenuity has
been wasted attempting to save the people from an
"ignorance" with which they are not in fact afflicted.
As Andrew Lang recognized long ago, concepts of physical-sexual and of spiritual procreation coexist in many
Australian societies - but there is no "obscuring" of
the former by the latter (unless it is for and by the
foreign inquirer!). Moreover, it is not true that concepts of spiritual procreation, so-called, "render a
knowledge of the [physical] facts quite superfluous"
(Montagu 1974: 249). For, again, the posited spiritual
and physical-sexual processes are complementary and do
not compete (intellectually or socially) with one anoth-

er.

Of course, it does not follow that because many reports of ignorance of physical paternity are demonstrably false therefore all such reports are false.[9] Even
so, considerable doubt has now been cast on the validity
of all such reports and they cannot be allowed to stand
in the way of recognition that relations of genealogical
connection, both maternal and paternal, are posited in
Australian cultures.

KINSHIP TERMS

The answer to the question "Do Australian societies
feature systems of kin classification?" depends in part
on the answer to the question "Do Australian cultures
posit the existence of relations of genealogical connection?" It is clear enough that relations of genealogical connection are posited in many Australian cultures,
and their wide distribution over the continent suggests
that such relations may be a universal feature of Australian cultures. Of course, it does not necessarily
follow from the fact that such relations are posited in
many (perhaps all) Australian cultures that these cultures also feature systems of kin classification. For
a culture to feature a system of kin classification it
must have in its language a set of expressions one of
whose functions is to designate egocentric classes of
persons assumed to be related by genealogical connections.

Anthropological opinion on whether or not Australian languages feature such expressions is sharply divided. One of the principal difficulties is that many of the Australian expressions that appear (to some anthropologists) to be kinship terms are used also to designate individuals as members of those "groups" or sets that comprise moiety, section, or subsection systems,[10] and they may also signify or connote kinds of moral, jural, and affective relationships (cf. Lang 1908: 141-2). Much controversy turns on different views of how these various "meanings" are related to one another.

The interpretation advocated by Fison and Howitt (1880) and Spencer and Gillen (1899, 1927), and by a number of more recent anthropologists, is that Australian "terms of relationship" signify relations between persons as members of moieties, sections, or subsections; each term has a single significatum (sense) and in no case does it correspond to the genealogical significatum of English words such as father, mother, brother, etc. According to Fison (1880: 77), "the ego is a group, not an individual; but each individual takes all the relationships which are taken by his group." Spencer and Gillen added (1899: 57): "In all the tribes with which we are acquainted, all the terms coincide, without exception, in the recognition of relationships, all of which are dependent upon the existence of a classificatory system [i.e., moieties, sections, or subsections] the fundamental idea of which is that women of certain groups marry the men of others."[11] Of course, they added, a term such as Aranda mia "does include the relationship which we call mother," but, they went on, "it includes a great deal more." In other words, in Spencer and Gillen's view, it just so happens that each person's genetrix is included in the category designated mia; the category consists of all women who belong to a specific section or subsection - the one from which one's father is "legally" obliged to take his wife - and the expression mia denotes one's mother only as a member of that category. In The Arunta, Spencer and Gillen (1927: 44) stated: "The whole classificatory system and social organization is based on the existence of . . . exogamous intermarrying groups, and in association with them there is a series of terms used that are expressive of the relationship that exists between each individual and every other member of the tribe. The fundamental feature of the terms used is that they are indicative of group relationship."

In brief, Spencer and Gillen's and Fison and Howitt's interpretation was that Australian "terms of relationship" signify relative positions within moiety, section, or subsection systems; as a consequence of the rules of intergroup marriage, and because membership in these sets is determined by one's parentage, it so happens that certain types of relatives always stand in the same "group relationship" vis-à-vis one another and, thus, always designate one another by a particular term or terms. Because of the consistency of application of certain terms to certain types of relatives (for example, anyone's mother is his or her mia), it may appear that the terms are kinship terms, but this would be an ethnocentric interpretation based on a misunderstanding of the Aranda language - or so they argued.

On one occasion, however, Spencer and Gillen (1927: 44) remarked that Aranda terms of relationship are indicative "sometimes, in a secondary way, of individual relationships." Unfortunately, they did not elaborate on this remark, but some indication of what they probably had in mind is given in the works of C. Strehlow (1913) and Schulze (1891) (see Lang 1908 and Howitt 1891 on some other cases).

Strehlow (1913: 65) took exception to Spencer and Gillen's statement (1899: 57, 106) that an Aranda man has no "name" by which to distinguish the woman he marries from other women of her subsection whom he does not marry, but any of whom he might lawfully marry. Strehlow noted that, while noa may be used to designate one's wife as well as her sisters and her "class sisters," one's "own wife" or "wife proper" alone may be designated as noatja, that is, noa + atja (also altja, iltja). Similarly, to distinguish "own father" from "class father," atja 'own', 'proper' is added to kata; thus kataltja designates one's own father as distinct from any other man or men who also may be designated as

kata. Of these same words, Schulze (1891: 224-5) notes
(in the context of a brief discussion of the terminological effects of marriage between kin): "Neither the
man nor the woman assume any new title, the connubial
relationship being only indicated by affixing the word
iltja to the names, while the affinity of the class is
denoted by the affix _lirra_. Thus _noa_ means spouse or
partner; _noa iltja_, real spouse, with whom he cohabits.
Again, _kata_ signifies the father of the class; _kata
iltja_, sexual father. Ordinarily they leave out the
words _iltja_ and _lirra_, and do not use them because they
all know among themselves who is personally related and
who is not. They are only used casually when conversing
with strangers, to whom they wish to explain their family relationships."

Strehlow and Schulze both took the same line as
Spencer and Gillen with regard to the structure of the
system of "terms of relationship." Thus it would seem
that in the view of all these men (if I may state their
arguments in terms they did not use), Aranda "terms of
relationship" designate categories defined by relations
of descent and intermarriage among subsections. These
categories, they argued, are the structurally primary
designata of the terms, but in addition the terms have
restricted, narrowed, or specialized senses and, when
so used, they designate the closest kinsman or kinsmen
in each of the broad categories. When used in these
specialized senses, they may be lexically marked by the

postfix _atja_ or _iltja_; the residual category of nonclos-
est kin in each section or subsection may be marked by
the postfix _lirra_ (according to Schulze). Clearly, this
is to imply that these words may be used as kinship
terms, that is, to designate egocentric kin categories.

Elsewhere (Scheffler and Lounsbury 1971: 60-1) it
has been pointed out that similar interpretations have
been placed on many non-Australian kinship terms and
that these interpretations are not altogether implausi-
ble. After all, the kind of polysemy[12] posited in these
interpretations is not uncommon. For example, in many
languages a man may denote his wife by a word ordinarily
used to designate any female human being as member of
the category "woman," and in other languages the term
for woman may be used to denote one's genetrix (mother).
In none of these cases do the words in question have the
senses of English mother (genetrix) or wife as their
structurally most basic senses. Our intuitions (which
are not misplaced here) tell us that these are specialized, narrowed senses of the terms: One's mother or
wife is a special case of a woman. This is indicated in
all of these languages in much the same way. To take
the case of French, because it is probably the most
familiar, the use of _femme_ in the specialized, narrowed
sense "wife" is lexically marked (Greenberg 1966; Lyons
1977: 305-311) by the obligatory cooccurrence of the
personal possessive pronoun (_ma_ _femme_, for example) or
by the possessive construction (_femme_ _de_ _X_) (see Figure

1.1). A wife, then, is a special kind of woman, <u>the</u>
woman to whom one is married and who is legally "pos-
sessed." The process whereby the designatum "woman"
is narrowed to "wife" consists in the addition of a
further stipulation or requirement for designation by
the term <u>femme</u>; the stipulation is that the denotatum
must be not only a female, adult human being but also
married to ego or the propositus. It so happens that
"woman" is also the historically prior of these senses
of <u>femme</u>, but that is not relevant here because we are
concerned with logical, synchronic semantic relationships that may or may not reflect historical, diachronic
relationships or processes (cf. Scheffler and Lounsbury
1971: 59-61; Lang 1908).

Thus it is conceivable that the structurally pri-
mary designata of Aranda <u>noa</u>, <u>kata</u>, <u>mia</u>, etc., might be

Figure 1.1. Relations among the senses of <u>femme</u>

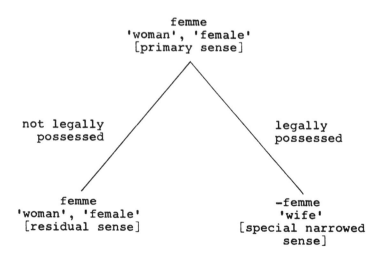

categories based on the subsection system of Aranda
society, but that they have in addition certain special,
narrower, or more restricted designata that, taken together, constitute a system of kin classes. This interpretation of the case of <u>kata</u> is illustrated in Figure
1.2.

On the basis of the evidence considered so far, it
would be premature to conclude that this interpretation
is correct. There is at least one other plausible interpretation, that advocated by Radcliffe-Brown (1913,
1930-31, 1951) and Elkin (1964 [1938]), and before them
by numerous authors such as Curr (1886), Roth (1897),
Mathew (1910), and Lang (1908). As they saw it, the
egocentric, genealogical designata of Australian "terms
of relationship" are their structurally primary designata; each term designates one or more category of kin

---

Figure 1.2. Spencer and Gillen's version of relations
among the senses of Aranda <u>kata</u>

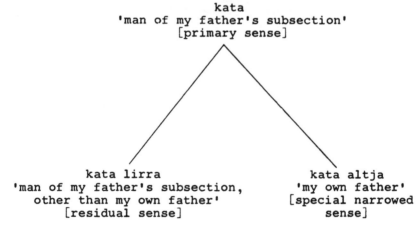

kata
'man of my father's subsection'
[primary sense]

kata lirra
'man of my father's subsection,
other than my own father'
[residual sense]

kata altja
'my own father'
[special narrowed
sense]

and may be _extended_ along section or subsection lines
(where such social divisions are present), or in other
ways. In other words, they argued that the kind of
polysemy we have here is the kind that results from
widening, that is, from the relaxing or weakening of
the conditions for designation by a term and, therefore,
for inclusion in a category.[13] To take the case of
Aranda _kata_, the argument is that 'father' (in the
ordinary English sense of male parent) is the primary
significatum of this term and that the term is extended
to certain male relatives of ego's father, ultimately
in some cases to many male members of ego's father's
section or subsection. From this perspective, the
postfix _atja_ (optionally) marks the use of a term in
its structurally primary or central sense, and the
postfix _lirra_ (optionally) marks its use in a derivative, structurally dependent, or "classificatory" sense.
This interpretation of the case of _kata_ is illustrated
in Figure 1.3.

The "kinship terms and extensions" interpretation
of the meanings of Aranda and other Australian "terms
of relationship" is the correct interpretation. There
are several serious defects in the alternative interpretation favored by Spencer and Gillen and others.

It is not true, as Spencer and Gillen (1899: 57)
asserted, that all Australian "terms of relationship"
are associated with "class" systems. Moiety, section,
or subsection systems are not present in many Australian

societies, and in other societies the systems of "terms
of relationship" are not congruent with the "class"
systems that do occur. For example, the Northern Aranda
have a subsection (eight-class) system and the Southern
Aranda have (or had) a section (four-class) system,
though apparently both have the same system of "terms
of relationship." Spencer and Gillen (1927: 42) argued
that the Aranda system of "terms of relationship" is
based on the subsection system and "as a matter of fact,
the eight subsections, though unnamed, exist in the
south." But the alleged evidence for the existence of
subsections in the South was only that in Southern
Aranda the "terms of relationship" have the same distribution over kintypes as they do in Northern Aranda.
Spencer and Gillen argued, in effect, that this distribution logically implies a subsection system, that is,

Figure 1.3. Actual relations between some of the
senses of Aranda <u>kata</u> 'father'

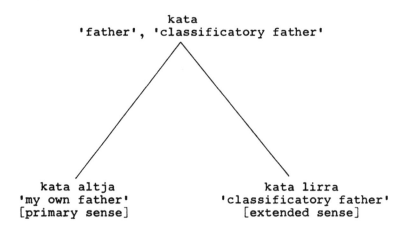

kata
'father', 'classificatory father'

kata altja
'my own father'
[primary sense]

kata lirra
'classificatory father'
[extended sense]

a further though unnamed or implicit division of each
of the four named sections. There is, however, an
essential difference between explicit (named) and implicit (unnamed) "classes" that makes it quite impossible to account for the distribution of terms over
kintypes in Southern Aranda in this way.

The theory maintains that any Northern Aranda
individual knows how to designate any other individual
by the appropriate "relationship term" if he knows the
subsections to which he and the other individual belong
and the terms that signify the "relationships" of descent and intermarriage between these subsections.
Further, any individual's subsection affiliation must
be identifiable independently of his or his parents'
designation by one of the "terms of relationship." If
his subsection affiliation cannot be identified independently of his or her parents' designation by one of
the "terms of relationship," such as by means of a subsection name (how else?), the theory offers no guide to
the application of the "terms of relationship." Of
course, where the alleged subsections lack names (in
other words, where the subsections do not exist), they
cannot function in this way. In short, the subsections
of the Southern Aranda are nothing more than hypothetical constructs, and Spencer and Gillen's argument reduces to the assertion that the "terms of relationship"
are distributed over kintypes in the way they would be
if the system of "relationship terms" were based on a

subsection system. This tells us nothing at all about
how it _is_ structured. Therefore, even if there were no
objections to their theory of the structure of the "relationship system" of the Northern Aranda, it would be
at best a special theory about a special case of systems
of the so-called Aranda type and not a useful theory
about the structure of Australian "relationship systems"
in general.

Spencer and Gillen's interpretation suffers from
yet another serious logical defect.[14] The theory maintains, in effect, that the Aranda expression _kata_ 'father' is used in two different though related senses; the
primary sense is "man of my father's subsection" and the
derivative sense is "my father" (cf. Spencer and Gillen
1899: 56). Note, however, if we take these glosses
seriously it is illogical to say that the latter sense
of the term is derived (by narrowing) from the former.
It is illogical because the category designated by the
term in its alleged primary sense ("man of my father's
subsection") not only includes ego's father but is defined by reference to ego's father; this definition
treats ego's father as _the_ relative by reference to
whom membership in the broader category is reckoned;
and the category that includes only ego's father is
logically prior to the more inclusive category defined
by reference to ego's father. Therefore, we may conclude that the narrow sense of _kata_ is its logically
prior or structurally primary sense, to which the broad

sense is related by widening or expansion.

This interpretation is confirmed by some of C. Strehlow's linguistic data. According to Strehlow (1913: 66-70, Table 2), the postfix larra (Schulze's lirra) is one of two that may be combined with a basic term to indicate a "class" relative rather than a "blood" relative of the designated category. The other is knara, which Strehlow glosses as 'big'. A kata knara, for example, is an older brother or classificatory 'brother' of one's 'own' or 'proper' father (i.e., one's kata atja); a kata larra is a younger brother or classificatory 'brother' of one's father; and similarly for most other Aranda "terms of relationship." These forms signify the relative age of the designated kinsman in relation to the closest kintype denotatum (or denotata) of the basic term, which is marked as the atja, 'own' or 'proper' designatum of the term. From all this it should be clear that the semantic function of atja is to specify the focal or logically primary designatum of a term, for it is in relation to this kintype that the terminological statuses of other kintypes as knara or larra are determined.

I have argued that the primary designatum of Aranda kata is ego's male parent. Ego's male parent is normally his mother's (female parent's) husband, and some anthropologists have argued that it is to this relationship (MH) or to some social relationship that expressions like Aranda kata primarily refer, rather than

to the genealogical relationship of male parent. Whatever the arrangement in other cases, this plainly is
not the arrangement in the Aranda case. T. Strehlow
(1971a: 595-7) reports that the Western Aranda form
tenama "can only be translated as 'beget'. It is used
only in connection with the father [kngeia = kata] and
is explained as katja mbarama 'to make a child'."

From these observations we must conclude that
Spencer and Gillen's interpretation of Australian "terms
of relationship" is seriously inadequate and misleading.
The evidence is quite clear that the so-called terms
of relationship designate egocentric, genealogically
defined categories and are polysemic; each term has a
structurally primary and specific sense and a derivative, expanded, or broader, sense (or senses). It may
well be that some kinship terms are sometimes used to
designate individuals according to their respective
section or subsection affiliations and the genealogical
and prescribed marital relationships between those
units. It should be virtually self-evident, however,
that not all applications of the terms beyond their
primary denotata are determined by considerations of
section or subsection affiliation; this would be quite
impossible in the case of the Southern Aranda, for example, and in other societies where no such units are
present.

It seems worth noting the reasons Spencer and Gillen were unable to present an adequate account of the

polysemy of Aranda and other Australian kinship terms.
One of the reasons was their implicit assumption of
monosemy, or "one word : one meaning" (one lexeme : one
significatum). This assumption made it difficult if
not impossible for them to deal adequately with their
own observations by forcing them to look for the "common denominator" in all uses of a term and to regard
that as the meaning of the term. Without this assumption, it is difficult to see what objection they could
have had to saying that some Australian words are
"equivalent" to English father, mother, brother, etc.,
in some but not all of their respective senses (especially in their respective structurally primary designata). Spencer and Gillen's quest for the "original"
meanings of Australian "terms of relationship" owed
much to their uncritical adherence to the assumption of
monosemy. Faced with the fact of current polysemy,
they were forced to offer a "historical" explanation,
to argue that the "group kinship" meanings of the terms
were their temporally prior and even "original" meanings, and to fall into the trap of self-contradiction
by sometimes writing of these alleged original or temporally prior meanings as though they were still the
only meanings of the words.

Unfortunately, Strehlow's published data on the use
of the postfixes knara and larra are not as extensive
as we might hope, and similar data for other Australian
systems of kin classification are scarce, aside from a

few other fairly well documented cases and the odd bit
of information here and there.[15] In addition, apparently similar terminological distinctions have been
reported but not documented for many other Australian
languages. The fairly well documented cases include
the Dieri of South Australia (Gason 1874; Curr 1886:
135-6; Fison and Howitt 1880: 61; Howitt 1891) and the
Mongkan of Cape York (Thomson 1935, 1936, 1946, 1972).
Neither the Dieri nor the Mongkan have section or subsection systems, and although the Dieri have matrimoieties there is no moiety division among the Mongkan.
Because the Dieri system of kin classification is like
that of the Aranda in many respects, there are many
aspects of this system that cannot be explained by a
reference to the more conspicuous descent-ordered units
of the society; a moiety or two-section system explanation would not take us very far in accounting for its
terminological equations and distinctions of kintypes,
and Korn's (1971) interpretation - a "four-line matrilineal symmetric prescriptive terminology" - takes no
account of the data on polysemy and terminological
extension or sense generalization.

In addition to making distinctions between focal
and nonfocal kintypes in much the same way as in the
Aranda system, the Dieri also employ expressions such
as 'my friend father' for father's brother (Gason, in
Fison and Howitt 1880: 61), the logic of which surely
implies that one's father's brother is regarded, termi-

nologically at least, as another and special kind of
'father' (apiri).[16] This is perhaps even more clearly
indicated in Fison and Howitt's (1880: 61-2) report
that in the Brabrolung (a Kurnai "clan") and Eildon
languages of Victoria one's father's brother may be
described as "my other father." A further interesting
and revealing aspect of Gason's data on the Dieri is
that he gives "descriptive" expressions for many kintypes, that is, mother's sister may be designated as
andrie 'mother', with or without a suffix indicating
her age relative to one's mother, or her relation to
ego may be described by the two-term relative product
andrie kakoonie (kakoo 'elder sister'). These descriptive expressions demonstrate that Dieri kinship terms
may be used in specific or in general (expanded) senses,
that is, that the terms are polysemic. If they were
monosemic or polysemous in the way claimed by Fison and
Howitt and Spencer and Gillen, these relative product
expressions could hardly serve, as clearly they are
intended, to specify a particular type of kinsman (see
also Scheffler 1977).

Social categories

So far the discussion has focused on the question
of how the more inclusive genealogical categories designated by certain expressions are related to the less
inclusive genealogical categories also designated by
those expressions. Some anthropologists might say that

I have, so far, avoided the central issue, for in their opinion the categories in question are "socially" rather than genealogically defined. Without denying that relationships of genealogical connection are posited in Australian cultures, some anthropologists deny that these relationships are directly relevant to classification under one of the expressions I have been calling kinship terms. These expressions, they argue, signify kinds of social relationships and the categories they designate are "social categories" and not kin classes.

Sometimes this phrasing is merely another way of stating the interpretation proposed by Spencer and Gillen, Fison and Howitt, and others. The social relationships allegedly signified by the terms are said to be those between persons as members of moieties (sometimes described as a pair of sections), sections, or subsections, and from this perspective Kariera-like systems of kin classification are sometimes described as "two-section systems of social classification." Perhaps needless to say, the minor changes in phrasing in no way save the argument from the defects that nullify Spencer and Gillen's and Fison and Howitt's interpretation.

Sometimes, however, when it is said that the terms designate "social categories" and not kin classes, it is meant that they signify distinctive sets of rights and duties or privileges and obligations or, in other words, distinctive "jural statuses." Thus the terms themselves are sometimes described as "jural categories." This argument may be traced back at least to McLennan and others in the 1870s, but by the early 1900s it was generally conceded that it is a mistake to pose the issue in either - or terms (cf. Rivers 1907; Lang 1908). One grain of truth in the "social" or "jural categories" argument is that membership in kin classes is typically a matter of some social significance. Distinctive kinds of rights and duties or privileges and obligations may be ascribed between kin of reciprocal categories, and the terms that designate those categories may also connote those statuses. In Goodenough's (1964) phrasing, to be a kinsman of one kind or another is to have a particular social identity in relation to some person; conversely, that person as a member of the reciprocal kin class has a particular social identity in relation to oneself. So the meanings of kinship terms, certainly in Australia (cf. Warner 1937; Meggitt 1962; Hiatt 1965; Thomson 1972), encompass both genealogical relationships and nongenealogical or social relationships. However, insofar as the social relations are ascribed between persons as members of kin classes or, in other words, predicated on the kin classes, the kin-class significata are structurally prior to the social-relationship connotations. As between the two kinds of meaning, the kin-class significata are the structurally central meanings and the social-relationship connotations are the structurally dependent meanings, consisting as they do of the social

implications of membership in the kin classes designated
by the terms.

It is important to remember that any one kinship
term may designate more than one kin class and that the
connotations (if any) of a kinship term are connected
with the kin class it designates rather than directly
with the term itself. Therefore, what a term connotes
may vary with the category it designates, and the connotations of a term are not necessarily uniform throughout its genealogical range. Moreover, some categories
of kin between whom distinctive kinds of social relationships are ascribed may not be distinguished by
simple kin terms; that is, social classification may
vary independently of terminological classification (cf.
Scheffler and Lounsbury 1971: Chapter 8). Therefore,
the structures of systems of kin classification are not
necessarily "mirrors" of social structure and they may
be poor reflections of it.

Another grain of truth in the social categories
argument is that kinship terms are sometimes used, especially in address, between persons who do not suppose
that they are related by birth or by marriage. Of
course, when so used, they do not designate kin classes
or signify genealogical relationships; instead they
signify (rather than connote) kin-like social relationships that are assumed (not ascribed) between the parties who so designate one another. That is, they may
be used metaphorically to designate classes that are

kin-like in that the parties who so designate one another treat one another as though they were kin of certain kinds. It would not be inaccurate to say that such
kin-like categories are socially rather than genealogically defined; they are, however, structurally dependent
on and derived from the kin classes designated by the
same terms (and from certain contingent or nondistinc-
tive features of those classes, rather than from their
distinctive genealogical features). At its most dogmat-
ic, the social categories interpretation tends to gloss
over this complex structure and to reduce the meanings
of kinship terms to that which they may normally connote
but sometimes, when used metaphorically, may indeed signify. In the process, a great deal of culture and
social structure is lost or analytically obscured.

Not the least thing that is lost is specification
of how the jural statuses are allocated (cf. Scheffler
1972b: 315; 1976c: 274). It is, for example, not very
instructive to say, as some anthropologists have, that
an expression like Aranda noa does not designate a kin
class but signifies only that the designated party is
one's spouse or is a person whom one might "lawfully"
marry. It may be that designation as noa does imply
that the designated party is one's spouse or is eligible
to be one's spouse, but this cannot be regarded as an
adequate definition of the designated category or, for
that matter, as a definitional statement. Such a specification of meaning describes the jural implications of

membership in a class and (obviously) not its defining
features. It would be nonsense to say that the condition of being regarded as potentially one's spouse is
the distinctive feature of the class, because this
statement fails to specify the conditions _for_ being
regarded as potentially one's spouse.

However the social relationships are allocated,
whether by reference to membership in major social
"groups" such as moieties, sections, or subsections and
the relations of descent (that is, genealogical connection) and marriage between them, or by reference to
relationships of egocentric genealogical connection,
the terms must also signify these relationships and
designate the categories constituted by them. It would
be difficult to argue that these categories are structurally derivative or that they are structurally insignificant artifacts. Insofar as the social relationships
are allocated by reference to them, they are, again,
structurally prior to the social relationships. As
meanings of the terms that allegedly signify those social relationships, they must be regarded as structurally prior meanings on which the alleged social-relationship significata are predicated.

It is hardly surprising, then, that one proponent
of the social categories interpretation of the meanings
of Australian kinship terms has found it necessary to
acknowledge that "such jural systems and their component
statuses can be genealogically defined" (Needham 1971:

4), by which he appears to mean that their structures
are such that they may be described as though they were
genealogical. He acknowledges that this is a curious
fact for which he can offer no explanation, yet he insists that it does not "mean that the relationships in
question are genealogical or that they are so conceived
by the actors." It should be clear by now, however,
that relationships of genealogical connection are posited in many if not all Australian cultures and that
these relationships are classified egocentrically and
signified by certain expressions which we may and must
describe as kinship terms. These expressions may, in
addition, connote social relationships ascribed between
kinds of kin, and they may at times (when used metaphorically) signify such relationships. Further still, when
used as vocatives or terms of address, they also have
certain pragmatic or indexical functions (cf. Silverstein 1976; Bean 1978).

It must be acknowledged that Australian kinship
terms have several different kinds of meanings, all of
which must eventually be specified in a global account
of their meanings. It does not follow that their kin-
class significata may not be analyzed and specified
independently of their social-status connotations, their
metaphorical significations, and their pragmatic or
indexical functions. To insist on doing otherwise would
be, in effect, to insist on conflating several quite
different kinds of meanings into one wholly artificial

kind that is quite alien to the object of analysis.
Moreover, because any social-relationship connotations,
metaphorical significations, or indexical functions
that the terms may have are structurally dependent on
their kin-class significata, and can be specified only
in terms of or by reference to their kin-class significata, the kin-class significata must be analyzed and
specified prior to any attempt to analyze and specify
how the other kinds of meanings are structured. This,
it seems to me, is the only ethnographically and theoretically defensible procedure, and it will be followed
in this study.

In what follows I will say very little about the
social-relationship connotations, the metaphoric significations, and the indexical functions of Australian
kinship terms. This, I should emphasize, is not because
I regard them as unimportant in any way. Far from it!
For many social and cultural anthropological purposes,
adequate comprehension of them is just as important
(that is, relevant and indispensable) as adequate comprehension of the kin-class significata. But for these
same purposes, I would argue, adequate comprehension of
the kin-class significata is the fundamental desidera-
tum. This is because the other kinds of meanings are
structurally dependent on the kin-class significata.
Therefore, adequate comprehension of the other kinds of
meanings must wait on adequate comprehension of the kin-
class significata.

CONCLUSION

In this chapter I have shown that many Australian languages and cultures include systems of kin classification. Because the presence of such a system has to be demonstrated for each language and culture individually, I have not argued that systems of kin classification are a universal feature of Australian societies, although I think there are few if any reasons to suspect that they are not and many good reasons to suspect that they are. Be this as it may, although such systems have been reported for all parts of Australia, the ethnographers have not always provided adequate evidence in support of their claims, and there are many systems of kin classification reported in the literature for which the documentation is deficient in many ways and about which additional information probably cannot now be acquired. For these reasons we may never be able to demonstrate conclusively that systems of kin classification are or were a universal feature of Australian societies, even if it is demonstrable that many classificatory systems alleged to be something other than systems of kin classification really are systems of kin classification after all.

In any event, the possible or probable universality of systems of kin classification in Australian societies is not the principal concern of this study. The aim here is to reveal the structural similarities and differences among certain Australian classificatory systems that may be assumed, with varying degrees of confidence, though usually beyond a reasonable doubt, to be systems of kin classification.

Chapter 2

TYPES AND VARIETIES

INTRODUCTION

This chapter reviews and critically evaluates
A. R. Radcliffe-Brown's theory of the structures and
the relations among the structures of Australian systems of kin classification.[1] His is the most compre-
hensive and realistic theory so far advanced on the
subject, and therefore the essential starting point for
any attempt to further our comprehension of it. Although seriously defective in a number of ways, both
semantically and sociologically, Radcliffe-Brown's
theory cannot be faulted (at least not fairly and sen-
sibly) for positing rules of terminological extension.
As we saw in Chapter 1, there is ample evidence that
polysemy by widening _is_ a feature of Australian systems
of kin classification. Serious questions may be raised,
however, about the formal adequacy of specific extension rules posited in the theory, and therefore about
the formal adequacy of the theory as a whole. Following a fairly detailed exposition of the theory (impor-

tant parts of which have not, I think, been understood
or appreciated by other critics), it is shown that the
rules of kin-class definition and terminological extension posited by Radcliffe-Brown, especially his rules
of interkin marriage, do not do what he claimed for
them. They must be replaced by other more adequate
rules, some of which must be common to a number of systems Radcliffe-Brown assigned to different typological
categories. This raises further questions about the
sociological adequacy of the theory.

RADCLIFFE-BROWN'S THEORY

In 1951 Radcliffe-Brown summarized the results of
his more than four decades of study of Australian "kinship systems." He said he had shown how, "starting
from the same base, namely an organization of small
patrilineal local groups," all Australian tribes have
found "some workable system in which every marriage is
a marriage between kin." They have produced "a number
of varieties" of "marriage systems" but "they are all
varieties of one general type" (1951: 55). He noted
that since 1910 he had been engaged, off and on, in
"determining and classifying these varieties," and he
discussed four of them - Kariera, Kumbaingeri, Aranda,
and Karadjeri. A fifth, Yaralde, may be added from
previous discussions (1918, 1930-31, 1952 [1941]). He
used these tribal or language names to denote types of
systems of kin classification as well as types of "mar-

riage systems" because it was an essential part of his argument that similarities and differences among Australian systems of kin classification are correlated with and may be accounted for by reference to similarities and differences in rules of interkin marriage.

Radcliffe-Brown's types

He described the five types (of marriage system and of kin classification) as follows.

Kariera. Two lines of descent are recognized, those of ego's FF and MF. MMB is designated by the same kin term as FF, and FMB by the same term as MF; similarly, FFZ and MM are designated by the same term, and so are MFZ and FM. The rule of marriage is that a man marries a MBD or FZD (both denoted by the same term) or a woman of their kin class, that is, a classificatory 'cousin'. (The gloss 'cross cousin' may be avoided because in no Australian system of kin classification are parallel and cross cousins classified together in opposition to siblings. Therefore, in this study 'cousin' stands for the kin term whose focal denotata are parents' opposite-sex siblings' children.)

Kumbaingeri. As in Kariera-type systems, two lines of descent are recognized; the system of kin classification differs little from Kariera-type systems. The marriage rule, however, is that a man marries the daugh-

ter of his father's classificatory 'sister' or the
daughter of his mother's classificatory 'brother',
that is, he marries a classificatory 'cousin'; marriage
between MBD and FZS or FZD and MBS is prohibited. Consequently, WM is not terminologically identified with
FZ.

Aranda. The number of lines of descent recognized in
these systems is either three or four, four in "fully
developed systems of the Aranda type"; MMB and FF are
designated by different terms and, in "fully developed"
systems of this type, so are FMB and MF. The "standard"
marriage is between the offspring of female cross cousins, for example, between a man and his MMBDD or his
MFZDD.

Karadjeri. These systems recognize three lines of descent: MMB and FF are designated by different terms
but FMB and MF are not. This results from the rule
that a man may marry a MBD but not a FZD and from the
correlative prohibition of sister-exchange (which is
permitted in all other systems).

Yaralde. These systems distinguish four lines of descent, those of FF, MF, MMB, and FMB. They differ from
Aranda-type systems, however, in that all members of
the MMB's and FMB's lines are terminologically identified with the grandkin (and their siblings) who initi-

ate those lines. (In Aranda-type systems, intergenerational terminological identifications occur only between alternate generations, e.g., MBS but not MB may
be designated by the same term as MF.) Correlated with
this mode of extension of the grandkin terms is the
rule that a man may not marry either a first or second
cross cousin (i.e., he may not marry MBD or FZD or the
daughters of his mother's female cross cousins). A man
must take his wife from some clan other than that of
FF, MF, MMB, or FMB.

In attempting to evaluate this typology (which was
not supposed to be exhaustive of the range of empirical
variation in Australian systems of kin classification),
we should first consider what Radcliffe-Brown meant by
"lines of descent" and how he viewed the relationship
between rules of interkin marriage and systems of kin
classification.

Descent lines

Barnes (1967: 31) observes that this is "probably
the least satisfactory of all the concepts introduced
into anthropology by Radcliffe-Brown" and points out
that, although he used the expression from at least
1918 onward, he did not provide a definition until 1951.
He then described a line of descent as "an arrangement
of the relatives of an individual which can be shown on
a chart of kinship terms" and added "there are always,
of course" patrilineal and matrilineal lines (1951: 43).

A few pages later (p. 46) he added, "The use of the idea of lines of descent is merely to facilitate descriptive analysis," and a few years later (1956: 364) he stated, "It has been my custom to speak of 'lines of descent' in reference to tables of kinship terms, the 'lines' of course being my own arrangement." It is doubtful, however, that Radcliffe-Brown intended to disavow the real-world structural significance of his posited patrilineal and matrilineal descent lines. His argument (1951: 49), that description of the Murngin and Yir Yoront systems in terms of three patrilineal and three matrilineal lines is "more correct" than description of them in terms of five or seven patrilineal lines, shows that he regarded these lines as much more than purely hypothetical constructs (cf. Barnes 1967: 34).

In most contexts where Radcliffe-Brown wrote of "patrilineal descent lines" it is fairly clear that he referred to egocentric kin classes, each composed of several lower-order kin classes (such as 'father's father', 'father', 'brother', 'son', and 'son's son', to mention only the male categories in ego's own line). He posited these conceptual entities because of his conviction that the similarities and differences among Australian "kinship systems" rest in large part on the universality of the patrilineal clan-horde. This may or may not be an accurate view of Australian local organization,[2] but, in any event, the patrilineal rather

than the <u>local</u> nature of such groups was the important
element in Radcliffe-Brown's theory. If we are to understand the "many different forms of social organization in Australia . . . as different varieties of a
single general type" (1930-31: 35), he argued, we have
first to locate one or more constant elements, and
this, for Radcliffe-Brown, was the element of patrilineality. Therefore he found it necessary to disagree
with those who described Dieri society as "matrilineal"
(1951: 40) and to show that the system of kin classification of the Wongaibon (who were organized in matrilineal totemic clans) could be analyzed into a number
of patrilineal descent lines, although he had no direct
evidence that the Wongaibon were organized in patrilineal clan-hordes (1923: 428).

Radcliffe-Brown argued (1930-31: 438-9, 446; 1952
[1941]: 83) that there is a tendency throughout Australia for patrilineal clans to be solidary social
units and for outsiders to treat them as undifferentiated wholes. A person tends to treat all the members
of each patrilineal group to which he is genealogically
related in much the same way as he treats his closest
relatives in that group and, conversely, all of them
tend to relate socially to him in much the same way.
Radcliffe-Brown assumed that this provides the social
motivation for extending the terms by which a person
designates his closest relatives in particular groups
to all other members of those groups; and of course the

genealogical criterion governing these extensions is
patrilineal or agnatic relationship to the primary
denotata of the terms (or so he argued). This tendency, he said, receives its fullest expression in
systems of kin classification of the Yaralde type.
However, in many Australian systems of kin classification not all agnatic descendants of one's MMB, for
example, are designated by the same term or terms as
MM and MMB; instead the agnatic descendants of MMB
are divided into two categories, those of MMB's generation and generations alternate to it, and those
of generations adjacent to these (G+2, G-2, and G=
versus G+1 and G-1). Radcliffe-Brown attributed
this division to the additional principle of "the
equivalence of alternate generations" and to the
correlative principle of the opposition of adjacent
generations (1930-31: 432-3). Also, the maximum
number of potentially discriminable types of grandparents is only four, but in many Australian systems
there are only two or three grandparent terms; this,
he argued, is because the form of the marriage rule
determines that some grandparent types (and their
siblings) will be "socially" and therefore terminologically equivalent to others (see below for further
discussion of this point).

Radcliffe-Brown allowed that in some instances
the number of descent lines recognized in a system
of kin classification might be the same as the number

of descent groups or categories of which some other system of classification is composed (but not the same as the number of patrilineal clans constituting the whole society). A simple and obvious example is a Kariera-type system of kin classification associated with a rule of cross-cousin marriage and a set of exogamous moieties. The system of kin classification recognizes two lines of descent in relation to any ego, and the moiety system divides all members of the society into two mutually exclusive patrilineal (or matrilineal) sets or categories. In a situation like this, any person and his or her same-generation patrilineal relatives denote all their relatives in either moiety by the same terms (except where considerations of relative age or relative sex may be relevant), or at least they would do so if the marriage rule were always followed. Such arrangements have led some observers to argue that the system of kin classification, so-called, is better understood as a system of "social classification" (the alleged kinship terms do not designate egocentric genealogical categories but, instead, relative positions in the system of moieties); and they have led other observers to argue that the sole rule of kin-term extension in an instance of this kind is moiety affiliation. In one way or another, according to these interpretations, the classification of kin is based on the moiety system. But Radcliffe-Brown thought otherwise.

He argued that the structural principles of a
Kariera-type marriage system, "if logically and consistently carried out, must inevitably result in the
formation of moieties" (1930-31: 440). "But," he added,
"the classificatory system of terminology does not and
cannot by itself produce any system of social segments
as absolute divisions of the society." In other words,
as a consequence of certain rules of kin classification
(among which he did not include moiety, section, or
subsection affiliation), it may so happen that certain
sets of individuals always denote other sets of individuals by the same kin terms, but this coincidence may
or may not be recognized by the people themselves. If
it is, they may reify or objectify the two sets and
their interrelationships, and the result would be a
moiety system. This moiety system would be a part of
the larger "kinship system" of the society, but it
would be based on and derived from the system of kin
classification, rather than the underlying basis of
that system. If the coincidence is not recognized, the
division of everyone's relatives into two mutually exclusive sets remains structurally insignificant; the
division remains merely implicit in the system of kin
classification and is not a structural principle of
that system or of the larger "kinship system" of the
society.

Unfortunately, Radcliffe-Brown often obscured this
argument by describing the coincidental division of

everyone's relatives into two (or four or eight) mutually exclusive sets as a moiety (or section or subsection) system. In his 1914 discussion of the Dieri case
he said, "the system of relationship proves the existence of four matrimonial classes in the Dieri tribe"
(1914: 56). But he did not claim that the Dieri system
of kin classification is based on a section system,
which the Dieri do not have. Probably all he meant to
imply was, again, that moiety, section, and subsection
systems are reifications or objectifications of mutually exclusive sets of relatives, which sets are themselves the products of certain rules of kin classification, and these mutually exclusive sets must exist
(given those rules of kin classification) even where
they have not been objectified and given proper names.

In this light, Radcliffe-Brown's seemingly cryptic
remarks about descent lines in his 1951 and 1956 papers
are not difficult to understand. Warner had diagrammed
the Murngin system of kin classification by means of
seven patrilineal descent lines. Lawrence and Murdock
(1949) had argued that these diagrammatic lines are
best understood as representing the number (or number
of types) of patrilineal descent groups ("patrilines")
into which the Murngin as a whole are divided, rather
than the number of descent lines terminologically distinguished within any person's kindred; and they had
argued that the real number is eight, not seven. As
Radcliffe-Brown (1951: 52) put it, they had inferred

the existence of a "semi-semi-moiety" system. He re-
plied that these "'patrilines' as social divisions were,
of course, purely imaginary" (1956: 364). That is, the
Murngin as a whole are not divided into eight patrilineal descent groups or categories. He argued also that
the Murngin system of kin classification recognizes
three "basic" patrilineal descent lines, but he emphasized that this was not to imply the Murngin as a whole
are divided into three patrilineal groups or categories;
the three lines in his diagram represent only the number of "basic" descent lines in the Murngin system of
kin classification, that is three agnatically constituted egocentric categories. His argument was not that
the concept of descent lines is a mere diagrammatic
convenience and nothing more, but that some observers
persisted in confusing the number of descent lines in a
diagram (and in a system of kin classification) with
the number (or number of types) of descent groups in a
society (see also Leach 1961 [1951]: 57) for much the
same point). He was objecting, in effect, to persist-
ent but ill-advised attempts to reduce "kinship systems"
to "descent systems" (cf. Scheffler 1973: 756-62), although his own language often facilitated that confusion.

In brief, for Radcliffe-Brown, patrilineal descent
lines were major kin classes, each composed of a variable number of lower-order kin categories. He argued
that all Australian systems of kin classification fea-

ture at least two or more such major kin classes and
this provides the principal basis for comparison of
them. As Radcliffe-Brown saw it, the exact number of
such lines "recognized" (or posited) in any system depends in large part on the nature of the rule of interkin marriage associated with that system. It is not
dependent on, but in fact varies widely in relation to,
social forms such as moiety, section, subsection systems and the like (cf. 1951: 41-2).

Marriage rules

Radcliffe-Brown often observed that particular
systems of kin classification are "based on" or "require" particular rules of interkin marriage. The Kariera system, for example, requires the rule of cross-
cousin marriage. In this context, "the rule of cross-
cousin marriage" does not refer to a rule that a man
must marry a woman (any woman) whom he designates as
'cousin' and may not marry a woman of any other kin
class (1913: 155; 1930-31: 48). It refers instead to
the rule that each man has a rightful first claim to
one of his cross cousins (MBD or FZD) as a wife for
himself (1913: 156; 1951: 41). Consistent with his
assumption that, in general, kinship-terminological
extensions are socially motivated and are based on or
follow jural-status extensions, Radcliffe-Brown assumed
that this marriage rule provides one of the jural bases
for extending kinship terms from their primary denotata

to more distant kin and relatives by marriage. In other words, he treated this marriage rule as, in effect,
a rule of kin-class expansion.

As noted above, Radcliffe-Brown argued that Australian systems of kin classification may be divided
into a number of types according to the number of lines
of descent that they discriminate terminologically. In
each case the form of the marriage rule determines
which ones of the potential four grandparental lines
are terminologically distinguished or merged (and thereby not "recognized" as distinct lines). It does this
by determining whether or not MMB is assimilated to the
jural status of FF, and whether or not FMB is assimilated to the jural status of MF. If ego's MMB is attributed the jural status of ego's FF (or one significantly similar to it), MMB is designated as 'father's
father'; if he is not attributed that jural status he
is designated by another term. Similarly, if ego's FMB
is attributed the jural status of ego's MF (or one significantly similar to it), FMB is designated as 'mother's father'; if he is not attributed that jural status
he is designated by another term.

In the Kariera case, Radcliffe-Brown argued, MMB
is attributed the jural and, therefore, the terminological status of FF; and FMB is attributed the jural and,
therefore, the terminological status of MF. These at-
tributions follow from a man's entitlement to claim a
MBD or FZD as his wife. Consistent with this right,

any person's MB is his or her potential or rightful
father-in-law; and, presumably, for the purpose of
reckoning the kin-class status of any relative's MB,
that MB is treated as though he were that relative's
father-in-law (whether or not he actually is). Therefore, MMB → (is structurally equivalent to) MHF or FF,
and FMB → FWF or MF (for FW → M and MH → F in the
Kariera system). It follows - or so Radcliffe-Brown
argued - that the agnatic descendants of ego's MMB are
terminologically identified with the agnatic descendants of ego's FF, and the agnatic descendants of ego's
FMB are terminologically identified with the agnatic
descendants of ego's MF. Similarly, the two lines of
descent of ego's spouse are jurally and terminologically identified with those of ego himself or herself
(WFF → MF, WMF → FF). The overall result is that the
Kariera system of kin classification "recognizes" only
two lines of descent - or so Radcliffe-Brown argued.
One consequence of all this is that certain more distant female relatives in ego's own generation - such
as MMBDD and FMBSD - are classified as though they were
ego's cross cousins, and these female relatives also
are marriageable. Indeed, they constitute the kin category (classificatory 'cousins') from which ego may
claim a wife, failing the availability of an actual
cross cousin (1913: 156; 1951: 41).

As noted above, Radcliffe-Brown (1918: 224; 1923:
436; 1951: 44) distinguished between "fully developed"

and not-fully-developed Aranda-type systems of kin classification. In the fully developed or "normal" or "typical" variety, four lines of descent are recognized. This follows, he argued, from the rule that a man's rightful spouse is the daughter of his mother's female cross cousin, and not his own cross cousin (1951: 42). Because a person's MB is not that person's rightful potential father-in-law, MMB is not attributed the jural and terminological statuses of FF. The marriage rule, however, still functions as a rule of kin-class expansion. It implies, for example, that the potentially discriminable lines of descent originating at higher generational levels and the four lines of descent of ego's spouse are jurally and terminologically assimilated to those originating in ego's four grandparents. Therefore, four lines of descent are recognized.

It is not clear, however, that Radcliffe-Brown regarded Kariera- and Aranda-type marriage rules as necessary and sufficient conditions for the recognition of two and four lines of descent respectively. In 1923 he reported on the Murawari and Wongaibon systems of kin classification, in which MMB and FF are not terminologically distinguished, and neither are FMB and MF, although in both societies the marriage rule is of the Aranda type (1923: 428, 431, 439). On the basis of the presence of this marriage rule, and the presence of some Aranda-like terminological equations and distinc-

tions in the first ascending and descending generations
and in ego's generation, he described these as Aranda-
type systems of kin classification, but he noted also
that they are "not as fully developed as the more typical systems" of that type (1923: 436). Earlier (1918:
224) he had written of "completely developed" Aranda-
type systems, referring to those systems in which four
lines of descent are recognized, but without specifying
any in which fewer are recognized.

Radcliffe-Brown nowhere elaborated on this distinction or explained in full what he meant by it. A
plausible interpretation, however, is that he assumed
that certain patterns of kin classification constitute
the logically fullest, most complete, or most consistent expressions of certain rules of interkin marriage.
He regarded the Kariera and Aranda systems themselves
(or the ethnographic data about them) as good examples
of the most complete realizations of the logical implications (but not logically necessary implications) of
the cross-cousin and second-cross-cousin marriage rules,
respectively (cf. 1930-31: 46-52). From this point of
view, the Murawari and Wongaibon systems may be regarded as "essentially" Aranda-type systems, because they
are associated with Aranda-type marriage rules, and
because they feature some of the terminological equations and distinctions that are characteristic of other
systems allegedly based on the same marriage rule, but
which are not found in Kariera-type systems (that is,

systems allegedly based on the rule of cross-cousin marriage). But Radcliffe-Brown was neither consistent nor clear about what he regarded as the "essential" features of an Aranda-type system of kin classification. In 1913 (p. 191) he stated, "the characteristic feature" of Aranda-type systems is that the term for MMB is different from that for FF (and in many cases it is the same as that for MM). But in 1923 (p. 427) he stated, "in essential features" the Wongaibon system "is similar to the systems of the Aranda and Dieri tribes," although in the Wongaibon system the term for MMB is not different from that for FF. And, finally, in 1951 (p. 44) he stated, "in tribes with marriage systems of the Aranda type it is essential to distinguish MMB and FF, since the daughter of the former is a possible mother-in-law, which the FFD is not."

Now, of course, if the Aranda-type marriage rule does not necessarily entail a "fully developed" Aranda-type system of kin classification, additional factors must be posited to account for the differences between fully and not-fully-developed Aranda-type systems. Radcliffe-Brown did not specify what these factors might be. Therefore, it is fair to say he did not account formally for much of the variation in Australian systems of kin classification. The formal limitations of Radcliffe-Brown's theory of the structures and the relations among the structures of Australian systems of kin classification are even more evident when

we consider some of the other types he distinguished,
although some of his most profound insights into these
structures are also evident in his interpretation of
one of these three other types, the Karadjeri type (in
which he included the Yir Yoront and Murngin systems).

Evaluation

Kumbaingeri-type systems. In his initial description
of Kumbaingeri-type systems, Radcliffe-Brown stated
"the classification of kindred is like that of the
Kariera type into two lines of descent," and "a man
marries the daughter of a man who is classified as
'mother's brother', but he may not marry the child of
a near 'mother's brother' or of a near 'father's sister'" (1930-31: 52). Later he stated "the classification of kin is to some extent carried out on the same
general principles as in the Kariera type" (1930-31:
326, emphasis added). Unfortunately, he never published his own data on the Kumbaingeri system itself
and, so far as I have been able to determine, there is
no analytically useful body of published data on this
system (cf. Smythe 1948). Even so, Radcliffe-Brown's
description gives no reasons to suppose that it differs
significantly from his Kariera-type systems in general.
The only distinguishing feature he ever noted is
this. In the Kumbaingeri system a man divides his
father's 'sisters' (that is the women classified as

'sister' by or in relation to his father) into two
kinds, "the genealogical near relatives whose daughters
he may not marry and whom he therefore does not have to
avoid, and those who are his possible mothers-in-law,
and whom he must avoid; and they are distinguished by
different terms. Each man or boy has his own set of
'mothers-in-law'" (Radcliffe-Brown 1951: 44). In
apparent contrast, in many systems of the so-called
Kariera type there is a single term for male and female
ego's FZ and it is extended to all kinswomen classified
as 'sister' by or in relation to ego's father. Ego's
spouse's mother also is denoted by this term, although
in some systems WM may be distinguished from other
kinswomen of the 'father's sister' class by the addition
of a suffix to the designation for that class (for some
examples see Chapter 4). Presumably, in the Kumbaingeri system a woman designates all her father's 'sisters'
by a single term. Her brother, however, designates
some of these women - those of his father's 'distant'
classificatory 'sisters' whom he regards as "possible"
mothers-in-law - by a different expression that he also
uses to denote his wife's mother. The remainder of his
father's 'sisters', including the 'distant' ones whom
he does not regard as "possible" mothers-in-law, he
designates as 'father's sister', the same as his sister
does. Apparently, the expression used by a man to
denote his WM and his father's 'distant sisters' whom
he regards as "possible" mothers-in-law is not merely

the designation for the father's sister class augmented
by a suffix.

To understand all this it is essential to know
that in many Australian societies the women who are
eventually to become a man's mothers-in-law are chosen
(by him or by those of his elder kinsmen who have the
right to do this) well in advance of his marriage.
Typically the choice is limited to women of a particular kin class, often his father's 'sisters' or a specified subset of them such as his father's 'distant'
classificatory 'sisters'. The set of kinswomen quali-
fied, genealogically or otherwise, to become a man's
mothers-in-law may be described as the set of his
potential WMs. Of course, a man may have a fairly
large number of potential WMs. But typically for each
man a few such women are singled out by some more or
less formal means to be regarded as his prospective
WMs. Such a woman is obliged to give him one or more
of her daughters to be his wife and he is obliged to
treat this woman more or less as if he were already
married to her daughter. He must ritually avoid her
and send gifts of food to her, and perhaps to one or
more of her brothers, his prospective WMBs. Again
typically, a man's prospective WMs are terminologically
distinguished from other members of their kin class.
If that is the class of women classified as 'sister' by
or in relation to his father, the designation for the
prospective WM class may be a lexically marked form of

the FZ term (that is, 'father's sister' plus a special
suffix) or it may be a wholly different expression. In
some systems, it seems, all of a man's potential WMs
(not only his prospective WMs) are terminologically
distinguished from the other members of their kin
class.[3]

In this light, the Kumbaingeri arrangement described by Radcliffe-Brown is easily understood by
supposing, as he must have realized, that the 'mother-
in-law' class is related to the expanded 'father's
sister' class by class inclusion. In this system, as
in Kariera-like systems in general, FZ is designated
by a distinct expression that is extended to all women
whom ego's father classifies as 'sister'. (It will be
convenient to represent this and other expanded kin
classes by means of upper-case letters. Henceforth
FATHER'S SISTER represents the class of all kintypes
denoted by the expression 'father's sister'.) But not
all members of the FATHER'S SISTER class are designated
only as 'father's sister'. Some of them may be designated also as 'prospective wife's mother'.[4] The women
so designated are those of a man's father's 'distant'
classificatory 'sisters' who have been singled out as
his prospective WMs. All members of the 'wife's-mother'
class are members also of the FATHER'S SISTER class,
but not vice versa.

Another way to describe this arrangement is to say
that the FATHER'S SISTER class is a superclass, that is,

a generic, higher-order, or more inclusive class, which consists of several subclasses.[5] One of these subclasses is the specially designated 'wife's mother' class; the members of this class include a man's prospective WMs. Another subclass of the FATHER'S SISTER superclass is the residual class composed of the kintypes left over from the FATHER'S SISTER class when the prospective WMs are singled out for special designation. The members of this residual class are designated as 'father's sister'; the members of the special prospective wife's mother class may be designated either as 'father's sister' or as 'wife's mother'. Because typically a man may not speak directly to his own or prospective WM, he does not address her by either term; but if he speaks about her he uses the expression 'wife's mother'. Presumably, other people speak of her as his 'father's sister' or as his 'wife's mother', depending on whether or not they wish to indicate that he is one of her sons-in-law or prospective sons-in-law.

One aspect of the structure of the FATHER'S SISTER class of the Kumbaingeri system - its structurally derivative nature - should be emphasized. Taxonomically, both the special wife's mother subclass and the residual subclass are derived from the FATHER'S SISTER class. One way to define the wife's mother class is to say that it consists of those members of the FATHER'S SISTER class who are in addition related to ego as prospective WM. The complementary residual subclass

may then be defined as consisting of those members of
the FATHER'S SISTER class who are <u>not</u> related to ego
as prospective WM. Both subclasses are derived from
the FATHER'S SISTER class because both are based on and
definable in terms of it; the FATHER'S SISTER class is
logically prior to either of the subclasses. But, of
course, the FATHER'S SISTER class itself is structurally derivative. It is not derived from either of the
subclasses already noted but from the kintype FZ, which
is the focal or structurally primary denotatum of the
expression 'father's sister'. The FATHER'S SISTER
class is generated from the FZ class by means of various rules of genealogical structural equivalence, which
determine that certain other kintypes also are to be
designated as 'father's sister'. The class consisting
of ego's FZ alone may be regarded as yet another subclass of the FATHER'S SISTER class, but it occupies a
rather special position as the focal class on which the
superclass as a whole, including its specially designated prospective wife's mother subclass, is based.

Because there is some confusion on the matter in
the anthropological literature (see for example Schneider 1968; Schneider and Smith 1973), it is important to
note that the most general sense of a polysemous expression is not necessarily its structurally central or
primary sense. As noted above, the superclass FATHER'S
SISTER is structurally prior to its two subclasses
distinguished by the opposition "prospective-WM versus

not-prospective-WM." Therefore, the most general (un-restricted) sense of 'father's sister' is logically or structurally prior to the restricted sense the term has ("FATHER'S SISTER and not-prospective-WM") when it is opposed to 'wife's mother' ("FATHER'S SISTER and pro-spective-WM"). But again, the FATHER'S SISTER super-class is not the primary designatum of 'father's sis-ter'. This superclass is itself a structurally deriva-tive class based on the kintype FZ, which is the primary designatum of 'father's sister'.

With this feature of the Kumbaingeri system now understood in a general theoretical context, we may note that, superficial appearances to the contrary, the arrangement in the Kariera system itself is not signif-icantly different. Here too a man designates his pro-spective WMs by the same term as he designates his WM, and this is quite different from the expression by which he designates his FZ and his father's classificatory 'sisters' (see Radcliffe-Brown 1913: 149). But, again, a man's prospective WMs are members of the FATHER's SISTER class. The difference is that a Kariera man may marry the daughters of any of his father's 'sisters', own, near, or distant; in the Kariera system any woman of the FATHER'S SISTER class qualifies for inclusion in the specially designated prospective WM subclass of the FATHER'S SISTER superclass. In both systems, then, the FATHER'S SISTER class is a superclass that focuses on the kintype FZ, and the other kintypes included in this

superclass are the same in both systems. We may suppose that the rules whereby these types are reckoned as
structurally equivalent to FZ also are the same in both
systems. The two systems differ only in how they par-
tition this superclass into subclasses. That is, they
differ only in certain subsidiary taxonomic rules.
These rules are subsidiary because their application is
contingent on the prior application of the primary defining rules (the rules that establish the definitions
of the primary senses of the principal or basic terms
of the system) and on the equivalence rules whereby the
expanded senses of the terms are generated (cf. also
Scheffler and Lounsbury 1971: 105-7).

Therefore, the differences between the Kumbaingeri
and Kariera systems are structurally quite minor. As
Radcliffe-Brown said, they share the same "general
principles." So it seems appropriate to regard the
Kumbaingeri system (and others like it) as constituting a subtype of Kariera-type systems. This subtype
differs from other subtypes not in class foci, primary
sense definitions, or equivalence rules, but only in
certain subsidiary rules of subclassification.

This procedure was not available to Radcliffe-
Brown because he based his typology on differences in
rules of interkin marriage, which he assumed are among
the structurally most basic features of Australian systems of kin classification. As we have seen, he supposed that rules of interkin marriage are expressed in

Australian systems of kin classification as rules of
kin-class expansion. In discussions of systems of the
Kariera-type he always insisted they are "based on and
imply the existence of . . . cross-cousin marriage"
(1930-31: 46). This rule accounts (in part) for the
extensions of the terms from their structurally primary
denotata to other more distantly related kintypes and
to relatives by marriage, and it determines that two
(rather than three or four) lines of descent are recognized - or so Radcliffe-Brown argued. In the case of
the Kumbaingeri, however, this type of marriage is prohibited. Therefore, if MMB is designated as 'father's
father' and FMB is designated as 'mother's father',
this cannot be because they might legitimately be MHF
(FF) and FWF (MF) respectively. Also, if the kintypes
FMBS and FFZS are designated as 'mother's brother' (as
they are in the Kariera system), this cannot be because
they might legitimately be FWB or MB. In this case another rule or rules must govern the collateral and in-
law extensions of the terms. Therefore, from Radcliffe-
Brown's perspective, the Kumbaingeri system cannot be
regarded as an exemplar of a subtype of Kariera-type
systems. It must be regarded as an exemplar of another
quite different type of system.

As Radcliffe-Brown saw it, that other type consists
of systems that somehow structurally incorporate the
rule of marriage to the daughter of a 'distant' classificatory 'mother's brother' or the daughter of a 'dis-

tant' classificatory 'father's sister'. It should be relatively obvious, however, that the collateral and in-law extensions of the terms of the Kumbaingeri system are not governed by such a rule. The marriage rule is specified in terms of and therefore presupposes the existence of the MOTHER'S BROTHER and FATHER'S SISTER classes, whose kintype compositions are (it seems) precisely the same as those of the corresponding categories of the Kariera system. The marriage rule does not govern inclusion in these categories. So the only feature of the Kumbaingeri system that is explainable by reference to the marriage rule is the composition of the specially designated prospective WM subclass of the FATHER'S SISTER class. The existence of such a subclass is not so explainable because similar subclasses, with slightly different compositions, occur also in systems of the so-called Kariera type such as the Kariera system itself.

In short, if the Kariera marriage rule is expressed in the Kariera system of kin classification in the form of a rule of kin-class expansion, the Kumbaingeri marriage rule is not so expressed in the Kumbaingeri system of kin classification. Radcliffe-Brown therefore failed to provide an adequate account of the structure of the Kumbaingeri system and of the similarities and differences between it and the Kariera system. These failures may be traced directly to his virtually unquestioned assumption that rules of interkin marriage are basic structural features of Australian systems of kin classification, wherein they take the form of rules of kin-class expansion. This assumption is brought seriously into question when we recognize that, aside from some structurally quite minor variation at the subclass level, the kintype compositions of the categories of the Kumbaingeri system and those of Kariera-like systems in general are precisely the same, despite differences in marriage rules. Recognition of this requires that we consider the possibility that, in their basic defining rules and in their rules of kin-class expansion, Radcliffe-Brown's Kumbaingeri- and Kariera-type systems are structurally identical. If they are, it must be that a rule of cross-cousin marriage is **not** one of their rules of kin-class expansion.

This conclusion cannot be avoided by supposing that the extensive similarities between Radcliffe-Brown's Kumbaingeri- and Kariera-type systems are merely fortuitous and do not reflect an underlying structural identity. To so argue and at the same time salvage Radcliffe-Brown's interpretation of the structure of his Kariera-type systems, we would have to suppose that the same overall pattern of kin classification (differences in subclassification aside) may be generated by two or more somewhat different sets of equivalence rules. We would have to suppose that where this pattern is found in association with a rule of cross-cousin marriage it is generated by a rule of kin-class expansion that ex-

presses this marriage rule; but where it is not associated with such a marriage rule it is generated by some
other rule or rules. It is shown in Chapter 4 (see
also Scheffler 1971b) that it _is_ possible to account
for the collateral and in-law extensions of the terms
of Kariera-like systems without positing a cross-cousin
marriage rule of kin-class expansion. However, in comparative perspective it is at best unparsimonious to
posit two different models to account for one and the
same overall pattern of kin classification. Considerations of parsimony require that we reject this hypothesis and suppose instead that the combination of a
Kariera-like system and a cross-cousin marriage rule
is merely fortuitous; this accidental association of
two variables in some instances has led some observers
to the fallacious conclusion that there is a causal or
structural relationship between them.

Karadjeri-type systems. Although Radcliffe-Brown
failed to appreciate the implications of the status of
the prospective WM classes of Kumbaingeri- and Kariera-
type systems as subclasses of their FATHER'SISTER
classes, he was not unaware of the existence of superclass-subclass relationships in Australian systems of
kin classification. Indeed, in his discussions of
Karadjeri-type systems he made exceptionally effective
use of this concept and was thereby able to demonstrate
the existence of significant structural similarities

among several systems that are superficially quite
different. He used this demonstration to support his
theory that the major source of structural variation
in Australian systems of kin classification is variation in the rules of interkin marriage associated with
them. But he stopped short of pursuing the evidence
for superclass-subclass relationships as far as he
should have. In the Karadjeri, Yir Yoront, and Murngin
systems, for example, there are many more such intercategory relationships than Radcliffe-Brown realized.
Their nature again casts serious doubt on the adequacy
of his interpretation of the structures of these systems.

Radcliffe-Brown's initial discussion of the Karadjeri system was quite brief, presumably because Elkin
had only recently studied it and had yet to publish his
data. He noted only that it "is a system based on or
implying a marriage rule whereby a man marries his MBD,
but may not marry his FZD" (1930-31: 353). He added,
"the system of terms differs from the Kariera type in
that FZ is distinguished from MBW, the former being
called by the same term as the father, while the latter
is actual or possible wife's mother" (1930-31: 341).
He did not state whether or not this is the only difference. In 1951 he added that among the Karadjeri a
man may not marry the daughter of any woman his father
designates as 'sister'; thus a man may marry a woman
of his mother's patrilineal clan, but he may not marry

the daughter of a woman of his own clan (1951: 42). In addition to the Karadjeri, he mentioned the Murngin, Yir Yoront, and Larakia as examples of "tribes" with "kinship systems" of the Karadjeri type. He had previously (1930-31: 53) described the Murngin system as "much more complicated" than the Karadjeri system because it recognizes "seven lines of descent," although he may have regarded it as a "variety" of the Karadjeri type, because it is "based on . . . matrilateral cross-cousin marriage" (but cf. 1930-31: 450, where he describes them as different types).

Radcliffe-Brown nowhere analyzed the Karadjeri system itself, but in his 1951 paper he dealt with the Yir Yoront and Murngin systems in some detail and described both as "based on three 'basic' patrilineal lines" (FF's, MF's, MMB's) and on the rule of matrilateral cross-cousin marriage. "The marriage system," he said, "requires that MMB should be distinguished from FF, but also requires that MF and FMB should be classified together, since a man's MF may actually be his FMB" (1951: 48).

The concept of "basic" as opposed to nonbasic or "secondary" lines of descent was introduced to deal with questions that had been raised by Sharp's (1934) five-line diagram of the Yir Yoront system and Warner's (1930, 1958 [1937]) seven-line diagram of the Murngin system. Radcliffe-Brown noted that in constructing diagrammatic representations of these systems the

ethnographers had, quite justifiably, made use of as
many "lines" as appeared necessary to fit all the terms
onto such diagrams (1951: 46-7). But, he argued, it is
a mistake to treat all such lines analytically as
though they were structurally on the same level. He
noted that some of these diagrammatic lines contain
only the reciprocals of terms found in other lines; any
two such lines logically imply one another and therefore must be regarded analytically as one line rather
than as two (see also Sharp 1934: 412). Further, some
of the lines are not terminologically distinct from
others; some diagrammatic lines share some of their
terms with other such lines. The implication appears
to be that such lines are not structurally distinct but
are related in some way, and this relationship must be
brought out through analysis. Radcliffe-Brown proposed
to explicate this relationship in terms of the difference between "basic" and "secondary" lines of descent.

"Basic" lines of descent, he argued, "are those
that have their origins in relatives of the second
ascending generation" and, therefore, "the first question that has to be asked in analyzing any Australian
system of kin classification is . . . how many kinds
of relatives does a man recognize in the generation of
his grandparents" (1951: 43). In the Yir Yoront system
(see Figure 8.1) there are only three: FF, MF (= FMB),
and MMB. (The other two lines in Sharp's diagram contain only the reciprocals of the terms of the MF and

MMB lines.) At first glance there appear to be four in
the Murngin system: FF, MF (= FMB), MMB, and WMMB (or
MMMBS) (see Figure 8.3). But, Radcliffe-Brown noted,
the designation for WMMB natchiwalker is morphologically derived (according to Warner) from the designation
for MF nati; presumably, more accurate orthographic
representations of the terms would be natji and natji-
walker (1951: 49). To account for this Radcliffe-Brown
introduced the concept of class inclusion or, in other
words, the concept of superclass-subclass relations.

He suggested that in the Murngin system the expression nati is extended from MF to WMMB, just as pa'a is
extended in the Yir Yoront system. But "the Murngin
distinguish certain relatives by marriage by special
terms within a more general category or class" (1951:
49). In this case the special term is natchiwalker
(nati + walker, a diminutive suffix) and it designates
"a special kind of nati," a classificatory 'mother's
father' who is potentially or actually one's WMMB. In
other words, natchiwalker designates a special subclass
of the MOTHER'S FATHER class (the expanded 'mother's
father' class), a superclass whose focus is MF and whose
designation is nati, the same as the designation of the
MF class.

Therefore, the question "how many kinds of relatives does a man recognize in the generation of his
grandparents?" may be answered in two ways. If we
attend only to the number of terms at this level and

disregard that morphologically one of them is derived from another, we may say there are four terms and therefore four categories. But if we take this morphological relationship into account and recognize that it indicates the presence of a lesser number of higher-order categories, we may say there are only three categories, one of which is a superclass consisting of two subclasses. (This was Radcliffe-Brown's conclusion, but it is not wholly accurate; see below.) While it may be that WMMB (or MMMBS) is the focus of a terminological class, this relationship has a different structural status than the other class foci, FF, MF, and MMB. It is the focus of a derivative class (a subclass) and not the focus of a more basic or principal class.

Radcliffe-Brown recognized, at least in this context, that intersystem differences in subclassification are superficial and may serve to mask and obscure more fundamental intersystem similarities. He realized also that analyses and comparisons that do not take into account that some terminological classes are subclasses (and do not have the same structural status as other terminological classes) must necessarily lead to fallacious theories. It was not his intention to deny that there are structural differences between the Yir Yoront and Murngin systems but only to put them in their proper perspective - by showing that the differences are not at the level of basic member definitions and equivalence rules but at the structurally much less significant level of differentiation <u>within</u> "generic" classes or superclasses. By discounting (but not ignoring) this low-level and structurally superficial variation, he was led to conclude that the Murngin system, like the Yir Yoront, contains only three "basic" patrilines. These originate in ego's FF, MF, and MMB, who are the foci of classes at the same structural level. Insofar as the line originating in ego's WMMB (or MMMBS) is terminologically at least partially distinct from the line originating in ego's MF, it may be counted as a fourth line, but it has to be described as a structurally derivative or "secondary" line whose existence is dependent on the recognition of a special subclass of the MOTHER'S FATHER class.

Unfortunately, Radcliffe-Brown did not follow this sound line of reasoning as far as he should have. If we must, for the reasons he adduced, regard the WMMB line as a "secondary" line related to the MF line, there are indications that we must also regard the MMB line as a "secondary" line related to the FF line. If it is, Radcliffe-Brown could have argued that the Murngin system contains only two "basic" lines. But this in turn would have cast doubt on the validity of his claim that the system is based on a rule of MBD-FZS marriage.

Because FF <u>marikmo</u> and MMB <u>mari</u> are designated by different terms they may be said to belong to different kin classes, but these expressions appear to be morphologically related (indeed, they are - see Chapter 8)

and this suggests that the two categories are related
as subclasses of a higher-order class or superclass.
Further discussion of the evidence for the existence of
this superclass and its structure must be postponed to
Chapter 8, but it may be noted here that it is entirely
consistent with this hypothesis, and with the two-line
hypothesis, that MMBD, a member of the MMB line, is
designated as mokul numeru. This is the expression by
which a man designates his WM and his prospective WMs,
but it is based on the FZ term mokul. Again, as in the
Kumbaingeri and Kariera systems, the prospective WM
class is a specially designated subclass of the FATHER'S
SISTER class. In Radcliffe-Brown's terms, kinswomen of
the MMB line are designated by a term derived from the
FF line. Kinswomen of the posited FF's and MMB's lines
are not radically distinct, categorically, at this generational level, and this suggests that the two posited
lines also are not radically distinct. This arrangement
is decidedly inconsistent with the hypothesis that this
is a three-line system based on MBD-FZS marriage, for
in such a system the FATHER'S SISTER and prospective
wife's mother classes would be wholly distinct. The
difficulty presented for Radcliffe-Brown's theory is
much the same as the difficulty presented by the Kumbaingeri system. As a rule of kin-class expansion the
MBD-FZS marriage rule does not account for the extension
of 'father's sister' to MMBD, a potential WM. One of
the implications of this rule is that MMBD is to be

regarded as structurally equivalent to MBW, but the rule
does not imply that MBW is to be regarded as structurally equivalent to FZ. Yet MBW is reportedly designated
as mokul numeru, that is, as a special kind of classificatory 'father's sister' who is also a potential or
prospective WM. The MBD-FZS marriage rule therefore
fails to account for the terminological facts.

As for the type-specimen, the Karadjeri system
itself, Elkin (1964: 70, 71) observes that a "special
term" is used for "the more distant 'aunt' ['father's
sister'], who can be wife's mother," and a man's WM
"must stand in the right kinship relationship to him -
a kind of 'aunt'." Again, one of the terminological
facts to be accounted for is that WM, MBW, and MMBD are
members of the FATHER'S SISTER class, even though none
of them may legally be FZ. And, again, Radcliffe-
Brown's theory does not account for this. The theory
can be modified and made to account for it only by elim-
inating one of the central tenets of the theory, the
assumption that terminological extensions are governed
principally by rules of interkin marriage.

The foregoing observations about superclass-
subclass relations in some Kariera-, Kumbaingeri-, and
Karadjeri-type systems demonstrate that many of the
differences among systems of these types are structurally superficial. Superclasses with the same foci and
the same overall composition occur in systems of all
three types; the rules that generate these superclasses

must also be the same or at least very similar. These
expansion rules cannot differ in the ways posited by
Radcliffe-Brown. Most especially, it cannot be that
the differences between systems of these several types
are attributable principally to differences in marriage
rules of kin-class expansion. The appearance, to the
contrary, results largely if not entirely from the
division of certain expanded kin classes into subclasses
in diverse ways that are more or less consistent with
variations in the associated local marriage rules.

This conclusion may be extended to many of the
systems classified by Radcliffe-Brown as Aranda-type
systems (and by Elkin as Nyulnyul-type systems). As
noted above, Radcliffe-Brown allowed that some Aranda-
type systems have only three "basic" lines of descent.
Following the line of reasoning he developed in dealing
with the Murngin system, he could (and should) have
argued that the Murawari and Wongaibon systems have
only two "basic" lines (see Chapter 7). The same thing
may be said of the Nyulnyul system.

Elkin (1964: 73) begins his discussion of the Nyulnyul system by arguing that "the prohibition of all
cross-cousin marriage" results in the four families of
origin of ego's grandparents being terminologically
distinguished from one another. Later (p. 75) he says,
"the prohibition of marriage with both kinds of cross-
cousins has also resulted in the differentiation of the
'uncles' (MBs) and 'aunts' (FZs) who can be parents-in-

law, from those who cannot. Thus, own father's sister
and mother's brother are yurmor and kaga respectively,
whereas wife's mother and father are yala and kaga
djaminir." The clear implication is that here, too,
the potential or prospective WM class is a specially
designated subclass of the FATHER'S SISTER class, and
the potential or prospective WF class is a specially
designated subclass of the MOTHER'S BROTHER class. The
superclass-subclass relationship is transparent in the
case of MOTHER'S BROTHER and prospective WF because the
expression for the special subclass is a lexically
marked form of the expression for the superclass (as
also in the case of Murngin natchiwalker versus nati).
In contrast, the expression for the prospective WM
class is not morphologically derived from the expression
for the FATHER'S SISTER class, but this should not mis-
lead us into thinking that the relationship between the
classes themselves is different. Just as one of the
qualifications for inclusion in the 'wife's father'
class is prior inclusion in the MOTHER'S BROTHER class,
one of the qualifications for inclusion in the 'wife's
mother' class is prior inclusion in the FATHER'S SISTER
class. Therefore, although the latter two classes are
designated by radically different expressions, they are
not radically different classes.

Further evidence for the superclass-subclass relationship between 'wife's mother' and FATHER'S SISTER is
that another "special term, ramba" is used for WMB "to

distinguish him from an ordinary 'father'" (Elkin 1964:
75). That is, a relative designated as ramba would be
designated as 'father' were it not that he is not only
ego's classificatory 'father' but also ego's WMB or
prospective WMB. So the kinsmen designated as ramba
(WMB, MMBS, MFZS) constitute a specially designated
subclass of the FATHER class. (In the Kariera system
itself the only designation for relatives of these types
is 'father'.) Their sisters constitute the 'wife's
mother' class, and as sisters of ego's classificatory
'fathers' they must be ego's classificatory 'father's
sisters'.

It should be fairly clear that the differences
between this system and Kariera-like systems in general,
attributed by Elkin and Radcliffe-Brown to differences
in rules of interkin marriage, are only differences in
subclassification. The Nyulnyul system contains more
specially designated subclasses than does the Kariera
system and, of course, even where the two systems have
the same subclasses their compositions may be somewhat
different. It may well be that these differences are
attributable to differences in rules of interkin marriage, but there are also certain structural similarities for which Radcliffe-Brown's and Elkin's theory does
not and cannot account.[6]

Kukata. The limitations of Radcliffe-Brown's descent-
lines and marriage-rules model are even more evident

when we consider the Kukata system. As he noted (1951: 43), this system is "exceptional" in recognizing only one (rather than two, three, or four) kinds of male kin in the second ascending generation. It is therefore inappropriate to describe this system and the many others like it in terms of the recognition of any number of lines of descent. Further, although this system has distinct FZ and MB terms, the collateral extensions of these terms are not the same as in Kariera- and Kumbaingeri-type systems. Here, mother's cross cousins are classified as 'mother' and 'mother's brother', and father's cross cousins are classified as 'father' and 'father's sister', just the opposite of the arrangement in the Kariera and Kumbaingeri systems. Yet the marriage rule is similar to that Radcliffe-Brown described for his Kumbaingeri-type systems. Again, the marriage rule cannot regulate the collateral extensions of the terms. It merely makes use of preexisting categories and does not determine their composition. Finally, this system has no distinct cross-cousin terms; cross cousins and parallel cousins are designated as 'brother' or 'sister'. The women whom a man may marry (the daughters of his mother's classificatory 'brothers' and of his father's classificatory 'sisters') are his own classificatory 'sisters'! At best, this is seriously inconsistent with Radcliffe-Brown's empirical claim (which he elevated to the status of a principal tenet of analysis) that, by and large, relatives of like jural status are designated by the same term and, conversely, relatives of dissimilar jural status are designated by different terms.

We have seen that Radcliffe-Brown's theory does not do what he claimed for it. There is no direct correlation between the number and types of descent lines apparently recognized in a system and the type of rule of interkin marriage associated with that system. Systems associated with Aranda-type marriage rules variously recognize two, three, or four lines of descent (or in more descriptive terms two, three, or four categories of kin in the second ascending generation); and there is much variation of the same sort within Radcliffe-Brown's categories of Kariera-, Karadjeri-, and Kumbaingeri-type systems. Radcliffe-Brown did attempt to deal with this difficulty for his theory, principally by distinguishing between "fully developed" and other varieties of his several types (especially the Aranda type), and by introducing the distinction between "basic" and "secondary" lines. But he stopped far short of applying these distinctions as thoroughly as he could have. Only in his comments on the Murngin and Yir Yoront systems did he show that the apparent discrepancy between theory (matrilateral cross-cousin marriage rules produce or require three lines of descent) and "fact" (Warner had produced a seven-line diagram of the Murngin system and Sharp had produced a five-line diagram of the Yir Yoront system) was only apparent - because both

systems may be analyzed into only three "basic" lines.
This was a very important point, but its significance
for our understanding of the Murngin and Yir Yoront
systems, much less for other Australian systems of kin
classification, was not fully developed. Had it been
developed Radcliffe-Brown might have come much closer
to accounting (formally if not sociologically) for much
of the apparent diversity among these systems - largely
by showing that much of the diversity is only superficial, that beneath their surfaces (on which most of the
tangled theoretical discussion has focused) many such
systems and "types" of systems are structurally identical and differ only in modes of subclassification.

To do this, however, he would have had to give up
his reliance on the concepts of descent lines and interkin marriage rules as the main structural principles of
Australian systems of kin classification. But this he
was not prepared to do, for it would have required a
radical alteration of his perspective on the analysis
of kinship systems and systems of kin classification.

In 1941 (1952: 61-3) he contrasted his perspective
on the study of systems of kin classification with that
of A. L. Kroeber. Kroeber, he argued, denied that
"there are important correspondences between kinship
nomenclature and social practices" and was concerned
only to isolate and describe the genealogical principles
of which various systems of kin classification are constructed, this in order to discover and define the

"historical relations of peoples by comparison of their systems of nomenclature." He added, "my own conception is that the nomenclature of kinship is an intrinsic part of a kinship system, just as it is also, of course, an intrinsic part of a language. The relations between the nomenclature and the rest of the system are relations within an ordered whole. . . . In the actual study of kinship systems the nomenclature is of the utmost importance. It affords the best possible approach to the investigation and analysis of the kinship system as a whole." In other words, Radcliffe-Brown saw systems of kin classification as essentially social phenomena and as convenient points of entry into the study of kinship systems as wholes - meaning by "kinship system" the totality of social relations normatively ascribed between persons who regard themselves as related by birth or by marriage.

On the matter of how systems of kin classification are related to "the rest of the [kinship] system," Radcliffe-Brown assumed that terminological equations and distinctions are socially motivated. He took it as well established that, by and large, terminological equations and distinctions of kintypes mirror or reflect equivalence and nonequivalence of jural status with respect to ego; or, in other words, that one "common feature" of kinship systems is the division of anyone's relatives into a number of jural categories, and kinship terms are the linguistic means of "recognizing" the jural categories.

Radcliffe-Brown realized it would be a serious methodological error to assume that "important correspondences between kinship nomenclature and social practices" must exist in every case; "they must be demonstrated by field work and comparative analysis" (1952: 61). But in practice he did not treat the proposition - that, structurally, systems of kin classification reflect systems of jural classification - as a hypothesis, as an arrangement known (or at least reported) to exist in some cases and to be looked for as a possibility in others. Instead he treated it as a principal tenet of analysis and used it not only to analyze systems of kin classification but also to analyze "kinship systems" as wholes. In the abstract he realized that an adequate test of the proposition would require comparison of the results of rigorous, independent structural analyses of whole terminological and jural-status systems. But in practice he never did this. In practice he fused the two systems (the terminological and the jural) a priori into a single system that he assumed had to be ordered in a certain way (terminological equations and distinctions reflect jural-status equations and distinctions). He then formulated structural principles phrased in sociological language to account for terminological equations and distinctions, sometimes where he had little data on jural classification to work with. He thus claimed to provide sociological explanations of systems of kin classification. But the models he produced of Australian systems of kin classification, and of the structural relations among them, are not formally adequate and the posited relationships between terminological and jural structures do not hold up when we look beyond the few cases he dealt with in some detail.

Although I have been critical of Radcliffe-Brown's fusion of sociological and structural-semantic analysis, it is not my intention to suggest (as he supposed Kroeber had) that there are no "important correspondences" between the structures of systems of kin classification and the larger social structures within which they occur and are used. My point is only that no good can come of assuming a priori that such correspondences must exist (much less that they must be of a particular kind) and from using this assumption as an excuse to shortcut the analytic process. Radcliffe-Brown's concern to discover the nature of "the relations between nomenclature and the rest of the [kinship] system" (1952: 61-2) was not unrealistic, but his methods were unsound insofar as they did not require him to analyze systems of kin classification rigorously and indepdendently of the social contexts within which they are used. This in turn severely limited his ability truly to accomplish his own ends (see Scheffler and Lounsbury 1971: 63-5 et passim for further discussion of this point).

CONCLUSION

Although much more could be said about Radcliffe-
Brown's theory of the structures and relations among
the structures of Australian systems of kin classification, enough has been said to show that the theory is
formally and sociologically inadequate. This is not to
say that the theory as a whole is without value. Far
from it! The most central feature of the theory is its
wholly realistic emphasis on the polysemy of Australian
kinship terms, and it is the one global theory of these
systems that is at all realistic in this respect. Also,
unlike most other theorists, Radcliffe-Brown (and Elkin)
perceived to some extent the indications that much of
the intersystem variation is at the subclass rather
than the superclass level and is therefore structurally
superficial. But his concern to relate the intersystem
variation directly to similarities and differences in
social structure, and in particular to rules of interkin marriage, led him to stress terminological differences at the expense of similarities and not to analyze
fully the indications that many of the differences are
indeed fairly superficial.

To deal more adequately with the less superficial
and not so readily apparent structural variation, it
appears necessary to consider more fully than Radcliffe-
Brown did the possibility that rules of interkin marriage are not the typologically most critical structural
variables in Australian systems of kin classification.

In the following chapters I will analyze the available
data on a number of Australian systems of kin classification and attempt to construct simple, adequate formal
semantic models of those systems. I will show that
these systems can be understood and related to one another in terms of a fairly small stock of structural
elements that may be variously combined to yield a
fairly wide empirical variety of systems of kin classification. In this scheme, rules having to do with interkin marriage have a relatively subordinate typological status. Although I hope to show that Radcliffe-
Brown was correct in claiming that there are structural
continuities throughout the continent, I do not attempt
to exhaust the known empirical variety of Australian
systems of kin classification, and I do not claim that
all such systems may be regarded as "varieties" of a
single, general structural "type." Even so, like Radcliffe-Brown, I would claim that much of the remaining
variation can be assimilated to the general scheme presented in the chapters that follow.

Chapter 3

PITJANTJARA

The focus of this chapter is the system of kin classification of the Pitjantjara dialect of the Western
Desert language family. In this system cross cousins,
as well as parallel cousins, are classified as 'siblings'; father's parallel and cross cousins are classified as 'father' and 'father's sister'; and mother's
parallel and cross cousins are classified as 'mother'
and 'mother's brother'. Although systems with this
feature are not uncommon in Australia, especially in
the Western Desert, they have received relatively little
analytic attention from anthropologists other than
Elkin, who calls them "Aluridja-type" systems, and who
has noted some of the similarities between them and the
"Hawaiian-type" systems of Oceania (cf. Elkin 1939:
214-5).

The data analyzed here were collected by Nancy Munn
at Areyonga settlement in the Northern Territory in
1964-65 and concern the Pitjantjara dialect spoken in
and around the Peterman Range (Munn 1965). Additional

information on the Pitjantjara system of kin classification has been recorded by Elkin (1939, 1940), Love
(Ms.), and Soravia (1969). Annette Hamilton provided
some important unpublished data on the Pitjantjara
spoken at Everard Park station in South Australia.

THE PITJANTJARA SYSTEM

Munn's data (1965: 7-9) are presented in Table 3.1.
Unfortunately the sources on the Pitjantjara dialect
provide little information on the lexical means whereby
structurally primary and nonprimary denotata may be
distinguished, but an apparently relevant bit of data
on these means is reported by Glass and Hackett (1970:
63) for the Nangatatjara dialect. They give the usual
form yuntal-pa for 'daughter' and the form yuntal-pa
kutjupa-nya, which they gloss as "the other daughter."
They do not explain this gloss, but according to A.
Hamilton (personal communication) the expression kutjupa
itself may be glossed as 'another', and in a kinship
context it is opposed to mulyapa 'truly'; yuntal-pa
mulyapa-nya designates one's true daughter or female
offspring. Thus yuntal-pa kutjupa-nya 'another daughter' is equivalent to "classificatory daughter," that
is, a 'daughter' by extension. According to A. Hamilton, the same expression is used with other Pitjantjara
kinship terms to distinguish nonprimary from primary
denotata (cf. Elkin 1939: 343). However, in the absence
of systematic data on the use of these expressions in

Table 3.1. Pitjantjara kin classification

	I Term	II English gloss	III Primary denotata	IV Other denotata
1.	tjamu	grandkinsman	FF, MF, SS, DS	all G2 males
2.	pakali	grandson	SS, DS	all G-2 males
3.	kami	grand kinswoman	FM, MM, SD, DD	all G2 females
4.	puliri	granddaughter	SD, DD	all G-2 females
5.	mama	father	F	FB, F's male Co., MZH
6.	kamaru	mother's brother	MB	M's male Co., FZH
7.	nguntju	mother	M	MZ, M's female Co., FBW
8.	kuntili	father's sister	FZ	F's female Co., MBW
9.	kuta	elder brother	B+	PSb+S, PCo+S
10.	kangkuru	elder sister	Z+	PSb+D, PCo+D
11.	malany-pa	younger sibling	Sb-	PSb-C, PCo-C
12.	watjira	distant sibling	C of distant 6 or 8 (beyond 2nd Co.)	
13.	yuntal-pa	daughter	D	mBD, wZD
14.	katja	son	S	mBS, wZS

15.	ukari	cross-nephew, niece	mZC, wBC	C of man's 'sister' or woman's 'brother'.
16.	waputju	wife's father	WF	WFB, WMB, mDH
17.	umari	wife's mother	WM	wDH
18.	minkayi	husband's parent	HF, HM	SW, HFZ, HMB
19.	inkani	cross sibling-in-law	mBW, WZ; wZH, HB	
20.	tjuwari	parallel sibling-in-law, female	wBW, HZ	
21.	marutju	parallel sibling-in-law, male	mZH, WB	
22.	inkilyi	co-parent-in-law	CSpP	SbCSpP, SbCSpPSb
23.	kuri	spouse	H, W	

relation to a large number of specific kintypes, we
must assume - on the basis of the limited available
information and on the basis of our knowledge of the
structure of systems of kin classification in general -
that the genealogically closest denotata of each term
constitute that term's structurally primary designatum.

The (presumed) primary denotata of the terms are
listed in the third column of Table 3.1 and the derivative denotata in the fourth column. An indented term
in the first column designates a subclass of the expanded class designated by the term listed immediately
above it. The evidence for these subclass relations
is as follows.

Super- and subclasses

The expression _tjamu_ may denote FF, MF, SS, or DS,
but SS and DS may also be designated as _pakali_. Similarly, _kami_ may denote FM, MM, SD, or DD, but SD and DD
may also be designated as _puliri_. This may be accounted
for by positing an underlying covert[1] grandkin class
(lineal kin two generations removed from ego, see Figure
3.1), which is divided into two sex-specific subclasses,
tjamu (male) and _kami_ (female). Each of these subclasses is further divided into a specially designated subclass consisting of kin of the second descending generation and a residual subclass consisting of kin not
of the second descending generation (and necessarily of
the second ascending generation because this is the only

Figure 3.1. Pitjantjara grandkin class

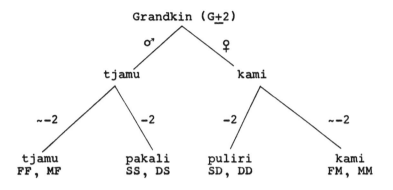

Figure 3.2. Pitjantjara PARENT-CHILD class

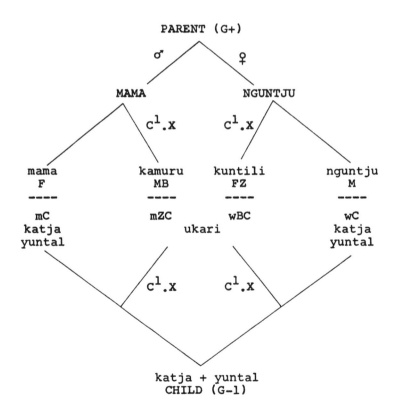

other generation level represented in the divided class). Therefore, as between <u>tjamu</u> and <u>pakali</u>, <u>pakali</u> is the marked term; it designates a special (G-2) subclass of the male grandkin class. <u>Tjamu</u>, the unmarked term, has two (at least two) possible senses: when not opposed to <u>pakali</u> it designates the male grandkin class, but when specifically opposed to <u>pakali</u> it designates grandfathers rather than grandsons. Between <u>kami</u> and <u>puliri</u> the arrangement is the same except that these terms signify that the designated kinsman is female rather than male.

As indicated in the table (see also Figure 3.2), the MB class (<u>kamaru</u>) is a subclass of the MAMA or FATHER (expanded 'father') class and the FZ class (<u>kuntili</u>) is a subclass of the NGUNTJU or MOTHER (ex- panded 'mother') class. Reciprocally, the mZC, wBC class (<u>ukari</u>) is a special subclass of the KATJA or SON class and of the YUNTAL or DAUGHTER class. The evidence for these relationships includes Annette Hamilton's report (personal communication, 1972) that at Everard Park any relative designated as <u>ukari</u> 'cross nephew, niece' may alternatively be designated as 'son' or 'daughter'. Some Pitjantjara groups do not use the expression <u>ukari</u> at all and designate cross nephews and nieces only as 'son' or 'daughter' (see Soravia 1969; and cf. Elkin 1939: 227 on the Kokata dialect, and Berndt and Berndt 1942-43: 149 on the closely related Antakirinya dialect). In addition Hamilton

reports that any 'brother' of one's mother, with the
exception of one's own MB, may be designated as 'father'; and any 'sister' of one's father, with the exception of one's own FZ, may be designated as 'mother'.[2]

To account for these data we may posit an underlying covert parent class (lineal kinsman of the first
ascending generation), which is divided into two lexically realized and sex-specific subclasses, designated
as nquntju 'mother' and mama 'father'. The reciprocal
child class (lineal kinsman of the first descending
generation) also is divided into two lexically realized
and sex-specific subclasses, designated as yuntal
'daughter' and katja 'son'. These classes are expanded
collaterally to include all kinsmen of the first ascending and first descending generations. But in addition
there are special terms for MB and FZ and these are
extended to mother's classificatory 'brothers' and
father's classificatory 'sisters'. MB is therefore
the focus and principal subclass of a class that is
itself a subclass of the FATHER class; and FZ is the
focus and principal subclass of a class that is itself
a subclass of the MOTHER class. Reciprocally, there is
a special term ukari for mZC and wBC and the child of
any kinsman designated as 'sister' by a man or as
'brother' by a woman. In some dialects this latter
special term is not used; the cross nephew and niece
subclass of the CHILD class is covert - although implied
by the reciprocal specially designated MB and FZ clas-

ses, it has no special designation of its own.

It is not inconsistent with this interpretation
that MB and FZ may not be designated as 'father' and
'mother'. As noted above, the MOTHER'S BROTHER and
FATHER'S SISTER subclasses are themselves composed of
two subclasses, own MB and FZ on the one hand and
classificatory 'mother's brothers' and 'father's sisters' on the other. Reciprocally, the CROSS NEPHEW and
NIECE subclass is composed of two subclasses, mZC and
wBC on the one hand and man's classificatory 'sister's'
child and woman's classificatory 'brother's' child on
the other. On the junior level it is permitted to
suspend the opposition between the CROSS NEPHEW and
NIECE class as a whole and the residual subclass of the
CHILD class. Thus, any relative designatable as ukari
is designatable also as 'son' or 'daughter'. In partial
contrast, the corresponding opposition on the senior
level is not fully suspendable. It is obligatory to
designate the foci of the senior subclasses (MB and FZ)
by the subclass labels; nonfocal members of the subclasses may be designated by the subclass or the superclass labels.

The sibling class of this system (also covert) is
divided first into two relative age classes (elder and
younger); the elder-sibling class is further divided
into two sex-specific classes (older brother and older
sister) but the younger-sibling class is not subdivided
(see Figure 3.3). This yields three terms, kuta 'older

brother', <u>kangkuru</u> 'older sister', and <u>malany</u> 'younger sibling'. These terms are extended collaterally to all kin of ego's generation. Alternatively, the children of mother's 'distant brothers' and father's 'distant sisters' may be designated as <u>watjira</u> (Munn 1965: 7, 7A). These 'cousins', but not the children of mother's 'close brothers' and father's 'close sisters', are marriageable. Mother's 'close brothers' include her own brothers and her male first cousins; father's 'close

---

Figure 3.3. Pitjantjara SIBLING class

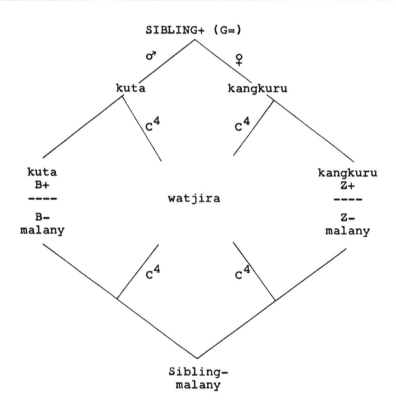

sisters' include his own sisters and his female first cousins (see also Elkin 1939: 210-2). It appears from this that watjira is a specifically cross-collateral designation, but it may be that it is not, at least not in all dialects of Pitjantjara. This possibility was suggested by a Pitjantjara speaker from the Docker River reserve who told me (in June 1972) that any 'distant' relative designated by one of the sibling terms could be designated also as watjira; it makes no difference, he said, whether the parents of the desig- nated relative are ego's classificatory 'father' and 'mother' or ego's 'mother's brother' and 'father's sister'. Either way, watjira designates a special subclass subtracted from the several expanded sibling classes.

As a final preliminary observation, it may be noted why the senior generation denotata of the self- reciprocal parent-in-law terms waputju, umari, and minkayi are specified (in Table 3.1) as their focal denotata, to the exclusion of their equally close junior generation denotata. An immediate reason is that this greatly simplifies the task of specifying componential definitions for the primary senses of the terms. Moreover, it simplifies the definitions them- selves. These simplifications are not methodologically or theoretically arbitrary. As Greenberg (1966: 103) has observed, it is a highly general (though by no means universal) feature of systems of kin classification that the designations of senior-generation kin classes are treated as unmarked in relation to the designations of the corresponding reciprocal junior-generation kin classes, whose members are no farther removed from ego than the members of the senior kin classes. Such an arrangement is especially obvious where a junior-generation class may be designated either by a distinct term or by the same term as its reciprocal senior class. We have dealt above with one case of this kind, in the relationships between the four grandkin terms, tjamu and pakali, and kami and puliri. It so happens that these terms are not self-reciprocal (at least not fully), although the three parent-in-law terms are. However, self-reciprocity is usually best (that is, most economically) interpreted in the same way, and we will have many occasions to so interpret it in the chapters and analyses to follow. Sometimes where a term is self-reciprocal the junior category has also another designation that may be used instead of the self-reciprocal term. This makes it more evident that the senior category is the focal designatum of the self-reciprocal term - at least more evident than it would be if the junior category did not have an alternative designation.[3]

Definitions and equivalence rules

The reciprocal sets of the Pitjantjara system of kin classification are illustrated in Figures 3.1, 3.2,

100

and 3.3. From these figures and the preceding discussion it should be evident that the following dimensions of opposition and their stated values are the ones that must be used in specifying the primary senses of the terms.

1. Kinsman (K) versus nonkinsman (~K). In this case, of course, most (perhaps all) in-laws[4] (I) are also kinsmen (waltja, Munn 1965: 8), because most (perhaps all) marriages are between kinsmen. Even so, it seems that no in-law category is a subclass of any particular consanguineal category. Waputju 'wife's father', for example, is not a subclass of kamaru 'mother's brother'. Although Munn (1965: 8) states that her informants "generally cited marriage with daughter of a distant kamaru (or kuntili) as the preferred type of marriage," there is no evidence that kamaru is an alternative designation for all wives' fathers.

2. Lineal (L) versus collateral (C) relationship. Colineal (Col.) relatives are a special case of first-degree collateral relatives, that is, those with whom ego shares both of his or her parents (full siblings) (see also Scheffler 1972a: 118-9).

3. Degree of generational removal: same generation as ego (G=) versus one generation removed (G1) versus two generations removed (G2).

4. Seniority: senior to ego (+) versus junior to ego (-).[5]

5. Sex of alter: male (♂) versus female (♀).

6. Relative sex: same (//) versus opposite (X).

7. Sex of ego: male (σ ego) versus female (φ ego).

The appropriate definitions are stated in Table 3.2. Kuri and inkilyi are left without definitions in this table; strictly speaking, neither is an in-law term (cf. Scheffler 1972a: 117).

The next step is to specify the rules for extending the terms to their additional, nonprimary denotata, or in other words for expanding the primary designata of the terms.

The sources contain no data on the classification of half-siblings and immediate stepkin (PSp and SpC), but it is likely that half-siblings are designated by sibling terms and immediate stepkin by parent and child terms - this is the general practice in Australian systems of kin classification. Furthermore, these rules are essential to account for numerous other classifications that are reported, for example, for the designation of FBW as 'mother', and for the designation of cousins as 'siblings'.

1. Half-sibling-merging rule:

(PC → Sb), self-reciprocal. Let anyone's parent's child be regarded as structurally equivalent to that person's sibling (or parents' child).

This rule expresses the neutralization of the distinction, made in the definition of the primary senses of the sibling terms, between colineal and other first-degree collateral kin of ego's own generation.

Table 3.2. Pitjantjara kin classification:
componential definitions of primary senses

	Term	Foci	Definition
1.	tjamu	FF, MF, SS, DS	K.L.G2. .♂
2.	pakali	SS, DS	K.L.G2.-.♂
3.	kami	FM, MM, SD, DD	K.L.G2. .♀
4.	puliri	SD, DD	K.L.G2.-.♀
5.	mama	F	K.L.G1.+.♂
6.	kamaru	MB	K.c¹.G1.+.♂.X
7.	nguntju	M	K.L.G1.+.♀
8.	kuntili	FZ	K.c¹.G1.+.♀.X
9.	kuta	B+	K.Col.G=.+.♂
10.	kangkuru	Z+	K.Col.G=.+.♀
11.	malany-pa	Sb-	K.Col.G=.-.
12.	watjira		K.c⁴.G=.
13.	yuntal-pa	D	K.L.G1.-.♀
14.	katja	S	K.L.G1.-.♂
15.	ukari	mZC, wBC	K.c¹.G1.-.X
16.	waputju	WF	I. .G1.+.♂.♂ ego
17.	umari	WM	I. .G1.♀.♂ ego
18.	minkayi	HF, HM	I. .G1.+. .♀ ego
19.	inkani	mBW, WZ	I. .G=. .X
		wZH, HB	
20.	tjuwari	wBW, HZ	I. .G=. ♀.//
21.	marutju	mZH, WB	I. .G=. ♂.//
22.	inkilyi	CSpP	
23.	kuri	H, W	

2. Stepkin-merging rule:

$(PSp \rightarrow P) = (SpC \rightarrow C)$. Let anyone's parent's spouse (who is not also his or her parent) be regarded as structurally equivalent to that person's parent; conversely, let anyone's spouse's child (who is not also that persons' own child) be regarded as structurally equivalent to that person's own child.

In addition, a lineal-collateral neutralization or sibling-merging rule is essential.

3. Sibling-merging rule:

$(PSb \rightarrow P) = (SbC \rightarrow C)$. Let anyone's parent's sibling be regarded as structurally equivalent to that person's parent; conversely, let anyone's sibling's child be regarded as structurally equivalent to that person's own child.

This rule, together with the half-sibling-merging rule, accounts for the classification of all cousins, parallel and cross, as 'siblings': If $PSb \rightarrow P$, then $PSbC \rightarrow PC$ (by rule 3), and $PC \rightarrow Sb$ (by rule 1). This rule (no. 3) does not necessarily imply that MB may be designated as mama 'father'; it implies only that MB is to be regarded as a member of the PARENT superclass, and we have already seen that kamaru designates a special subclass of the FATHER (expanded 'father') class, which is one of the subclasses of the PARENT class. Similarly, the rule does not necessarily imply that FZ may be designated as nguntju 'mother'; it implies only that FZ is to be re-

garded as a member of the PARENT superclass, and we have
already seen that _kuntili_ designates a special subclass
of the MOTHER (expanded 'mother') class, another subclass
of the PARENT class. These special subclasses of the
FATHER and MOTHER classes consist of parents' opposite-
sex siblings, and their designations are extended to
all cross collaterals of one's parents. That is to say,
the subclass statuses of one's parents' cousins are
determined in the same way as the subclass statuses of
one's parents' siblings - by simple comparison of the
sex of the linking parent with that of the designated
relative. Consider, for example, the case of MMBD. The
kintype MMBD is a special case of the type PPSbC. Via
rule 3 this type reduces to PPC; via rule 1 this type
reduces to PSb; and via rule 3 again this type reduces
to P. That is, MMBD is to be regarded as a kind of
PARENT. There are two kinds of female PARENT, MOTHER
and FATHER'S SISTER, and MMBD must belong to the MOTHER
subclass because the linking relative is one's own
mother and MMBD is of the same sex as one's own mother.

The parent terms and the subclass designations
'mother's brother' and 'father's sister' are extended
also to parents' siblings' spouses who are not collateral relatives (but who are the spouses of collateral
relatives). In the classification of these relatives
the comparison is not between the linking parent and
the designated relative but is between the linking
consanguineals themselves. Consider, for example, the

case of MBW. If the comparison were between the sex of
the linking parent and the sex of the designated relative, MBW would be classified as 'mother', but she is
classified as 'father's sister'. This is because M and
MB are relatives of opposite sex and because MBW's
status as a parallel or cross relative is determined by
the parallel-cross status of her husband, ego's MB.
Similarly, FBW is classified as 'mother' because F and
FB are of the same sex and FBW's status as a parallel
or cross relative is determined by that of her husband,
ego's FB.

Rule 3 accounts also for the extension of the in-
law terms to the collateral relatives of their respective foci. Waputju, for example, is extended from WF
to WFB and WMB. This extension is governed by rule 3
because an in-law is, by definition, a lineal or colineal kinsman of one's spouse (though of course not all
lineal kin of one's spouse are regarded as in-laws, for
example, one's spouse's children who may or may not be
one's own children as well).

Finally, several of the in-law terms are extended
by the rule of self-reciprocity (no. 4). This is a
special rule, rather than a general equivalence rule
(cf. Lounsbury 1965: 151); it applies only to certain
specified terms: the parent-in-law terms, waputju (WF),
umari (WM), and minkayi (HP). It extends these terms
from their respective foci to the reciprocals of those
foci. The (presumed) primary denotatum of waputju, for

example, is WF; the reciprocal of this relationship is
mDH, and the rule of self-reciprocity extends waputju
to this relationship. Note that the rule does not
apply to the sibling-in-law terms, which are self-
reciprocal by definition.

These four rules (listed in Table 3.3) are necessary and sufficient to account for the extensions of
Pitjantjara kinship terms listed in Table 3.1.

Marriage rules and kin classification

As noted above, Munn (1965: 8) reports that her
informants "generally cited marriage with the daughter
of a distant kamaru [MB] (or kuntili [FZ]) as the
preferred type of marriage." The definition of 'close'
and 'distant' kamaru and kuntili precludes marriage
between ego and any of his or her first and second
cousins. Munn does not elaborate on the nature of this
"preference," but it is clear from her account that
numerous other kinds of marriage between kinds of kin
are not considered 'wrong' or improper. The Pitjantjara

Table 3.3. Pitjantjara equivalence rules

---

1. Half-sibling rule, (PC → Sb)

2. Stepkin-merging rule, (PSp → P) = (SpC → C)

3. Sibling-merging rule, (PSb → P) = (SbC → C)

4. Special self-reciprocal rule, applies only to
   the three parent-in-law terms

---

divide themselves into endogamous generational "moieties" (Munn 1965: 7),[6] and they appear to regard as
improper or (as they say) 'wrong' only those marriages
that are between persons of different "moieties." Munn
(1965: 8) states, "Members of the opposite moiety are
regarded as 'the place of shame' (kunta-ngka) and marriages between members of adjacent generations [two
instances were encountered at Areyonga] were remarked
upon with disgust. These were the only marriages ever
pointed out to me by informants as being clearly
'wrong'. They considered it unkind to refer to the
misalliances of these two couples, but their general
prestige and acceptance did not otherwise suffer and
both were, to my knowledge, well-regarded in the camp."
Presumably, marriage between the offspring of siblings
or 'close' classificatory 'siblings' (whether of the
same or opposite sex) would be disapproved also. Munn
(1965: 8) reports that her informants specifically
approved of marriages between men and their 'distant
granddaughters' (puliri), and some informants approved
of marriages between men and the daughters of their
'mothers' - presumably 'distant' classificatory 'mothers' is intended.

Elkin reports much the same rules for the "Northern
Aluridja" in general, in which group he includes the
Pitjantjara. He states (1931b: 69): "The general
marriage rule is that a man marries the daughter of a
'mother's brother' from 'long-way' and that usually a

man does not marry his own second cousin." Also (1939:
218-9): "The difference between right and wrong marriage, granted that the blood relationship is more distant than second-cousin, consists of the observance or
non-observance of local exogamy and exogamy of the
alternate generation 'lines' or levels. That is, marriage should not take place within one's own horde or
between successive generations."

Of course, because of the "preference" for marriage
between persons related as the children of 'distant'
classificatory opposite-sex 'siblings', it must be that
in many instances WF waputju is a man designated as
'mother's brother' prior to marriage to his daughter;
WM umari is a woman designated as 'father's sister'
prior to marriage to her daughter; WMB waputju (or
umari, according to Elkin) is a man designated as 'father' prior to ego's marriage to his ZD; and so on. As
Elkin (1939: 217) puts it in relation to the latter
relationship, "Normally, [WMB] would be a kind of 'father," that is, an umari mama 'tabu father' or kadu mama
'like my father' (a classificatory 'father'). "Normally" in this context must refer only to a relative fre-
quency, to the more-or-less common coincidence of
particular kinds of in-law and genealogical relationships, which is brought about by observance of the rule
that a man may marry a 'mother's brother's' daughter
or a 'father's sister's' daughter. But, as we have
seen, men may also marry classificatory 'granddaughters'

and daughters of classificatory 'mothers'. Elkin (1939: 216) notes that in certain circumstances a man may marry the daughter of a classificatory 'father' and 'mother', in particular his WMBD (and, perhaps, his WFZD). This is possible, Elkin says, if the first wife is the daughter of a 'mother's brother' and 'father's sister'. Then ego's WMB would be his 'tabu father' and his WMBW would be his 'tabu mother' and their daughter would be his 'sister' and also the 'sister' of his wife (who would be his 'sister', too). Apparently it is reasoned that, because WMBD is wife's 'sister', she is like a wife kuri to ego and so he may marry her. In an instance of this sort, ego's second wife's father is his classificatory 'father', and not his classificatory 'mother's brother'. Elkin does not report how ego then designates his second wife's father - whether he retains the designation he had prior to ego's marriage to his daughter, umari and 'father', or whether he is designated as umari and 'mother's brother'.

The latter designation would be expectable if the Pitjantjara and similar systems of kin classification feature, in addition to the equivalence rules already noted, a spouse-equation rule of kin-class expansion.[7] If these systems do feature such a rule, that rule does not specify that a person's cross cousin is to be regarded as structurally equivalent to his or her spouse or spouse's sibling, either as a designated or linking relative, or both. First, marriage between actual cross cousins is prohibited, and so the use of such a rule would be culturally and analytically inappropriate. Second, such a rule would result in the classification of mother's cross cousins as 'father' and 'father's sister' and of father's cross cousins as 'mother' and 'mother's brother' - but the arrangement in these systems is just the opposite. Instead, the rule (if any) would have to specify that one's spouse is to be regarded as structurally equivalent to a cross cousin, or rather to a classificatory 'cousin' - because marriage between actual cross cousins is prohibited. That is to say, the directionality of the rule would have to be just the opposite of the rule mentioned above. Such a rule might specify, for example, that a man's WF is to be regarded as structurally equivalent to a classificatory 'mother's brother' (that is, as a member of the MOTHER'S BROTHER class) and, therefore, is to be designated as 'mother's brother' even if designated as, say, 'father' prior to ego's marriage to his daughter. This rule would imply also that WMB is to be regarded as structurally equivalent to a classificatory 'father' (a member of the FATHER class) even if designated as, say, 'mother's brother' prior to ego's marriage to his ZD. The available evidence strongly suggests, however, that this is not what is done among the Pitjantjara and closely related peoples with similar systems of kin classification. Elkin (1939: 217) cites an instance (among either Pitjantjara or Jangkundjara speakers), in

which a man's WMB was also his mother's 'brother' (they
were from the same 'country') and ego designated him as
'mother's brother' or 'wife's father' but not as 'father'. Ego's WM was therefore his mother's classificatory
'sister' and ego designated her as 'mother' or 'wife's
mother' but not as 'father's sister'. (Cf. also Tindale
in Elkin 1939: 334.) It seems relatively certain from
this that, in the case of marriage between persons related in some way other than as the offspring of 'distant' classificatory opposite-sex 'siblings', no effort
is made systematically to reorder the classification of
in-laws to make it appear as if the marriage were between persons related as the offspring of 'distant'
classificatory opposite-sex 'siblings'. If so, these
systems do not contain spouse-equation rules of kin-
class expansion of either of the two kinds noted above.

SOME SIMILAR SYSTEMS

To all appearances (see Elkin 1939: 240 ff.), the
systems of kin classification of many (but not all)
other dialects of the Southwest group of the Pama-
Nyungan language family (Oates and Oates 1970) are
similar to the Pitjantjara system. It may be that
similar systems occur also in other languages of the
Pama-Nyungan family. If Elkin's assessment of the
evidence is correct, "with local variations, this type
of kinship system also prevails in the east coast dis-
tricts of the continent from Gippsland (Victoria, the

Kurnai tribe) north to the Queensland border" (1962: 17). There is, however, relatively little published data on the systems of kin classification of this area and most of the data are too limited to be of analytic value.

The best data are Howitt's (Ms.; 1878; 1904: 169, 270; Fison and Howitt 1880: 236-7) on Kurnai. All first and second cousins are designated as 'sibling' and, consistent with this, all 'sisters' of one's father are designated as 'father's sister' and all 'brothers' of one's mother are designated as 'mother's brother'. The most notable difference between this system and the Pitjantjara system is that it features four rather than only two grandparent categories, but this is a relatively minor difference. The Kurnai arrangement is readily understood as based on a distinction between parallel and cross kin of the second ascending generation, supplemented by the distinction between male and female alter. Ego's FF and MM are his parallel kin, because they are of the same sex as the linking parents; in contrast ego's MF and FM are his cross kin because they and ego's linking parents are of opposite sex. There are two kinds of parallel kin and two kinds of cross kin, male and female.

SUMMARY

The focus of this chapter has been the Pitjantjara system of kin classification. The most distinctive

features of this system are the sibling-merging (or
lineal versus collateral neutralization) rule of kin-
class expansion, and inclusion of the FZ and MB classes
in the PARENT (expanded 'parent') class - FZ is a special kind of MOTHER and MB is a special kind of FATHER.
The rule for determining the designations of more distant collaterals considers the relative sex of the
linking parent and the designated kinsman. Thus, for
example, father's male kinsmen of his own generation
are designated as 'father' and mother's male kinsmen of
her own generation are designated as 'mother's brother'.
To all appearances, this system does not feature a
spouse-equation rule of kin-class expansion.

Similar systems of kin classification occur in a
number of dialects more or less closely related to
Pitjantjara. All of these dialects are (or were)
located in the western half of South Australia and
adjacent Western Australia and the Northern Territory.
Outside of this area, the system of kin classification
of the Kurnai of southern Victoria also features the
general sibling-merging rule; its father's sister and
mother's brother classes are subclasses of expanded
PARENT classes. It may be that systems with these
features were characteristic of coastal southeast
Australia.

ADDENDUM: PITJANTJARA AND IROQUOIS-LIKE SYSTEMS
There are some striking similarities between the

Pitjantjara system and systems of kin classification conventionally described as Iroquois- or Dakota-type systems. But the similarities are largely superficial and should not be misinterpreted as reflecting an underlying structural identity.[8]

In systems of the so-called Iroquois type FB and MZ are designated as 'father' and 'mother' respectively, and their reciprocals mBC and wZC are designated as 'child'. These systems feature distinct MB and FZ categories and terms, and distinct cross-nephew and -niece categories and terms. Moreover, in these systems, father's parallel and cross cousins are designated as 'mother' and 'mother's brother'. The most readily noticeable difference is that these systems have distinct cross-cousin categories and terms, whereas in the Pitjantjara system cross cousins, like parallel cousins, are classified as 'siblings'. This difference may seem relatively minor, but it is a reflection of a significant structural difference.

The presence, in so-called Iroquois-type systems, of distinct cross-cousin categories and terms indicates that the 'mother's brother' and 'father's sister' categories of these systems are not subclasses of the extended PARENT classes. For this shows that MB is not treated as structurally equivalent to FB (who is structurally equivalent to F), either as a designated or linking kinsman; and similarly for FZ and MZ. Thus, these systems do not feature the rule (PSb → P) = (SbC → C).

The extension of their parent terms to parents' same-sex
siblings is governed by a more limited sibling-merging
rule, one that establishes a limited structural equivalence between same-sex siblings only. The rule is:

(mB... → m...) = (...mB → ...m)

(wZ... → w...) = (...wZ → ...w), that is, let
anyone's sibling of the same sex as himself or
herself, when considered as a linking kinsman,
be regarded as structurally equivalent to that
person himself or herself; conversely, let any
linking kinsman's sibling of the same sex as
himself or herself be regarded as structurally
equivalent to that linking hinsman himself or
herself.

This rule expresses neutralization of the opposition
between lineal (and colineal) and parallel collateral
relatives (whereas the sibling-merging rule of the
Pitjantjara system neutralizes the opposition between
lineal [and colineal] and collateral relatives in
general). As a consequence of this rule parallel
cousins are classified as 'siblings' but cross cousins
are not; they are not affected by the rule and remain
to be designated by other terms.

Another similarity between the Pitjantjara and the
so-called Iroquois-type systems is that all grandparents' siblings are designated by one of the two sex-
specific grandparent terms, and all siblings' grandchildren are designated as 'grandchild'. The Pitjant-

jara rule (PSb → P) = (SbC → C) accounts for this, but the same-sex sibling-merging rule accounts only for the classification of grandparents' same-sex siblings as 'grandfather' or 'grandmother', and (reciprocally) for the classification of the grandchildren of one's same-sex siblings as one's 'grandchildren'. The opposite-sex siblings of one's grandparents and the grandchildren of one's own opposite-sex siblings are not affected by this rule. Moreover, the same-sex sibling-merging rule accounts for the classifications of parents' parallel cousins but not for the classifications of parents' cross cousins; these, too, are not affected by the rule.

The residual kintypes whose classifications are not accounted for by the same-sex sibling-merging rule (and, of course, by the half-sibling- and stepkin-merging rules, which these systems also feature) may be accounted for by positing the rule:

> Let one's parent's parent's opposite-sex sibling be regarded as structurally equivalent to that person's parent's parent's same-sex sibling; conversely, let the child's child of one's opposite-sex sibling be regarded as structurally equivalent to the child's child of that person's same-sex sibling.

This rule neutralizes the opposition (made in the definitions of the primary senses of the terms of these systems) between parallel and cross collaterals, but it does so only in the second ascending and descending

generations, and the rule applies whether the grandparent's opposite-sex sibling is considered as a linking
or as a designated kinsman. The result is that the
categorical opposition between 'siblings' (including
ego's parallel cousins) and 'cousins' (cross cousins)
is neutralized in the classification of ego's parents'
cousins. That is, although ego distinguishes terminologically between his or her own siblings and parallel
cousins (as 'siblings') and his or her own cross cousins
(as 'cousins'), he or she does not distinguish terminologically between his or her father's siblings, parallel
cousins, and cross cousins (they are all 'father' or
'father's sister'), nor does he or she distinguish
terminologically between his or her mother's siblings,
parallel cousins, and cross cousins (they are all
'mother' or 'mother's brother').

The principal differences between the Pitjantjara-
and Iroquois-like systems are differences in their
equivalence rules. Both kinds of system feature the
half-sibling- and stepkin-merging rules. Therefore,
the most distinctive equivalence rule of the Pitjantjara
system is the sibling-merging rule. In contrast, the
most distinctive equivalence rules of Iroquois-like
systems are the same-sex sibling-merging rule and the
parallel-cross neutralization rule (limited to G2).

So far as I have been able to determine, there are
no Iroquois-like systems in Australia.[9] There are,
however, a great many systems of kin classification in

Australia whose structures are readily understandable
as simple permutations of the structure of Iroquois-
like systems. Some of these are considered in the
next chapter.

Chapter 4

KARIERA-LIKE SYSTEMS

The initial focus of this chapter is the system of kin
classification of the Mari'ngar or Maranunggu of the
Northern Territory, as described by Falkenberg (1962:
38-40 et passim). The distribution of terms over kintypes in this system is much the same as in the Kariera
system itself except that the Mari'ngar system has only
two (not four) grandparent terms, but this difference
is structurally quite minor. I do not begin this chapter with an analysis of the Kariera system itself,
despite its anthropological historical interest, because
the available data appear to be incomplete and in some
respects confused and contradictory. The difficulties
may be resolvable, but to compose the differences among
the numerous published and unpublished sources[1] would
require an extended discussion that would prove tedious
to most readers, and in the end the conclusions could
be only tentative.

Table 4.1. Mari'ngar kin classification

Term	Gloss	Primary denotata	Other denotata
1. tjan'angga	parallel grandparent	FF, MM	FFSb, MMSb, FMBW, MFZH, mSC, wDC, ZDC, BSC, HZSC, WBDC
2. tamie	cross grandparent	MF, FM	MFSb, FMSb, FFZH, MMBW, mDC, wSC, ZSC, BDC, HZDC, WBSC
3. it:a	father	F	FB, FFBS, FMZS, MMBS, MFZS, SpMB
it:a il:ingeri	wife's mother's brother	WMB	
4. ngaia	father's sister	FZ	FFBD, FMZD, MMBD, MFZD, MBW, SpM
ngaia il:ingeri	wife's mother	WM	
5. kela	mother	M	MZ, MFBS, MMZD, FMBD, FFZD, FBW, SpFZ
6. kaka	mother's brother	MB	MFBS, MMZS, FMBS, FFZS, FZH, SpF
kaka boi	wife's father	WF	
7. ngauwe	brother	B	S of (3), (5)
8. ngaka	sister	Z	D of (3), (5)
9. manggen	female cousin	MBD, FZD	D of (4), (6), WZ, BW, HZ
10. do'goli	male cousin	MBS, FZS	S of (4), (6), HB, WB, ZH

11.	bange boi	wife's brother	WB	
12.	nea	man's child	mC	C of (7), (9), wSW, wDH
13.	mulugu	woman's daughter	wD	C of (8), (10), mSW
14.	magu	woman's son	wS	S of (8), (10), mDH
15.	wam:a	any relative of the third ascending or descending generation		

Source: Falkenberg 1962: 38-40.

122

MARI'NGAR

Falkenberg's (1962: 38-40) data on the Mari'ngar
system of kin classification are presented here in Table
4.1, and the system is represented diagrammatically in
Figure 4.1. The diagram must not be read as implying
that the system is based somehow on a rule of bilateral
cross-cousin marriage, although the distribution of
terms over kintypes would be entirely consistent with

Figure 4.1. Mari'ngar kin classification
(after Falkenberg 1962: 39)

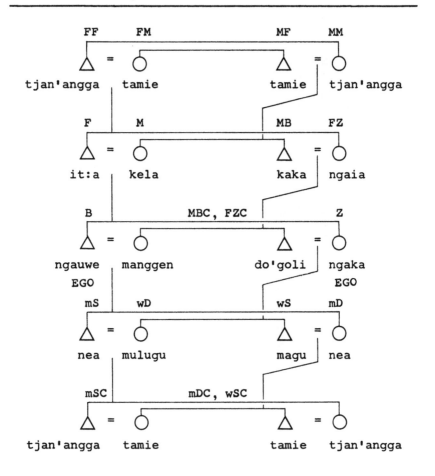

such a rule. Falkenberg (1962: 35) observes "bilateral cross-cousin marriage and 'sister exchange' are the common marriage forms," but he qualifies this by noting (1962: 45-6) that a man "can marry his cross cousins belonging to his own mother's local clan, the only exception being his own mother's full-brother's daughter." In other words, a man may marry any woman whom he designates as 'cousin', except his MBD. It follows from this that the collateral extensions of the terms of this system are not governed by a rule of bilateral cross-cousin marriage (expressed as a rule of kin-class expansion), for there is no such rule in Mari'ngar culture. Therefore, the diagram must be understood only as a convenient representation of the distribution of terms over kintypes; it must not be understood as implying any particular rule or set of rules whereby this distribution is determined.

Super- and subclasses

In Table 4.1 some terms are indented either two or four spaces. These terms designate subclasses of the expanded classes designated by the terms listed immediately above them. It:a il:ingeri (WMB) designates a subclass of the expanded it:a or FATHER class, and ngaia il:ingeri (WM) designates a subclass of the expanded ngaia or FATHER'S SISTER class. The only relative designated as it:a il:ingeri is WMB. If a man marries a kinswoman whom he designates as 'cousin', her MB is one of his 'fathers' (possibly his own father, or his MMBS, or MFZS, etc., because FZD marriage is permitted), but as WMB this relative becomes a special kind of 'father', the 'father' who is also ego's WMB, and as such is designated as it:a il:ingeri. Similarly, if a man marries one of his 'cousins', her mother is one of his 'father's sisters' (possibly his own FZ, or his MMBD, or MFZD, etc.), but as WM this 'father's sister' is designated as ngaia il:ingeri. Il:ingeri signifies the additional in-law relationship. Note, however, that WF is distinguished from other classifi-catory 'mother's brothers' kaka by the addition of boi rather than il:ingeri. (WFZ is not distinguished in this way from other classificatory 'mothers' kela.) That is, il:ingeri does not distinguish in-laws in general from other relatives of particular kin classes who are not related also as in-laws; it distinguishes only WM and WMB from other FATHER'S SISTERS and FATHERS.

The FATHER'S SISTER class is a subclass of the FATHER class and the MOTHER'S BROTHER class is a sub-class of the MOTHER class (just the opposite of the arrangement in the Pitjantjara system!). This is indicated by certain reciprocal relationships. In this system there are two child classes, man's child and woman's child (if we omit consideration of the division of the latter class into two sex-specific subclasses). This is a fairly common feature of Australian systems of kin classification (cf. Scheffler

1971a), and so is the manner in which these categories are expanded. A man designates his own child and his BC by the same term and this term is used also by a woman to designate her BC. Similarly, a woman designates her own child and her ZC by the same term that is used also by a man to denote his ZC. Thus 'man's child' is the common reciprocal of 'father' and 'father's sister', and 'woman's child' is the common reciprocal of 'mother' and 'mother's brother'. (Again, this contrasts with the Pitjantjara system in which 'father' and 'mother' have a common reciprocal term, and so do 'father's sister' and 'mother's brother'.) The implied superclass-subclass relationships are indicated in Figures 4.2 and 4.3. These figures are constructed on the assumption that parents are opposed to their siblings as lineal versus collateral kin and that the collateral kin are divided into two classes, parallel and cross. Further, FZ is identified (covertly) with FB (rather than with F directly) and MB is identified (covertly) with MZ (rather than with M directly). Presumably, the opposition between parallel and cross collaterals is neutralized and cross collaterals are made structurally equivalent (in limited contexts) to parallel collaterals, who are structurally equivalent to lineal kin - this via the same-sex sibling-merging rule (cf. the discussion of parallel-cross neutralization, in the Addendum to Chapter 3).

That this must be the structure we have to deal

126

with here is evident if we consider that the extensions of 'man's child' to mBC and to wBC cannot be governed by the same rule, nor can the extensions of 'woman's child' to wZC and to mZC. The single rule in question could only be (PSb → P) = (SbC → C), but this would imply wBC → wC and mZC → mC, which is not the case in the Mari'ngar system. Also the structural equivalence between same-sex siblings is more general than is the

---

Figure 4.2. Mari'ngar FATHER class

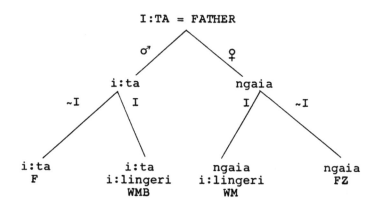

Figure 4.3. Mari'ngar MOTHER class

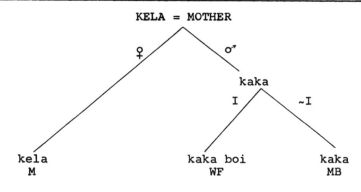

I = in-law, ~I = not in-law

structural equivalence between opposite-sex siblings;
parallel cousins are classified as 'brother' and
'sister' but cross cousins are not.

Although the status of FZ as a kind of FATHER is
covert (FZ is not, so far as we know from Falkenberg's
report, designated as 'father'), the status of wBC as
a kind of 'man's child' is overt. Similarly, although
the status of MB as a kind of MOTHER is covert, the
status of mZC as a kind of 'woman's child' is overt.
In some other Australian systems of kin classification
where a similar superclass-subclass structure is present, FZ's status as a kind of FATHER and MB's status
as a kind of MOTHER may be made overt. The terminological oppositions between FB and FZ, and between MB
and MZ, may be neutralized, and FZ may be designated
as 'father' (the same as FB) or as 'female father', and
MB may be designated as 'mother' (the same as MZ) or as
'male mother' (Hernandez 1941; Thomson 1972: 11-13).
There is no evidence for (or against) this possibility
in the Mari'ngar system. Even so, the evidence afforded
by the relations among the reciprocals is decisive - FZ
is a special kind of FATHER and MB is a special kind of
MOTHER.

It is important to note that this system features
no structurally independent in-law categories; all in-
laws (with the possible exception of WB, bange boi) are
designated by consanguineal terms (cf. Chapter 3, note
4). Certain in-laws of male ego may be denoted by

special modified forms of consanguineal terms, and so
certain in-law classes are recognized, but only as subclasses of certain consanguineal classes. A woman's
parents-in-law and their siblings are not terminologically distinguished, even as special subclasses, nor is
a man's WFZ terminologically distinguished from his
other classificatory 'mothers', although presumably
there are linguistic means (such as "descriptive" or
relative product expressions such as 'my wife's father's
sister' for WFZ) by which to make such distinctions
when required. Children-in-law are terminologically
identified with the children of opposite-sex siblings
by both sexes of ego, and siblings-in-law are terminologically identified with cross cousins.

In Table 4.1 cross cousins are listed as the foci
or primary denotata of manggen and do'goli, to the
exclusion of the spouse and sibling-in-law relationships. The assumption is that these are 'cousin' terms
that are extended to spouses and siblings-in-law.
Falkenberg glosses manggen as 'wife', but it is improbable that wife (rather than MBD, FZD) is the primary
designatum, because manggen is used by women as well as
by men, and do'goli is used by men as well as by women.
If we suppose that 'wife' and 'husband' are the primary
senses of manggen and do'goli, we must suppose also
that a woman designates her female siblings-in-law and
her female cross cousins as 'wife', and that a man
designates his male siblings-in-law and his male cross

cousins as 'husband'. Such an arrangement may not be impossible, but it would be very difficult to account for formally. These difficulties may be avoided, and the whole analysis greatly simplified, by supposing that manggen and do'goli are 'cousin' terms that are extended to spouses and siblings-in-law (by the spouse-equation rule specified below).

Definitions and equivalence rules

We may now consider the dimensions and their values relevant to the definitions of the primary senses of Mari'ngar kinship terms. They are:

1. Kinsman (K) versus nonkinsman (~K). All the categories of this system have the feature K. The specially designated in-law classes are subcategories of kin categories and so must possess this feature. Of course, if a man marries a woman to whose parents he is not otherwise related, his in-laws are not also his kin; but he designates them as though they were. Thus, the in-law categories are subclasses of kin classes, even though, in the rare instance, particular in-laws may not be kin also.

2. Lineal (L) versus collateral (C) relationship. Again, colineals (Col) are a special case of first-degree collaterals. Degree of collateral removal is indicated by a superscript numeral.

3. Degree of generational removal: G=, G1, G2.

4. Seniority: generationally senior (+) versus

130

generationally junior (-).

   5. Sex of alter: male (σ) versus female (φ).

   6. Relative sex: same or parallel (//) versus different or cross (X).

   7. Sex of ego: male ego (σE) versus female ego (φ E).

The appropriate definitions are stated in Table 4.2. In the cases of 'parallel grandparent' and 'cross

Table 4.2. Definitions of primary senses of Mari'ngar kin terms

Term	Primary denotata	Definition
1. tjan'angga	FF, MM	K.L.G2.+.//
2. tamie	MF, FM	K.L.G2.+.X
3. it:a	F	K.L.G1.+.σ
4. ngaia	FZ	K.C^1.G1.+.X.φ
5. kela	M	K.L.G1.+.φ
6. kaka	MB	K.C^1.G1.+.X.σ
7. ngauwe	B	K.Col.G=.σ
8. ngaka	Z	K.Col.G=.φ
9. manggen	MBD, FZD	K.C^2.G=.X.φ
10. do'goli	MBS, FZS	K.C^2.G=.X.σ
11. bange boi	WB	
12. nea	mC	K.L.G1.-.σ ego
13. mulugu	wD	K.L.G1.-.φ ego.φ
14. magu	wS	K.L.G1.-.φ ego.σ
15. wam:a		K.L.G3

grandparent' the comparison of sex is between the linking and the designated relative, as also in the cases
of 'father's sister' and 'mother's brother', but in the
case of 'cousin' the comparison is between the linking
kinsmen (ego's and alter's parents). The definitions
stated in Table 4.2 specify the primary or structurally
most prior senses of the terms. The definition of
'parallel grandparent', for example, applies only to
FF and MM, and not to MMB or FFZ, who are collateral
rather than lineal kin. That they are, indeed, cross-
collateral kintypes is irrelevant to the problem of
specifying definitions for the primary sense of 'parallel grandparent', for their inclusion in one of the
categories designated by this term is <u>by extension</u>, and
not by definition of the primary sense of the term.

Two of the equivalence rules of this system are
the same as two of the equivalence rules of the Pitjantjara system. These are: (1) the half-sibling-merging
rule and (2) the stepkin-merging rule. In contrast
with the simple sibling-merging rule of the Pitjantjara
system, this system features (3) the same-sex sibling-
merging rule (see Addendum, Chapter 3). These rules,
taken together, determine that all parallel collaterals
are merged terminologically with lineal or colineal kin.
Thus, for example,

FB → F, by rule 3.

FFB → FF, also by rule 3.

FBS → FS, by rule 3, and FS → B, by rule 1.

132

FFBS → FFS, by rule 3, and FFS → FB, by rule 1;

by rule 3 again FB → F.

The rule that must be posited to account for the designation of wBC as 'man's child' and for the designation of mZC as 'woman's child' (and, conversely, for the subclass statuses of FZ and MB) is a parallel-cross neutralization rule whose scope is somewhat different from the scope of the parallel-cross neutralization rule of Iroquois-like systems (see Addendum, Chapter 3). It is:

4. Parallel-cross neutralization rule:

(FZ. → FB.) = (.wBC → .mBC) and

(MB. → MZ.) = (.mZC → .wZC); that is, let any person's FZ as a terminus (this restriction is indicated by the single postfixed dot) be regarded as structurally equivalent to that person's FB as a terminus; conversely, let female ego's brother's child (the restriction to the locus of ego is indicated by the single prefixed dot) be regarded as structurally equivalent to male ego's BC; and similarly in the case of anyone's MB as a terminal kintype, etc.

The specified context restrictions are necessary because FZ as a linking kinswoman is not structurally equivalent to FB, nor is MB as a linking kinsman structurally equivalent to MZ - cross cousins are not designated as 'siblings'.

It should be noted that this rule does not necessarily imply that FZ is or may be designated by the same term as FB, or MB by the same term as MZ. It implies only that FZ belongs to the same kin class as FB, and this is a superclass. FZ is a member of a specially designated subclass of this superclass and, so far as we know, it is mandatory to designate her by the subclass label. The case of MB is similar. In the case of the reciprocals, however, there are no specially designated subclasses and so wBC is designated as 'man's child' and mZC as 'woman's child'.

An additional implication of this rule is .wBCC → .mBCC, that is, a woman's brother's grandchildren belong to the same kin classes as do a man's brother's grandchildren. In the Mari'ngar system both wBSC and mBSC are designated as 'parallel grandparent', and wBDC and mBDC are designated as 'cross grandparent'. Similarly mZDC and wZDC are designated as 'parallel grandparent' and mZSC and wZSC are designated as 'cross grandparent'. Of course, this is consistent with the arrangement among the reciprocals, the siblings of grandparents. But in the grandparental generation, in contrast to the parental generation, there are no specially designated cross-collateral subclasses. Thus, for example, FFZ is terminologically identified with FFB (who is identified with FF by the same-sex sibling-merging rule); both are 'parallel grandparent'. Similarly, FMB is terminologically identified with FMZ (who is identified with FM); both are 'cross grandparent'.

One implication of the context restrictions on parallel-cross neutralization in this system is that the relationships parent's cross cousin and, reciprocally, cross cousin's child are not affected by the rule. Although FFZ, for example, is structurally and terminologically identified with FFB, FFZC are <u>not</u> structurally or terminologically identified with FFBC (and, hence, FFC or FSb, via the same-sex sibling- and half-sibling-merging rules). Instead, FFZC are terminologically identified with MSb; they are designated as 'mother' and 'mother's brother'. However, because parents' cross cousins and, reciprocally, cross cousins' children are designated by the same terms as parents' siblings and siblings' children, we may say that these kintypes are attributed parallel and cross statuses. This attribution must be <u>by</u> <u>extension</u>, rather than by definition of the parallel-cross opposition, for that opposition is between kinds of sibling relationships (or between parent and child in the case of the grandparent categories).[2] Therefore, ego's parents' cousins (parallel and cross) are not discriminable along these lines without some rule or rules whereby the parallel-cross opposition is generalized beyond the range of first-degree collaterals (in the first ascending generation, but beyond second-degree collaterals in ego's own generation). Certain second-degree collaterals (in G+1) are attributed the terminological statuses of the parallel collaterals (FB and MZ), whereas others are attributed the terminological statuses of the cross collaterals (FZ and MB). FFBS, for example, is a collateral kinsman who is designated as 'father'; this must be because he is attributed (by extension) the status of a parallel-collateral kinsman, FB. This attribution is governed by the same-sex sibling- and half-sibling-merging rules. Similarly, FFBD is a collateral kinswoman who is designated as 'father's sister'; this must be because she is attributed (by extension) the status of a cross-collateral kinswoman, FZ. Again, this attribution is governed by the same-sex sibling- and half-sibling-merging rules. These rules, however, do not affect parents' cross cousins, who are invariant also to the parallel-cross neutralization rule of this system. What, then, is the rule or rules by which these kintypes are attributed parallel or cross status?

The conventional account of Kariera-like systems (see Chapter 2) maintains that the designations of parents' cross cousins, cross cousins' children, and second cousins are based on a rule of cross-cousin marriage. The argument is that FMBS, for example, is designated as 'mother's brother' because he is or legally could be ego's FWB who is or legally could be ego's MB; conversely, FZSC is designated as 'woman's child' because he (or she) is or legally could be ego's ZHC or ZC. If we formulate the underlying or covert structural equivalence posited in this theory in kintype

notation, we may say that the theory posits, for
example, the extension rule (mMBS... → mWB...) and
(...mFZS → ...mZH), that is, let a man's MBS, when
considered as a linking kinsman, be regarded as
structurally equivalent to his WB, when considered as
a linking kinsman, and so on. (The theory must posit
also the stepkin- and same-sex sibling-merging rules.)
The structural equivalences posited in this rule are
not inconsistent with the assumption that FMBS is
attributed the status of a cross-collateral kinsman
(MB), but it is inconsistent with the ethnographic
facts of the Mari'ngar case (and many others like it),
because of the prohibition on marriage between a man
and his MBD. That is, the structural equivalences
specified in the rule are contrary to the jural rules
of Mari'ngar society, which prohibit a man from marrying
his MBD and thus from making her structurally equivalent
to his W, BW, or WZ, and from making her brother, his
MBS, equivalent to his WB.

The difficulties for the "marriage rule" interpretation cannot be resolved by positing a rule of kin-
class expansion that specifies that a man's <u>classifica-
tory</u> 'cousin' (rather than his MBD or FZD) is to be
regarded as structurally equivalent to his W, BW, or WZ
as a linking kinswoman. Although this would permit us
to account for the terminological statuses of the
children of ego's classificatory 'cousins' and the
classificatory 'cousins' of ego's parents, we would

still have to account for the children of actual cross
cousins and the actual cross cousins of ego's parents,
and, more important, for the composition of the classificatory 'cousin' category itself. Such an equivalence
rule cannot account for the composition of that class,
because the rule presupposes the existence and composition of that class. In short, we cannot account for
the collateral extensions of the terms of this system
by means of any kind of "cross-cousin marriage rule."

The problematic kintypes are parents' cross cousins
and, reciprocally, cross cousins' children, and the
problem is to specify a rule whereby these kintypes may
be attributed parallel and cross statuses. The appropriate rule for the Mari'ngar system (and many others
similar to it) may be brought out by considering first
the rule that is most characteristic of Iroquois-like
systems. In Iroquois-like systems the parallel-cross
opposition is fully neutralized in G+2; the opposite-sex
siblings of grandparents are treated as structurally
equivalent to the same-sex siblings of grandparents,
both as designated and as linking kin. Therefore,
parents' cross cousins are treated as though they were
parents' parallel cousins. In other words, the terminological opposition between 'cousins' and 'siblings'
is neutralized when ego's cross cousins are regarded
as linking kin or when the cross cousins of linking kin
are considered as designated kin. In effect, then, the
parallel-cross statuses of ego's parents' cross cousins

are dependent only on whether the designated kinsman
is of the same sex or of different sex in relation to
the linking parent. In contrast, in the Mari'ngar
system the parallel-cross opposition is neutralized
for grandparents' opposite-sex siblings only when they
are regarded as designated kin; their cross-collateral
statuses remain relevant to the determination of the
parallel-cross statuses of their offspring, and the
parallel-cross statuses of their offspring may be determined as a function of the parallel-cross relationships
between the linking kin in both immediate ascending
generations. The appropriate formula is:

Relationship of linking siblings in G+2	X	X
Relationship of cross cousins in G+1	//	X
Then category in G+1 (and G=)	X	//

Of course, the converse applies in the case of the
child of ego's cross cousin.

This formula may be expressed also in kintype
notation. The appropriate expression is:

5. Parallel-cross status-extension rule:

$(FPSb_xC \rightarrow MSb) = (PSb_xSC \rightarrow ZC)$,

$(MPSb_xC \rightarrow FSb) = (PSb_xDC \rightarrow BC)$,

that is, let anyone's father's cross cousin be
regarded as structurally equivalent to that
person's mother's sibling, and conversely, let
one's male cross cousin's child be regarded as

structurally equivalent to one's sister's child;
also, let anyone's mother's cross cousin be
regarded as structurally equivalent to that
person's father's sibling, and, conversely, let
one's female cross cousin's child be regarded as
structurally equivalent to one's brother's child.
Some implications of this rule are shown in Table 4.3.

Table 4.3. Mari'ngar classification of parents'
cross cousins and cross cousins' children

---

1. FFZS, rule 5 → MSb, male, therefore 'mother's
   brother'.

2. FMBS, see (1).

3. FFZD, rule 5 → MSb, female, therefore like MZ,
   rule 3 → M, 'mother'.

4. FMBD, see (3).

5. MFZS, rule 5 → FSb, male, therefore like FB,
   rule 3 → F, 'father'.

6. MMBS, see (5).

7. MFZD, rule 5 → FSb, female, therefore 'father's
   sister'.

8. MMBD, see (7).

9. MBSC, rule 5 → ZC: for female ego, rule 3 → wC,
                       'woman's child'.

                       for male ego, rule 4 → wZC,
                       rule 3 → wC, 'woman's child'.

10. FZSC, see (9).

11. MBDC, rule 5 → BC: for male ego, rule 3 → mC,
                        'man's child'.

                        for female ego, rule 4 → mBC,
                        rule 3 → mC, 'man's child'.

12. FZDC, see (11).

Note that this rule determines the parallel-cross statuses of more distant collateral kin, but not the terminological statuses of their spouses. Note also that although this is an equivalence rule, it is not a neutralization rule. Unlike the same-sex sibling-merging rule, for example, it does not neutralize an opposition made at the level of the definitions of the primary senses of the terms. It does not express parallel-cross neutralization (as does rule 4) but determines the parallel-cross statuses of certain more remote collateral kintypes by generalizing the parallel-cross opposition itself.

The only extensions within the consanguineal domain not yet accounted for are those of the grandparent terms to grandchildren. These terms are self-reciprocal and their senior designata may be regarded as their primary designata; the terms are extended to the reciprocal junior categories by the rule of self-reciprocity, which is a special rather than a general rule - it applies only to the grandkin terms. The only other way to account (formally) for the self-reciprocity of the grandkin terms would be to suppose that the lineal G+2 and G-2 denotata of each term are equally focal, and that the grandkin terms are self-reciprocal by definition, rather than by extension. This is possible in the Mari'ngar case because there are only two grandkin terms that are not sex specific. Therefore, it is possible to regard FF, MM, mSC, and wDC as equally focal or primary denotata of 'parallel grandparent' and to define the class as (K.L.G2.//), in which case the dimensions of seniority, sex of ego, and sex of alter are not relevant to its definition. In comparative perspective, however, this is not the most economical analysis. Many Kariera-like systems of kin classification have three or four grandkin classes rather than just two, and in the analysis of these systems it is necessary to suppose that the senior denotata of the terms are their focal denotata. This is necessary if we are to avoid specifying disjunctive definitions for the relevant classes.

Another rule that must be posited for this system may be described as the cross-stepkin rule. It is:

6. Cross-stepkin rule:

(...wBW → ...wHZ) = (wHZ... → wBW...),

(...mZH → ...mWB) = (mWB... → mZH...), that is, let a linking kinswoman's BW be regarded as structurally equivalent to that kinswoman's HZ and, conversely, let any woman's HZ when considered as a linking relative, be regarded as structurally equivalent to that woman's BW as a linking relative; and similarly for a linking kinsman and his ZH, etc.

This rule combines with the stepkin-merging rule to determine that the spouses of cross-collateral kin in senior generations are designated as though they themselves were cross-collateral kin of ego or the propositus. For example, by this rule MBW is structurally equivalent to MHZ, and by the stepkin-merging rule (no. 2) MHZ is structurally equivalent to FZ; therefore MBW is designated as 'father's sister'. The effect is to attribute cross-collateral status to the spouses of cross collaterals in senior generations, just as the same-sex sibling-merging rule combines with the stepkin-merging rule to attribute parallel-collateral status to the spouses of parallel collaterals in senior generations. Of course, the terminological effects of this rule are consistent with the Mari'ngar practice of sister exchange (Falkenberg 1962: 35) or, in other words, with the possibility that two men may marry one another's sisters.

We may now consider the equivalence rule that accounts for the designations of the various in-law types. According to Falkenberg (1962: 35) the "ortho-dox" marriage is between a man and one of his manggen 'cousins', although marriage between MBD and FZS is prohibited. Falkenberg's account makes it clear that marriage between 'cousins' is not jurally prescribed; not all other forms of interkin marriage are regarded as illegitimate or "wrong" and negatively sanctioned. But terminologically this kind of interkin marriage is prescribed, at least insofar as the Mari'ngar extend kinship terms to in-laws according to the rule (and its corollaries) that a person's spouse is to be regarded as a COUSIN, regardless of his or her kin-class status (if any) prior to marriage.

Falkenberg reports that a person designates his or her spouse and spouse's siblings as manggen and do'goli 'cousins'. Consistent with this, spouse's mother is designated as 'father's sister' and spouse's father is designated as 'mother's brother'; also WFZ, HFZ are designated as 'mother' and WMB, HMB as 'father'. Conversely, a man designates his children-in-law as 'sister's child' (i.e., 'woman's child') and a woman designates her children-in-law as 'brother's child' (i.e., 'man's child'), and so on. Now, of course, if a man marries a kinswoman of the 'cousin' class, his WM is already his 'father's sister' and his WF is already his 'mother's brother', and so on. Moreover, it is clear from some of the examples recorded by Falkenberg that, in the event that a man does not marry a woman of the 'cousin' class, he too may designate his WM and WF as 'father's sister' (+il:ingeri) and 'mother's brother' (+boi), and he may designate his wife and her siblings as 'cousin', regardless of their kin-class statuses prior to the marriage. So, it must be that the Mari' ngar system features a rule of kin-class expansion that specifies that a person's spouse is to be regarded as his or her COUSIN, for the purpose of designating him or her and his or her immediate relatives. Even so, a person who marries "irregularly" does not lose sight of the fact that his in-laws are related to him in other ways, and their classification as in-laws does not nec-

essarily affect their statuses as linking kin (i.e., it
does not necessarily affect the terminological statuses
of any and all persons to whom ego is related through
them).[3]

Again, we need not (and may not) suppose that the
spouse-equation rule of this system specifies that a
person's spouse and his or her father and mother, and
so on, are to be regarded as structurally equivalent to
specific kintypes, such as FZ and MB in the cases of
WM and WF. Mari'ngar pragmatic marriage rules prohibit
MBD-FZS marriage and, therefore, prohibit WF from being
MB and WM from being MBW. So it would be analytically
(because culturally) inappropriate to specify that WF
is to be regarded as structurally equivalent to MB, and
so on. All we need to specify is that a person's spouse
is to be regarded as a member of the COUSIN class. We
may not specify that a person's spouse is to be regarded
as a classificatory 'cousin', because he may marry his
own FZD.

The spouse-equation rule (no. 7) and its corollaries (i.e., logical implications in other genealogical
contexts) are listed in Table 4.4, along with the other
equivalence rules of the Mari'ngar system. The spouse-
equation rule must be ordered in relation to all other
rules; all other rules must have priority. Thus, for
example, in the case of MBW the cross-stepkin rule (no.
6) has priority over the spouse-equation rule (no. 7)
and MBW → MHZ → FZ, rather than MBW → M's COUSIN (sic.).

Table 4.4 Equivalence rules in Mari'ngar
kin classification

---

1. Half-sibling-merging rule
   (PC → Sb), self-reciprocal

2. Stepkin-merging rule
   (PSp → P) = (SpC → C)

3. Same-sex sibling-merging rule
   (...m/wSb$_{//}$ → ...m/w) = (m/wSb$_{//}$... → m/w...)

4. Parallel-cross neutralization rule
   (FZ. → FB.) = (.wBC → .mBC)
   (MB. → MZ.) = (.mZC → .wZC)

5. Parallel-cross status-extension rule
   (FPSbxC → MSb) = (PSbxSC → ZC)
   (MPSbxC → FSb) = (PSbxDC → BC)

6. Cross-stepkin rule
   (...wBW → ...wHZ) = (wHZ... → wBW...)
   (...mZH → ...mWB) = (mWB... → mZH...)
   corollary
   (wBWP → HP) = (DHZ → SW)
   (mZHP → WP) = (SWB → DH)

7. Spouse-equation rule, subordinate to all the above
   a. (Sp → COUSIN), self-reciprocal
   b. (SpF → MO.BRO.) = (mCSp → man's SIS.CH.)
   c. (SpM → FA.SIS.) = (wCSp → woman's BRO.CH.)
   d. (SpFZ → MOTHER) = (wBCSp → woman's CHILD)
   e. (SpMB → FATHER) = (mZCSp → man's CHILD)
   f. (SpSb → COUSIN) = (SbSp → COUSIN)
   g. (COUSIN Sp → Sb) = (Sp COUSIN → Sb)
   h. (Sp FF → MO.FA.) = (mSCSp → man's DAU.CH.)
   i. (SpMF → FA.FA.) = (mDCSp → man's SON'S CH.)
   j. (SpMM → FA.MO.) = (wDCSp → woman's SON'S CH.)
   k. (SpFM → MO.MO.) = (wSCSp → woman's DAU.CH.)

Without this ordering, the latter reduction would be permissible, but then it would be necessary to posit yet another rule specifying that a parent's COUSIN is to be regarded as a SIBLING of the other parent. However, this rule would be redundant and, therefore, should not be posited.

Marriage and kin classification

As noted above, the "orthodox" marriage among Mari'ngar is between a man and a woman whom he classifies as 'cousin'; the only 'cousin' whom a man may not marry is his MBD. But men may marry women other than those they designate as 'cousin'. Falkenberg (1962: 38) notes "a man has a right to marry his sister's son's daughter" (and presumably other women who are son's daughters of his classificatory 'sisters'). If a man marries such a kinswoman, he designates her and her brother as 'cousin'. He designates his ZS (or 'sister's' son) who is his father-in-law as <u>kaka boi</u> (WF), and he designates his ZSW (or 'sister's' SW) who is his mother-in-law as <u>ngaia il:ingeri</u> (WM). These designations are consistent with the spouse-equation rule posited above, but it is unknown how far the corollaries are allowed to apply in an instance of this kind.

In addition, a man may marry kinswomen of other classes, provided they are not closely related and that certain other proprieties are observed. The principal consideration is suggested in a passage in which Falken-

berg comments on the case of a man who married a classificatory 'sister' who belonged to his own clan. The
marriage, Falkenberg says, was considered "wrong" but
the people ignored this because the woman was previously
married to two men of other clans. He explains: "The
point here is that if a woman has been married before,
in the normal way, outside the clan, it is not considered to be such a serious matter if she later marries
irregularly. It is thought to be wrong, but it can be
accepted. A woman's worth declines sharply with age so
that if no man who stands in the right kinship relationship to such a widow demands her in marriage, it is not
considered to be a serious offence even if she breaks
the marriage rules" (1962: 42-3).

These observations suggest that it is not so much
that a man must marry a woman of the 'cousin' class as
it is that men have marital rights in respect of women
who are their 'cousins' (as against men who are not the
'cousins' of those women). Men have no rights in
respect of women who are not their 'cousins', but this
does not prevent them from marrying such women, provided
that they are distantly related, and provided that any
men who do have rights in respect of the women do not
wish to claim them. Elderly widows are not regarded as
desirable and so they are relatively available to men
who are not their 'cousins'. Also, when a man attempts
to marry a woman to whom he has no right but to whom
another man has a right, the offense is not against the

moral sensibilities of the community so much as it is
against the rights of the other man (or men). Falkenberg notes at least one case in which a Mari'ngar man
killed another who attempted to marry a woman "who
stood in the correct dyadic relationship" to the killer
(1962: 43).

SOME OTHER KARIERA-LIKE SYSTEMS

Many Kariera-like systems of kin classification
differ from the Mari'ngar system in some respects, most
of which are relatively minor insofar as they are differences in superclass-subclass relationships. Even so,
it may be useful to note and comment on a few of the
differences because they bear on the theory that rules
of interkin marriage are critical determinants of similarities and differences among Australian systems of
kin classification.

Murinbata

The Murinbata are neighbors of the Mari'ngar and,
prior to about 1930, they too had a relatively simple
Kariera-like system of kin classification (Stanner 1936;
Falkenberg 1962). Radcliffe-Brown (1951) regarded the
Murinbata system as one of his Kumbaingeri type, because
cross-cousin marriage is prohibited and MBC and FZC
(designated as kale nan 'little mother' and kaka nan
'little mother's brother') are distinguished terminologically from the children of classificatory 'mother's

brothers' and classificatory 'father's sisters' (designated as <u>purima</u> and <u>nanggun</u>). But this is a relatively
minor structural difference, readily accounted for by
positing a covert COUSIN class that is divided into two
subclasses, cross cousin and classificatory cross cousin, each again divided into two sex-specific overt subclasses (see Figure 4.4). The COUSIN class as a whole
has the same composition as the COUSIN class of the
Mari'ngar system and is generated by the same equivalence rules.

We may suppose that the spouse-equation rule of the
Murinbata system is the same as that of the Mari'ngar
system, even though the associated interkin marriage
rules are slightly different. A Mari'ngar man has a
right to any COUSIN other than his own MBD, but this
kinswoman is not distinguished terminologically from
FZD and classificatory 'cousins' in general. A Murinbata man has a right to any COUSIN other than his MBD

Figure 4.4. Murinbata COUSIN class

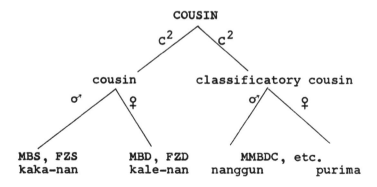

150

and FZD, but these women <u>are</u> distinguished terminologically from other COUSINS whom he may marry. But this difference does not necessarily imply a different spouse-equation rule. The only other rule that could be posited is that a person's spouse is to be regarded specifically as a member of the 'classificatory cousin' subclass of the COUSIN class (rather than as a member of the COUSIN class in general). But it is not necessary to specify this restriction on the spouse-equation rule. Spouse's designation as 'classificatory cousin' follows from the prohibition on marriage between cross cousins and from the structure of the COUSIN class. A person's spouse cannot (legally) be his or her cross cousin and so cannot qualify for designation as <u>kale</u> <u>nan</u> or <u>kaka</u> <u>nan</u>.

The child class is overtly expressed in Murinbata. Falkenberg (1962: 214) notes that the designation for the 'man's child' class <u>wakal</u> may be postfixed to the designations for the 'woman's child' classes <u>mulok</u> and <u>nawoi</u>. This indicates that <u>wakal</u> has a general as well as a specific sense. In its general sense it designates the child class; the woman's child terms designate special (marked) subclasses of this class, leaving the residual man's child class also to be designated as <u>wakal</u>.

Ompela and Mongkan

The systems of kin classification of the Ompela

(Umpila) and Mongkan of Cape York differ from most other
Kariera-like systems in making more extensive use of
the criterion of relative age in definitions of basic
classes and subclasses (Thomson 1935, 1955, 1972; McConnell 1934, 1940, 1950; McKnight 1971). For example, in
the Ompela system the FATHER class includes two specially designated subclasses, 'father's elder sibling' and
'father's younger sister'. FB- is designated as 'father'. Within the reciprocal MAN'S CHILD class there is
only one specially designated subclass 'younger brother's child', whose designation is derived from 'father's
elder sibling' by addition of a diminutive suffix. The
MOTHER class is similarly subdivided. Within the COUSIN
class there are covert close and distant subclasses:
The close COUSINS are divided into elder and younger,
according to the relative ages of the linking parents;
the distant COUSINS are divided into females and males,
males again by relative age. For second-degree and
more distant collaterals, classification as elder or
younger depends on the relative ages of the apical
linking siblings, not on the relative ages of ego and
alter. In general, distant COUSINS are marriageable,
close COUSINS are not.

　　With some minor differences, the Mongkan arrangement is the same. The Mongkan COUSIN class is not subdivided, but there is a special expression, <u>kutth</u> (or
<u>kort</u>), by which a man addresses his prospective wife
and wife's siblings.

Thomson and McConnell appear to disagree on Mongkan marriage rules and how they affect kin classification. Thomson says the Mongkan have a "second-cousin marriage system," but McConnell claims that the orthodox or preferred marriage is between MB-D and FZ+S. She argues that the Mongkan marriage system should be classified with those of the Karadjeri and Yir Yoront "as variations of a unilateral cross-cousin marriage, as distinct from the bilateral Kariera cross-cousin marriage" (1940: 343-5). The apparent contradiction is resolved when it is realized, first, that there is some local variation in rules of interkin marriage and, second, that Thomson's account emphasizes a man's relationship to his prospective WMs, whereas McConnell's account considerably overemphasizes his possible relationship to his WF. According to Thomson, in orthodox marriage by betrothal, one or more of a man's classificatory 'father's mothers' (FM's or MF's 'sister') is chosen to be his prospective WMM, her daughters become his prospective WMs, and their children become his prospective wives and siblings-inlaw. A man's prospective WMs may be married to 'mother's younger brothers' or to 'mother's elder brothers'. From McConnell's accounts it appears that most groups of Mongkan prohibit a man from marrying his MB+D but not his MB-D, provided she is not also his FZD. Therefore, in these groups a man's prospective wives must be at least his second cousins through their mothers, but they may be his first cousins through their fathers (see also Needham 1963).

McConnell's account is especially misleading in
that she frequently gives the impression that the designation for MB-C as such is <u>kutth</u> (or <u>kort</u>), while
other cross cousins are designated as <u>moiya</u>. This
accords with her interpretations that Mongkan kin classification is based on "unilateral" cross-cousin marriage or on a "junior marriage system," but these interpretations are obviated by recognition that <u>kutth</u>
(or <u>kort</u>) designates a special prospective in-law
subclass of the COUSIN class. The collateral extensions
of Mongkan kin classes and their designations are not
governed by a rule of marriage between MB-D and FZ+S.

Another interesting and theoretically significant
aspect of the Mongkan system is that the kintypes FFZ
and MMB are subject to alternative designations (Thomson
1972: 18; McConnell 1934, 1940). The parallel-grandparent class is divided into two sex-specific subclasses,
FF <u>pola</u> and MM <u>kema</u>. Usually FFZ is designated as
'mother's mother' and MMB is designated as 'father's
father', but alternatively FFZ may be designated as
'father's father' and MMB as 'mother's mother'. Similar
alternatives are possible for FMB and MFZ in the Ompela
system (Thomson 1972: 6) but not in the Mongkan system,
where the cross-grandparent class has no subclasses.
Thomson and McConnell do not report the conditions that
govern these alternative designations, but a linguist,
Barbara Sayers, made additional inquiries in November

1971. She reports (see Thomson 1972: 18, 43-4) that
these alternative usages are conditioned by the sex of
the speaker and by 'tabu' prospective and actual marital
relationships. In general FFZ may be designated in
either way, but female speakers tend to use 'mother's
mother'; and MMB may be designated in either way, but
male speakers tend to use 'father's father'. However,
when speaking to a 'tabu' prospective or actual in-law,
it is obligatory to designate any member of the
PARALLEL-GRANDPARENT class as one's 'father's father'.
Thus, for example, a woman ordinarily refers to or
addresses her DC as 'little mother's mother' and they
refer to or address her as 'mother's mother'; but if a
woman is speaking to her 'brother's son' who is also
her prospective or actual DH, she must refer to her DC
as 'little father's father' and to herself as their
'female father's father'.

Clearly pola, in addition to being the designation
of the FF class, is also the cover term for the parallel-grandparent class as a whole. As between pola and
kema, pola is the unmarked and kema the marked term,
and the relevant opposition is "not-female versus
female." In 'tabu' contexts this opposition is suspended and all members of the parallel-grandparent class
are designated by the expression for the class as a
whole, which is 'father's father'. In ordinary contexts, however, grandparent's opposite-sex siblings may
be designated by the term appropriate to their sex

(i.e., FFZ as 'mother's mother' and MMB as 'father's father'), or by the term for the subclass to which they are assimilated by the parallel-cross neutralization rule (FFZ → FFB → FF, and MMB → MMZ → MM). Because it is not improbable that this pattern of alternative designations for grandparents' opposite-sex siblings is fairly common in Kariera-like systems, it is unlikely that patterns of kin classification in the second ascending and descending generations are significantly related to rules of interkin marriage.

Another aspect of the Mongkan system that should be noted is how its terms are extended to kin of the third ascending and descending generations. These extensions receive little or no attention in McConnell's first published account (1934). In her second account (1940: diagrams 1-4) she reports that the two child terms are used to denote kin of the third <u>ascending</u> generation, whereas their reciprocals, the parent and parent's sibling terms, are used to denote the recip- rocal kintypes in the third <u>descending</u> generation. Later, however, McConnell (1950: 121) produced a dia- gram showing a quite different arrangement: Kin of the third ascending generation are designated as 'father's older sibling' and 'mother's older sibling', and kin of the third descending generation are designated by the terms appropriate to parents' younger siblings. She adds, the child terms "would seem to be used also in the third ascending generation" (1950: 124). She gives

no indication of the circumstances in which a person
would designate a G+3 kinsman by one of the child terms
rather than by one of the parent's elder-sibling terms;
nor does she comment on the fact that this latter
account strangely suggests that the parent's sibling
terms may be used reciprocally - for example, a man may
designate his FFF as 'father's elder brother' and be
designated as 'father' (the same as FB-) in return.
McConnell's diagram (1950: 122) of the Kandyu system
shows the same arrangement, and she says, "but the terms
older brother's and sister's children . . . are apparently used also for the third ascending generation"
(1950: 126).

Confusing as it may seem, it is not difficult to
account for all of this. First it must be noted that
in Mongkan, as in many other Australian languages, the
term used to designate a particular relative may be
allowed to depend on his or her age relative to ego.
This is true not only of sibling terms, which often
designate categories defined in part by the criterion
of relative age, but also of parent, child, and parent's
sibling and sibling's child terms. For example, if a
man A should have a classificatory 'sister' whose son B
is older than A, A may designate B as 'mother's brother'
rather than as 'sister's son', and B may designate A as
'sister's son' rather than as 'mother's brother'. The
semantic basis of this arrangement is discussed in the
following section on the Kariera system. At this point

we need only note that this possibility accounts for the fact that kin of G+3 may be designated by G-1 or G+1 terms. Presumably, the underlying pattern here is that kin of G-3 are designated by G+1 terms (parents' junior siblings) and kin of G+3 are designated by G-1 terms (senior siblings' children), but kin of G+3 are likely to be older than ego and so may be designated by the appropriate senior terms (parents' senior siblings). The problem, then, is to account for how it is that G+3 kin are designated by G-1 terms and, conversely, G-3 kin are designated by G+1 terms.

This cannot be done by positing a general equivalence between G+3 and G-1 and, conversely, between G-3 and G+1, because we have to account also for the fact that some PPP are designated as 'man's child' whereas others are designated as 'woman's child' and, conversely some CCC are designated as 'father's sibling' whereas others are designated as 'mother's sibling'. The simplest solution is to suppose that the designation of kin in G-3 is a function of the designation of kin in G-2 and, conversely, the designation of kin in G+3 is a function of the designation of kin in G+2. For example (see Figure 4.5), the classification of mSSS depends on the classification of mSS, and reciprocally the classification of mFFF depends on the classification of mFF. But the designation assigned to mSS depends on the designation of FF, because the designation of mSS is derived from the designation of FF by the rule of self-reciprocity. That is, FF is designated as <u>pola</u> and the term is extended by the rule of self-reciprocity to the reciprocals wSS and mSD. Thus, a man's SSS is the son of a kinsman whom ego designates as <u>pola</u>, and this is the term ego uses to denote his own FF. If we leave aside the possibility of designating the son of a junior <u>pola</u> (SS) by any term, we may say that normally the son of a <u>pola</u> (FF) is classified as some kind of FATHER, depending on his age relative to one's own father. Analogously mSSS may be classified as a kind of FATHER, for he too is the son of a <u>pola</u> (but a junior one). He

---

Figure 4.5. Mongkan G+3 classifications

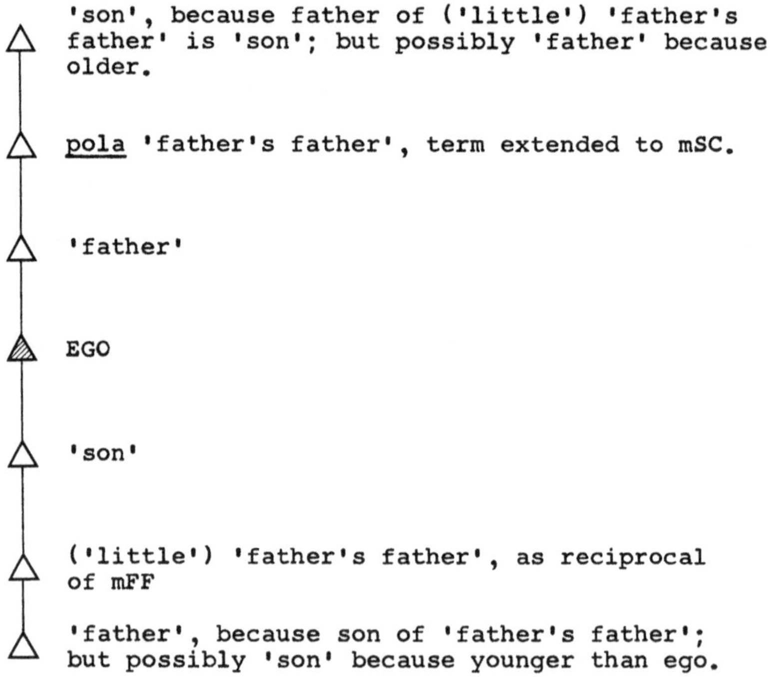

'son', because father of ('little') 'father's father' is 'son'; but possibly 'father' because older.

<u>pola</u> 'father's father', term extended to mSC.

'father'

EGO

'son'

('little') 'father's father', as reciprocal of mFF

'father', because son of 'father's father'; but possibly 'son' because younger than ego.

is necessarily younger than ego's own father, and so
the appropriate designation is the same as that for a
member of the FATHER class who is younger than ego's
father, which is 'father'. Conversely, male ego's FFF
is the father of a man whom ego designates as <u>pola</u>.
Leaving aside the possibility of designating FFF by any
term, we may say that the father of <u>pola</u> (SS) is normally a relative classified as some kind of MAN'S CHILD,
for within G+2 - G-2 range the only <u>pola</u> whose fathers
have designations are male ego's SC (or kintypes structurally equivalent to mSC). So, analogously, FFF may

'sister's son', because father of ('little')
'mother's father', i.e., mDC is 'sister's son';
but possibly 'mother's brother', because older.

'mother's father', term extended to mDC

'mother'

EGO

'son'

'father's father' (w) or 'mother's mother'

'mother's younger brother', because as son of
(w) 'father's father' or 'mother's mother' is
like FFZC or MMC; but possibly 'sister's son'
because younger

be classified as some kind of MAN'S CHILD. As a senior
relative, he is appropriately placed in the senior subclass of the MAN'S CHILD class, which is 'man's child'.
But he is also senior in age to ego and ego's father
and so may be designated by the senior term of the reciprocal FATHER class, that is as 'father's older sibling'.

Similarly, a man's SDS is the son of ego's SD or
female pola. Normally the son of a female pola (FFZ or
MM) is designated as some kind of MOTHER'S BROTHER,
depending on his age relative to ego's mother. Male
ego's SDS is necessarily younger than ego or his mother,
ego's SD, and so may be classified as 'mother's younger
brother' kala. Conversely, mMFF is the father of a man
whom ego classifies as ngatja, and normally the father
of ngatja is 'sister's son' tuwa (Z+C) or mukaiya (Z-C).
As a senior relative MFF is appropriately designated as
tuwa, or by the senior term of the reciprocal MOTHER
class, which is muka 'mother's older brother'.

In analogous fashion it is possible to account for
all the G+3 and G-3 designations reported by McConnell:
We need not go through the whole list here. McConnell
says nothing about the possible social motivations for
assigning kin terms to G+3 and G-3 kin, but we may
speculate that there are at least two possible reasons.
First, because of the relatively young age at which
women marry and bear children, it may be that it is not
uncommon for a person to have living relatives in his or

her third descending and ascending generations. Second,
it is forbidden to use the personal names of deceased
relatives; yet it may be useful, if not essential, to
have some means by which to refer to them, as it may be
necessary to do when, for example, marriages are being
arranged and kin-class statuses are an important consideration.

As a final observation on the Mongkan system, we
may note Thomson's (1946: 58-9) report of a full set of
"mourning terms" that are in effect special kinship
terms used to address the kin of a deceased person or
to refer to the deceased. Some of these terms are:

1.a. aimi   bereaved M, MZ, MB

   b. wall'wala   deceased wC, wZC, mZC

2.a. maki   bereaved F, FB, FZ

   b. oimp'watjaman deceased mC, mBC, wBC

3.a. oma   bereaved wC, wZC, mZC

   b. ?   deceased M, MZ, MB

4.a. ant   bereaved mC, mBC, wBC

   b. pinyawun   deceased F, FB, FZ

5.a. koi'yomo   bereaved Sb-

   b. mant   bereaved Sb+

6. ankalamwet   bereaved spouse or sibling-in-law

As Thomson remarks, some of these terms "throw light on
the kinship organization," for "they group together
under a single name a number of relatives who are distinguished normally by separate terms." The first four
pairs of terms are especially interesting because they

are in effect special designations for the MOTHER,
FATHER, WOMAN'S CHILD, and MAN'S CHILD superclasses of
the ordinary system of kin classification (see Figures
4.2 and 4.3). That is, these "mourning" classes are
created by neutralizing all the terminological distinctions made within those superclasses.

Kariera

The Kariera system itself need not be discussed in
detail, but a few aspects of Radcliffe-Brown's (1913)
data on this system are especially worthy of comment.

It was noted above that relative age is a classificatory criterion of the Ompela and Mongkan systems;
it operates also in the Kariera system, but in a somewhat different way. Although the Kariera system distinguishes four sibling classes, B+, Z+, B-, and Z-, it
does not distinguish among parents' siblings by relative
age (though it may distinguish two FZ classes by sex of
ego). Even so, relative age _in relation to ego_, rather
than as between alter and the linking kinsman, is taken
into account in designating collateral kin in the first
ascending and descending generations. Radcliffe-Brown
(1913: 154) gives the following example: "It may happen
that a man B is by genealogy the 'father' of a man A,
but is younger than A [see Fig. 4.6]. In such a case
A calls B not 'father', but 'son', and B calls A 'father', although by genealogy he is his 'son'. The same
thing may occur in the case of a _kaga_, a _nganga_, or a

toa [MB, M, FZ]. In one case I found three men, A, B,
and C, aged about 65, 63, and 60, respectively. The
father of A and C, who were brothers, was the 'elder
brother' of B, and therefore, both A and C were by
genealogy, the 'sons' of B. He called C his 'son', but
as A was older than himself, he called him not 'son',
but 'father', thus reversing the genealogical relation."
He found that the Mardudhunera may do much the same
thing (1913: 184). This may be accounted for as follows
(cf. Maybury-Lewis 1967).

In the Mari'ngar and Murinbata systems siblings
and parents' siblings are not distinguished by relative
age; so it is not necessary to posit a relative age
dimension when definitions are specified for the primary
senses of the kinship terms. It is necessary, however,
to posit a seniority dimension, which has to do with
generational status. In the case of a lineal kinsman

Figure 4.6 Kariera relative age example

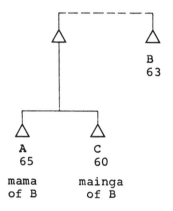

seniority of generation necessarily coincides with seniority of age, and juniority of generation necessarily
coincides with juniority of age. But in the case of a
collateral kinsman there may be a discrepancy between
generational status and relative age, such that a man
might designate an older man as, say, 'sister's son'.
For the Ompela and Mongkan systems, however, it is necessary to posit a relative age dimension, as well as a
seniority dimension that has to do with generational
status, because siblings and parents' siblings are
distinguished terminologically by relative age. (Or,
alternatively, we may omit the seniority dimension and
treat the generational dimension as five-valued, that
is, as having the values G+2, G+1, G=, G-1, and G-2,
rather than as three-valued, G2, G1, G=, as was done
for the Murinbata system.) In all these systems when
the terms are extended collaterally natural generational
status is respected. In the Ompela and Mongkan systems
this occurs because the relative age status of a collateral kinsman is determined by the relative ages of the
apical linking siblings, and not by the relative ages
of ego and alter or by the relative ages of some more
immediate linking kinsman and alter (e.g., ego's F and
alter in the case of FMB+S, for example).

The Kariera case is slightly different. Siblings
are distinguished terminologically by relative age but
parents' siblings are not. Also, as the parent and
parent's sibling terms and their reciprocals are ex-

tended collaterally, natural generational status is not respected but the relative age status of ego and alter is respected. This may be accounted for by assuming that in the Kariera system, as in the Mari'ngar and Murinbata systems, the generational dimension is three-valued, G2, G1, and G=; but whereas the Mari'ngar and Murinbata systems feature a seniority dimension (having to do with generational status) but not a relative age dimension, the Kariera system features a relative age dimension but not a seniority dimension. Therefore, in the Kariera system as the parent and parent's sibling terms and their reciprocals are extended collaterally they may become "inverted." That is, the definition of Kariera mama 'father' in its primary sense may be specified as $(K.L.G1.+.\sigma^{\gamma})$, where "+" signifies relative age, and ego's own father is necessarily older than ego. Similarly, Kariera mainga 'son' (which is not specific for sex of ego) may be defined as $(K.L.G1.-.\sigma^{\gamma})$, where "-" signifies relative age, and ego's son is necessarily younger than ego. Note, then, that aside from the fact that mama signifies "elder" and mainga signifies "younger," the terms have identical values on the several dimensions of contrast that enter into their definitions. As the terms are extended collaterally they no longer signify "lineality" of relationship but they remain semantically identical except for their signification of relative age. Now ego's father's 'brother' may be older or younger than ego himself (or herself), yet both

166

are collateral kinsmen, one generation removed, and
male. Therefore, if designation as 'father' or as 'son'
depends on relative age rather than on generational
status (direction of generational removal), it follows
that ego's father's 'brother' who is younger than ego
is to be designated as 'son' and, conversely, ego's
'brother's' son who is older than ego is to be designated as 'father'.

One bit of Radcliffe-Brown's data on the Kariera
system may seem to contradict this interpretation. He
reports that there are no special terms for relatives
in the third ascending and descending generations, but
he was told that a man would designate his FFF as 'son'
and his FFM as 'sister's daughter' (or 'son's wife').
Goodenough (1970: 133-4) has argued that the Kariera
system of kin classification does not extend beyond two
generations from ego and that Radcliffe-Brown's report
to the contrary was based on a misunderstanding. Goodenough notes that _maeli_ (FF) is self-reciprocal and
also denotes mSS. Therefore, says Goodenough, if asked
what he calls the 'father' of his _maeli_ (FF or SS), a
man may think that the reference of _maeli_ is to his SS
and reply 'son' (S). Similarly, if asked what he calls
'mother' of his _maeli_ (FF or SS), he may again think
that the reference is to his SS and reply 'sister's
child'. Goodenough concludes that the system does not
go on alternating throughout the generations as Radcliffe-Brown claimed it did. The major difficulty for

Goodenough's argument is that, according to his own report, Radcliffe-Brown did not fall into the trap of gathering his data in the way that Goodenough's argument suggests he did. Radcliffe-Brown explicitly states that he obtained the _name_ of his informant's FFF and asked what this man would be called, and similarly for the case of FFM. Moreover, this same extension of first descending generation terms to lineal kin in the third ascending generation is extensively documented for a great many Australian societies. I see no reason to suppose that all or even most ethnographers who have reported this have misled themselves in the way Goodenough's argument would suggest.

In short, this seems to be a real enough fact of Kariera kin classification. Therefore, it is particularly unfortunate that Radcliffe-Brown's data on this aspect of the Kariera system are too limited to justify much further comment on them. About all that may be said with confidence is that, probably, all kin of the third ascending generation may be designated by first descending generation terms, and all kin of the third descending generation may be designated by first ascending generation terms. The reported designation of FFF by a junior term is not necessarily inconsistent with the fact that FFF is necessarily older than ego, because it may be that, as in the Mongkan system, in direct address such a kinsman would be designated by the reciprocal term more appropriate to his relative age.

SUMMARY

This chapter has dealt with several Kariera-like systems of kin classification. The structural differences among them are relatively minor and consist for the most part of differences in subclass structures. They share all their equivalence rules, including the parallel-cross status-extension rule and the spouse-equation rule, which specifies that a person's spouse is to be regarded as his or her COUSIN. This latter rule, together with its corollaries, governs the classifications of ego's immediate in-laws and specifies that they are to be classified as though they were members of cross-collateral categories. It does not govern the classifications of ego's more remote collateral kin. These classifications are governed by the parallel-cross status-extension rule (and by other rules). These two rules are the most distinctive rules of these systems.

This analysis of the structure of Kariera-like systems differs from the analysis offered by Radcliffe-Brown principally in that it accounts for the collateral and in-law extensions of their terms by two different rules, rather than by a single rule (and its corollaries) that specifies structural equivalences between specific consanguineal relationships and specific in-law relationships (such as WF → MB; W → MBD, FZD, and so on). It was pointed out in Chapter 2 that Radcliffe-Brown's model is inadequate because it requires the

possibility of marriage between MBD and FZS and FZD and
MBS (not just any persons who designate each other as
'cousin'), but Kariera-like systems of kin classification are found in association with various prohibitions
on marriage between cross cousins. Among the Kariera,
men are permitted to marry MBDs and FZDs; but in none
of the other cases dealt with here is marriage permitted
between cross cousins in general.

Aside from the possibility that the subclass distinctions within the COUSIN categories of the Murinbata
and Ompela systems may be motivated by the associated
prohibitions on marriage between cross cousins, none of
this variation in practical marriage rules is reflected
in these systems of kin classification, certainly not
in their rules of kin-class expansion. In all these
systems father's cross cousins are designated as 'mother' and 'mother's brother' and mother's cross cousins
are designated as 'father' and 'father's sister'. If
we attempt to account for this by rules of interkin
marriage, the only rule that will "work" is one that
permits (but does not necessarily require) marriage
between cross cousins in general. But, again, in most
of the cases dealt with here that rule is absent. The
marriage rules that are present in these cases, if
treated also as rules of kin-class expansion, would
yield quite different results. For example, the Mari'
ngar permit a man to marry his FZD but not his MBD.
This implies that a person's FFZD may "legally" be his

or her father's wife (or wife's sister) and perhaps his
or her mother, and so it may seem feasible to account
for the Mari'ngar designation of FFZD as 'mother' by
reference to this marriage rule. But in the Mari'ngar
system FMBD also is designated as 'mother', although
she may not "legally" be ego's father's wife and perhaps
ego's mother. Positing an asymmetric marriage rule
(with regard to cross cousins) would leave us without
a means to account for this aspect of the Mari'ngar
system.

The theory that the collateral extensions of terms
are governed by rules of interkin marriage cannot be
salvaged by supposing that the rules deal with whole
categories rather than with specific kintypes and
specific kinds of in-law relationships. In all these
cases a man is free to marry any woman of the COUSIN
category, sometimes with the exception of certain
cross cousins. But a rule of kin-class expansion based
on this practical marriage rule cannot be used to
account for the collateral extensions of the terms of
these systems. The purpose of positing such a rule
would be to account for the kintype compositions of
the expanded kin classes, but such a rule is incapable
of doing this because it presupposes the compositions
of the expanded kin classes. The same objection does
not apply, however, to the hypothesis (presented in
this chapter) that such a rule does govern the extensions of kin terms to ego's in-laws.

It may be, as Radcliffe-Brown (and Elkin) sometimes
argued, (although not in these terms) that subclass
distinctions within the COUSIN class are motivated by
marital prohibitions; that is (for example), that
'distant' kin within the COUSIN class are distinguished
from 'close' kin because the former are marriageable
and the latter are not. Of course the marital prohibition does not require the terminological distinction,
for there need be no difficulty in distinguishing
socially between 'close' and 'distant' COUSINS even if
they are not terminologically distinguished. There are
plenty of Australian societies in which this is done.
But, even so, where marriage between relatively close
kin is prohibited, it may be thought inappropriate to
designate siblings-in-law as 'cousins' and additional
expressions may be introduced to distinguish the 'distant' COUSINS who are potential or prospective spouses
and siblings-in-law from the 'close' COUSINS who are
not. This addition of a special subclass to the COUSIN
class has no effect on the basic structure of the system
as a whole.

Chapter 5

NYULNYUL AND MARDUDHUNERA

This chapter focuses on the Nyulnyul and Mardudhunera
systems of kin classification. Both have been described
(by Elkin and Radcliffe-Brown respectively) as Aranda-
type systems, largely because they are associated with
prohibitions on marriage between cross cousins and close
classificatory 'cousins', and because they classify
grandparents and their siblings in a manner that seems
consistent with this prohibition and with preference
for marriage between the offspring of same-sex cross
cousins (e.g., between a man and his MMBDD). Radcliffe-
Brown divided his Aranda-type systems not only into
"fully developed" and not-fully-developed "varieties,"
but also into two other "varieties" according to whether
or not certain kintypes of ego's own generation (e.g.,
MMBSC) are terminologically identified with certain
kintypes of the second ascending and descending generations (e.g., MM, MMB, wDC, ZDC). Where they are he
attributed this to the principle of equivalence of agnatic kin of alternate generations. According to Rad-

cliffe-Brown the Mardudhunera system features this
principle, and from Elkin's published accounts it would
seem that the Nyulnyul system does, too.

It may be shown, however, that there are no basic
structural differences between the Nyulnyul and Mardudhunera systems and the Kariera-like systems discussed
in Chapter 4. The Nyulnyul and Mardudhunera systems do
not incorporate rules of structural equivalence of agnatic kin of alternate generations; their most distinctive equivalence rules are, again, the parallel-cross
neutralization rule and the parallel-cross status-
extension rule. Appearances to the contrary in the
published accounts are attributable to insufficient
recognition of superclass-subclass relationships and to
misinterpretation of certain practical consequences of
a form of intergenerational marriage as though they
were "basic" structural features of these systems.

It should be emphasized that the argument of this
chapter is _not_ that there are no significant structural
differences between so-called Aranda-type systems and
Kariera-like systems in general. There are significant
and basic structural differences between the Nyulnyul
and Mardudhunera systems and many other systems of the
so-called Aranda type, which do incorporate rules of
structural equivalence of agnatic kin of alternate
generations (as well as rules of structural equivalence
of uterine kin of alternate generations; see Chapter 9).
Analytical preoccupation with the possibility of

accounting for similarities and differences among Australian systems of kin classification by relating them
to similarities and differences in rules of interkin
marriage has tended to obscure these basic structural
differences, while highlighting superficial similarities. Moreover, some of these superficial similarities
are not even structural, but are entirely contingent on
one of the practical consequences of the possibility of
intergenerational marriage, which is that two individuals may be related in more than one way.

NYULNYUL

The Nyulnyul data presented in Table 5.1 and elsewhere in the text are taken from published accounts by
Elkin (1932: 300-1; 1964: 74) and Jolly and Rose (1943-
45, 1966), and from an unpublished field report (Elkin
1928b) and an extensive collection of genealogies that
Elkin kindly permitted me to consult. In most respects
Elkin's and Jolly and Rose's accounts are in agreement;
the differences center mainly on the classifications of
spouses, siblings-in-law, and other in-laws of a woman.
These differences are not of an order that would materi-
ally affect an analysis of the principal structural
features of this system, so I pass over them without
further comment. There are, however, a few differences
between the data presented in Table 5.1 and Elkin's
published accounts. These should be explained.

Elkin's published diagram (reproduced here as Fig-

ure 5.1) is potentially misleading in two ways. First,
it represents the system as though it were based on a
rule of marriage between the offspring of same-sex
cross cousins, but Elkin (1932: 307) reports that this
form of marriage "is rare and disparaged." Indeed, he
recorded no instances of it (1928b). Jolly and Rose
(1966: 103) state that one Nyulnyul marriage rule is,
"I can marry djalel ['cousin'] as long as she is more
distant than second parallel cousin to my own djalel."
That is, the closest 'cousin' a man may marry is a
third cousin (or a fourth-degree collateral relative).
It is clear that the children of same-sex cross cousins
stand in no special potential marital relationship, and
we may safely infer that MMBD and MMBS are not designated as yala 'prospective WM' and ramba 'prospective
WMB'. Because, as noted in Chapter 2 (see Karadjeri-
type systems), Elkin provides evidence to show that the
prospective WM class is a subclass of the FATHER'S
SISTER class, and that the prospective WMB class is a
subclass of the FATHER class, we may infer also that
MMBD is classified as 'father's sister' and MMBS as
'father'. Conversely, the designation for MBDC and FZDC
is 'man's child' wal, not yala and ramba as shown in
Elkin's diagram. If yala 'prospective WM' designates a
special kind of FATHER'S SISTER, and a prospective or
actual WF is a classificatory 'mother's brother', it
follows that mala 'prospective wife or sister-in-law'
designates a special kind of djalel 'cousin', as does

Table 5.1. Nyulnyul kin classification

Term	Gloss	Primary denotata	Other denotata
1. kalod	father's father	FF	FFSb, FMBW, mSC, BSC, HZSC
2. kamad	mother's mother	MM	MMSb, MFZH, wDC, ZDC, WBDC
3. djam	mother's father	MF	MFSb, MMBW, mDC, BDC, HZDC
4. kabil	father's mother	FM	FMSb, FFZH, wSC, ZSC, WBSC
5. ibal	father	F	FB, FFBS, FMZS, MMBS, MFZS, MZH, SpMB
ramba	prospective WMB, HMB		any ibal more than two degrees of collaterality removed from ego.
6. yirmor	father's sister	FZ	FFBD, FMZD, MMBD, MFZD, MBW, SpM
yala	prospective WM		any yirmor more than two degrees of collaterality removed from ego.
7. berai	mother	M	MZ, MFBD, MMZD, FMBD, FFZD, FBW, SpFZ
berai djaminir	prospective WFZ		any berai more than two degrees of collaterality removed from male ego.
8. kaga	mother's brother	MB	MFBS, MMZS, FMBS, FFZS, FZH, SpF

Term	Gloss	Symbol	Definition
kaga djaminir	prospective WF		any kaga more than two degrees of collaterality removed from male ego.
9. kaga rangen	prospective HF		any kaga more than two degrees of collaterality removed from female ego.
10. babal	brother	B	S of (5), (7)
11. marer	sister	Z	D of (5), (7)
12. djalel	cousin	MBC, FZC	C of (6), (8)
mala	prospective wife or sister-in-law		any female djalel more than three degrees of collaterality removed from ego.
malp	prospective husband or brother-in-law		any male djalel more than three degrees of collaterality removed from ego.
14. wal	man's child	mC	C of (10), mala (under 12)
yala			reciprocals of yala under (6) for female ego.
ramba			reciprocals of yala and ramba under (5) and (6) for male ego.
15. bap	woman's child	wC	C of (11), malp (under 12)
rangen			reciprocals of kaga rangen under (8)
16. wainman			prospective SpMP, DCSp

178

Figure 5.1. Nyulnyul kin classification
(after Elkin 1964: 74)

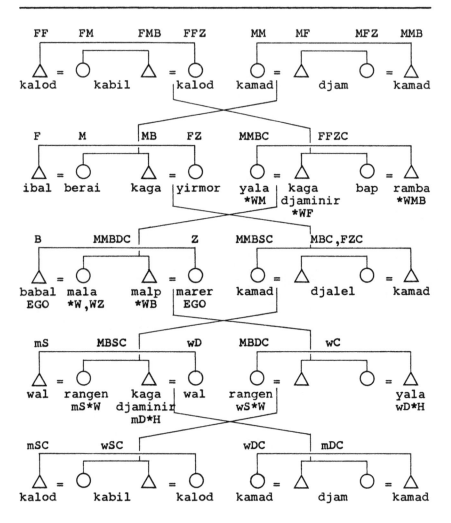

<u>malp</u> 'prospective husband or brother-in-law'.[1]

To account for these classifications we must posit
the same rules of kin-class expansion as were posited
for the Kariera-like systems in Chapter 4. Under these
rules, however, MMBSC, MFZSC, FFZDC, and FMBDC would be
classified as 'brother' and 'sister', and not as 'mother's mother' or 'sister's (woman's) daughter's child',
as indicated in Elkin's diagram. Again, the diagram is
potentially misleading. It illustrates some of the
terminological consequences of a contingent marital
relationship as though they were structural features
of the system of classification itself.

Elkin (1928b) reports that a man has a rightful
say in the betrothal of his ZD and it is considered
appropriate for two men to arrange to betroth their
ZDs to one another. This does not necessarily entail
marriage between ZSD and FMB, but it may do so if a
man's ZDHZ (X in Figure 5.2) has married his own ZS
(by "sister exchange," which also is favored), in which
case ZDHZD would be ZSD also. The two men would be
both WMB and ZDH to one another, and their respective
WMBWs would be their respective ZDs. In contrast, if
ego's WMB marries within his own generation to one of
his classificatory 'cousins', WMBW would be ego's
classificatory 'mother'. Thus WMBW may be ego's
'sister's daughter' or 'mother', depending on the kind
of kinswoman ego's WMB has married.

Similarly, ego's MMBS may be married to a woman

otherwise related to ego as a classificatory 'mother',
or to a woman otherwise related to ego as a classificatory 'sister's daughter'. If MMBSW is 'mother', the
only designations available for MMBSC are 'brother' and
'sister' – for the children of 'fathers' and 'mothers'
are 'brother' and 'sister'. (These designations frequently appear in the genealogies recorded by Elkin.)
But if MMBSW is 'sister's daughter', there is an alternative possibility, for ego's MMBS's children are also
his 'sister's daughter's' children. Thus, the designation for MMBSC (and MFZSC, etc.) as such is 'brother'
or 'sister', but a person who is ego's MMBSC may be
related also in another way, through his or her mother,
as a 'sister's daughter's child'. More generally, many
people may have dual kin-class statuses in relation to
one another, but for each kintype there is only one
appropriate classification – the same as in the Kariera-
like systems discussed in Chapter 4.

The Nyulnyul system differs from the systems dis-

Figure 5.2. ZD exchange

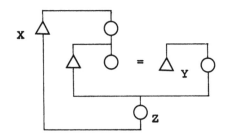

cussed in Chapter 4 in that it has more specially designated prospective in-law subclasses, and in the kintype compositions of these subclasses. But its equivalence rules must be the same as those listed in Table
4.4, because the PARENT and CHILD superclasses of this
system have the same kintype compositions as the PARENT
and CHILD classes of the Mari'ngar and other Kariera-
like systems.

The superclass-subclass relations indicated for
the PARENT and CHILD classes of the Nyulnyul system are

Figure 5.3 Nyulnyul FATHER-MAN'S CHILD class

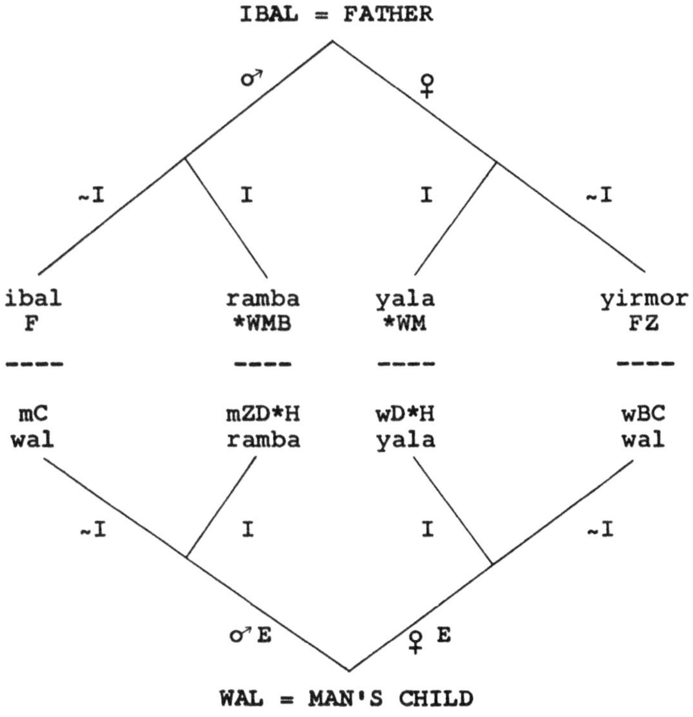

IBAL = FATHER

♂	♀

~I	I	I	~I

| ibal | ramba | yala | yirmor |
F	*WMB	*WM	FZ
mC	mZD*H	wD*H	wBC
wal	ramba	yala	wal

~I	I	I	~I

♂ E   ♀ E

WAL = MAN'S CHILD

I = prospective in-law
~I = not prospective in-law

182

shown in Figures 5.3 and 5.4. The kintypes shown in
the tables represent the foci of the various subclasses.
*WM, it should be emphasized, does not represent ego's
actual WM but ego's prospective WMs; therefore, it
represents a fairly broad range of kintypes, no one of
which may be regarded as focal to the exclusion of the
others. Ego's actual WM cannot be regarded as the focal

Figure 5.4. Nyulnyul MOTHER-WOMAN'S CHILD class

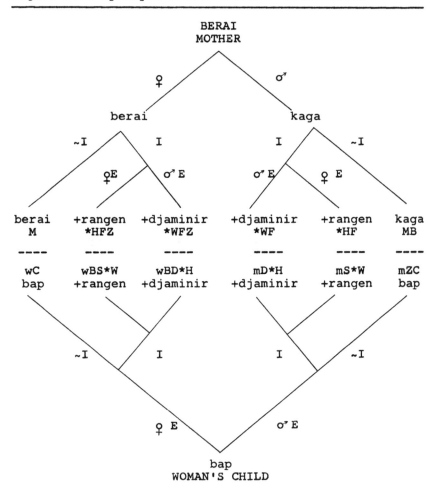

member of the yala class because this class and the
other specially designated prospective in-law classes
are genealogically defined. The evidence for this
assertion is as follows.

Note that there are no designations for HM and wSW
other than 'father's sister' and 'man's child' (which
includes a woman's 'brother's' children). Therefore,
if (as it occasionally happens, see Jolly and Rose 1966:
103) a woman marries a kinsman other than one of her
distant 'cousins', she still designates her HM as though
she were her classificatory 'father's sister'. That
is, a woman's HM who is not already her classificatory
'father's sister' is assimilated to the FATHER'S SISTER
class. The same is true of a man's WF who is not
already his classificatory 'mother's brother', for the
only designation available for such a WF is kaga
djaminir, which is merely a lexically marked form of
the MB term. The cases of HF kaga rangen and WFZ berai
djaminir are similar. It is difficult to imagine that
the rules of kin-class expansion that determine these
classifications do not also determine that a WM who is
not already ego's classificatory 'father's sister' is
assimilated to the FATHER'S SISTER class. Therefore,
it is highly probable that a WM who is not already ego's
classificatory 'father's sister' is assimilated not only
to the yala class but also to the FATHER'S SISTER class.
Of course, it may be that she is never spoken of as
ego's 'father's sister', but this would not be defini-

tive evidence against her inclusion in the FATHER'S SISTER class. So we may say that inclusion in the yala class, whether as prospective WM or as actual WM, implies inclusion in the FATHER'S SISTER class, and the yala class is a subclass of the FATHER'S SISTER class that is genealogically defined. Therefore, a genealogical relationship of some specified kind is a necessary condition for inclusion in the yala class, and to this extent that class is genealogically defined. The inclusion of a WM who is not so related (although she may be genealogically related in some other way) is not evidence against this interpretation, for such a WM is a yala only by extension, and not by definition.

It is not clear from the available ethnographic accounts just how distant genealogically two relatives must be if they are to qualify as potential in-laws. But if Jolly and Rose's report is correct (as assumed in Table 5.1), the rule is that only those FATHER'S SISTERS who are at least three degrees of collaterality distant from ego qualify as potential WMs, and their children, who are four degrees of collaterality removed from ego, qualify as potential spouses or siblings-in-law. But, assuming that the specially designated subclasses are prospective in-law classes (and not, more generally, potential in-law classes), it is not fully adequate to say they are genealogically defined. Membership in a prospective in-law class depends not only on membership in a certain kinship superclass and on a

certain degree of genealogical distance, but also on
social relationships. If the Nyulnyul are, in this
respect, like many other Australian peoples, a man must
ceremoniously avoid and send gifts to the women whom he
would have regard themselves as his prospective WMs.
If he does not do these things he is not entitled to
speak of them as his 'wife's mother'. Therefore, if
the yala class is a prospective WM class (and not, more
generally, a potential WM class), it is "socially" as
well as genealogically defined; but it is none the less
a kin class (and not simply an in-law class).

As already indicated, the rules of kin-class expansion of this system must be the same as those of the
Kariera-like systems discussed in Chapter 4. This
includes the spouse-equation rule, although Elkin's
diagrammatic representation of this system may appear
to indicate the presence of some other rule. Elkin's
diagram implies the following:

WFM	kalod	(= FF or FMBW),
WMF	kamad	(= MMB or MFZH),
WMM	djam	(= MFZ or MMBW),
WFF	kabil	(= FMB or FFZH).

But, again, the diagram - although perhaps useful as a
didactic device, especially in the context of a brief,
nontechnical discussion - is misleading. In his field
report (1928b), Elkin records the following:

wainman   a special self-reciprocal term for
          spouse's mother's parents and their
          siblings

kalod	HMF	(= FF, FFB, FFZ),
kamad	WFM, HFM, WMF	(= MM, MMZ, MMB),
djam	WFF, HFF	(= MF, MFB, MFZ),
kabil	WMM, HMM	(= FM, FMZ, FMB).

There is only one item on this list that is consistent
with Elkin's diagram, that is, WMF kamad. Otherwise,
these classifications are much more like what we would
expect, on the basis of Radcliffe-Brown's and Elkin's
theories, to find in a "Kariera-type" system. Of
course, they indicate the presence of the spouse-
equation rule posited in Chapter 4 (rule 7, Table 4.4).

These classifications are not inconsistent with
the prohibition of cross-cousin marriage or with the
classifications of grandparents' opposite-sex siblings
shown on Elkin's diagram. In most Kariera-like systems,
FM and MF are classificatory 'siblings' of one another,
and so are FM and MFZ. In the Nyulnyul case, also,
they would be classificatory 'siblings' of one another,
given a sequence of orthodox intragenerational marriages, and probably they are in principle as well. It
is not a necessary implication of this 'sibling' relationship, between the 'father's mother' and 'mother's
father' categories, that FM and MFZ should be designated
by the same term. MFZ's terminological status is
dependent on the equivalence rules and the rules of
subclassification of the system, whereas FM's terminological status depends only on the rules that define

the primary senses of the terms. Individuals of these
two kintypes may designate one another as 'sister' and
yet not be designated by the same term by ego, if the
equivalence and subclassification rules of the system
require that MFZ should be designated as 'mother's
father' rather than as 'father's mother'. Of course,
if the rule governing the terminological status of MFZ
were a spouse-equation rule that specifies that certain
kintypes are to be regarded as structurally equivalent
to certain in-law types (e.g., FZ → HM), we would expect
MFZ to be designated as 'father's mother' (for the rule
would require MFZ → MHM → FM). But there is no such
rule in the Nyulnyul system, or in other Kariera-like
systems wherein MFZ is designated as 'father's mother'.
In all these systems MFZ's terminological status is
governed by the parallel-cross neutralization rule and
by certain rules of subclassification. In some systems
the rules of subclassification determine that MFZ is
designated as 'father's mother'; in others they determine that she is designated as 'mother's father'. Thus,
comparing the Nyulnyul system with other Kariera-like
systems, it is clear that the differences among them in
the designations of grandparents' opposite-sex siblings
are attributable entirely to different rules of subclassification, and not to different rules of interkin marriage or to different spouse-equation rules. The Nyulnyul classification of MFZ, for example, as 'mother's
father' is <u>not</u> a function of the Nyulnyul prohibition

of marriage between actual cross cousins (for so do the
Murinbata); nor is it a function of a preference for
marriage between the children of same-sex cross cousins
or classificatory 'cousins' (for there is no such
preference).

Presumably, then, a man's classificatory 'father's
sisters' who are his potential WMs include not only
daughters of classificatory djam (MF's classificatory
'sisters') but also daughters of classificatory kabil
(FM's classificatory 'sisters'). Because Elkin does
not report that WMM may be designated as kabil or as
djam, it seems that the designation for WMM as such is
kabil (or wainman). If a man's WMM happens to be one
of his djam, when he refers to her in her status as his
WMM he designates her as kabil (or as wainman) - as
specified by the spouse-equation rule.

The possibility of ZSD-FMB marriage receives no
formal expression in the Nyulnyul system of kin classification, certainly not in its equivalence rules. Although a man may designate his ZSD as mala 'prospective
wife or sister-in-law', rather than or in addition to
kabil, this is contingent on whether or not his ZSW is
assigned the status of prospective WM. It is not the
case that mZS is regarded as structurally equivalent to
WF, either as a designated or as a linking kinsman, and
that ZSD therefore is structurally equivalent to WFD or
WZ or W. Nor are there any reasons to suppose that the
Nyulnyul system features the rule (mWF → mZS), etc.

Certainly, a man's WF and WFZ may be kinsmen he designated as 'sister's child' prior to his marriage, but
there is no evidence that a man designates his WF or
WFZ as 'sister's child' regardless of their premarital
kin-class statuses - indeed, there is much evidence to
the contrary. Even so, possibility of ZSD-FMB marriage
does have practical implications for the designations
of particular kinsmen (not kintypes) who, because of
this possibility, may be related to one another in two
or more ways that do not have the same designation.

We may ask why it is that ZSD-FMB (and other classificatory kabil), in addition to distant 'cousins',
are singled out from the various classes of kin as
potential spouses. At first glance, the Nyulnyul four-
section system may seem to provide an explanation, for
in this system of classification, as in all four-section
systems, a man's ZSD normally belongs to the same section as his cross cousins. But membership in a particular section is not sufficient to account for ZSD's
status as a potential wife. Not all women of this section qualify as potential spouses; among the COUSINS
only distant 'cousins' qualify; and, in addition to
'cousins' and 'sister's son's daughters', a man has
'daughter's daughters' in this section, none of whom
are marriageable.

For a more satisfactory explanation we must take
some additional information into account. Elkin (1928b)
reports that rights of bestowal of women in marriage

rest with their mother's brothers, and reciprocity is expected if not required. Therefore, if one man, X (see Figure 5.2), gives his ZD to another, Y, the other, Y, is expected to give his ZD to X in return. It seems that X has a right to claim the ZD of Y, that is, X's own ZDHZD. This woman Z may or may not be X's own ZSD. She will be X's own ZSD if and only if X's ZS has married the sister of Y, that is, if and only if "sister exchange" has also occurred, and this would require a conventional intergenerational marriage between classificatory 'cousins'. So, it seems, a man has no marital rights in respect of his ZSD as such; his right is to his ZDHZD via his ZDH (provided he was instrumental in arranging the marriage of his ZD), and it is based on reciprocity in the 'giving' or bestowing of women as wives, rather than on kinship. In this respect it is unlike the rights that men have in respect of their classificatory 'cousins', which are rights as against other men who are not reciprocally the 'cousins' of those women. But this latter right must be compromised if it is to operate in conjunction with the rights of men to bestow their ZDs as wives and with the expectation of reciprocity in bestowal. It is not necessarily compromised very much, however, for it seems probable that the only 'sister's son's daughter' whom a man may marry by right is his ZDHZD (who may be also his ZSD).

If this interpretation of the possibility of ZSD-FMB marriage is correct, we can see why it is that men are not permitted (by right, anyway) to marry their 'daughter's daughters' djam, even though these kinswomen belong to the same section as their 'cousins' and their 'sisters' sons's daughters', some of whom they may marry. The reason is that marriageability is not contingent on section membership and intersection relationships, but on kin-class membership. Orthodox marriage is between distant classificatory 'cousins', but men may claim their ZDHZDs in return for their own ZDs. In all Kariera-like systems of kin classification ZDHZD is designated by the same term as ZSD. This follows from the cross-stepkin rule and its corollaries, one of which specifies that a woman's DHZ is structurally equivalent to her SW (i.e., ZDHZD → ZSWD by corollary of rule 6, Table 4.4; and ZSWD → ZSD by the ordinary stepkin-merging rule). In many Kariera-like systems, no terminological distinction is made between mZSD and mDD, because no distinction is made between mDD and wSD. But in the Nyulnyul system the latter distinction is made, and so mZSD is designated as kabil, and not as djam. Therefore, ZDHZD is not a djam, and so no djam is marriageable, at least not by right.

MARDUDHUNERA

Radcliffe-Brown (1913: 177) described the Mardud-hunera system of kin classification (see Figure 5.5) as "very different" from the Kariera system and as based on four descent lines and preference for marriage

192

Figure 5.5. Mardudhunera kin classification
(after Radcliffe-Brown 1913: 177-83

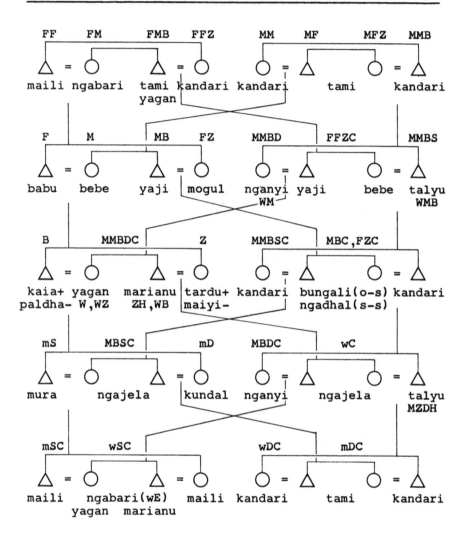

between the offspring of same-sex cross cousins (or persons classified as such). Again, the description is seriously misleading. There is ample evidence to show that, in its basic structural features, this system also is identical to the systems discussed in Chapter 4. The differences emphasized by Radcliffe-Brown are at the subclass level.

The relevant evidence need not be discussed in full. To establish the point beyond a reasonable doubt it will be sufficient to note some of the indications that this system also features the parallel-cross status-extension rule.

We may begin by noting that father's cross cousins, WF and WFZ, are classified as 'mother' and 'mother's brother', and their reciprocals are classified as 'sister's child' (or 'woman's child'). This indicates the probable presence of the parallel-cross status-extension rule. However, this rule would determine also that mother's cross cousins are classified as 'father' babu and 'father's sister' moqul; but Radcliffe-Brown reports that they are designated as nganyi and talyu, the same as WM, WMB. The apparent contradiction may be resolved by demonstrating that nganyi and talyu are special designations for potential or prospective in-law sub-classes of the FATHER and FATHER'S SISTER classes - and, because the terms are self-reciprocal, special potential or prospective in-law subclasses of the MAN'S CHILD class.

The first thing to note is that the husband of nganyi is yaji 'mother's brother', the same as the husband of mogul 'father's sister'. Also, the wife of talyu is bebe 'mother', the same as the wife of babu 'father'. This indicates that, at some level of classification, kinswomen designated as nganyi are equivalent to kinswomen designated as 'father's sister', and kinsmen designated as talyu are equivalent to kinsmen designated as 'father'. In comparative perspective it is easy to see that the relevant level of classification is the superclass level. That is, nganyi designates a subclass of the FATHER'S SISTER class and talyu designates a subclass of the FATHER class. These relationships are confirmed by Bates's (Ms.) report that WMZ (who surely belongs to the same kin class as WM herself) may be designated as 'father's sister'.

The relatives singled out of the FATHER and FATHER'S SISTER classes (and their reciprocals) and designated as nganyi or talyu have a special prospective in-law status, which is the basis for their designation as nganyi or talyu. Radcliffe-Brown (1913: 184-5) reports that it is conventional for a boy's father to arrange for his (the father's) WMBD or WFZD to be the boy's WM, and the two are thereby "made nganyi" to one another. This may be done before the man has a son or the woman a daughter. Every boy is "made nganyi" to several women (and vice versa), "so that he may have the better chance of ultimately obtaining a wife." To

keep "alive" his marital claim to the daughter of a
particular woman to whom he has been "made nganyi," a
man must ritually avoid her and he must visit and give
presents to her husband, who is normally his classificatory 'mother's brother', and to her brother (who is
his talyu). This gives a man the "first right" to the
daughter of this nganyi. He has "only a secondary
right" to the daughters of other nganyi.

This analysis accounts for several observations
that Radcliffe-Brown found puzzling. He reports (1913:
183-4) that in one case a man's male kandari 'mother's
mother's brother' married a woman who was that man's
ngabari 'father's mother' and had five children, three
sons and two daughters. The man told Radcliffe-Brown
that he called the two eldest brothers babu 'father',
and the other brother talyu; he called the two sisters
nganyi. Radcliffe-Brown was unable to determine why
his informant differentiated among the brothers, but
the informant said that all of them were "really half
babu ['father'] half talyu." It is easy to see why he
said this. Regardless of the kin class to which his
mother belongs, any man designated as talyu is a special kind of classificatory 'father'. As for why the
informant differentiated among the three brothers, it
seems that a man does not necessarily designate all
brothers of his prospective WM as talyu. Perhaps, if
a woman has several brothers, one of them is singled
out as the one who will have the greatest authority in

regard to the disposition of her daughters in marriage.
He would be the one whom the woman's prospective DHs
would regard as their prospective WMB and the one to
whom they would give gifts in anticipation of receiving
his sister's daughter as a wife. If a woman had several
brothers and all of them had the same rights with regard
to the disposition of her daughters in marriage, this
arrangement could only lead to conflict within the
sibling set. Therefore, we may suppose that the Mardudhunera worked out an arrangement whereby the potential
for such conflicts was minimized. Also, if a woman had
several brothers and her prospective DHs had to make
prestations to all of them, more or less equally, this
might place an undue burden on the prospective DHs
(depending of course on the nature of the gifts in
question).

Radcliffe-Brown says (1913: 188) that the Mardudhunera system differs from the Aranda-like systems of
Central Australia only in that it "is not used consistently." As an example, he notes that A may call B
'brother' and C 'father'; "if the system were used
consistently," he says, B also would call C 'father',
but in some instances B calls C talyu (WMB) instead of
'father'. This, however, is not a logical or structural
inconsistency, for talyu (WMBs) are special kinds of
FATHERS. The talyu kind of FATHER is the brother (or
one of the brothers) of ego's prospective WM, but not
all men who classify one another as 'brother' have the

same prospective WMs. We have no reasons (from Radcliffe-Brown's report) to suppose that it is not possible for two brothers to have different prospective WMs
and, therefore, to differentiate terminologically among
their (the WM's) brothers. Indeed, it is highly probable that this often happens.

It follows from what has been said so far that a
person does not designate all of his mother's 'cousins',
own or classificatory, as nganyi (WM) and talyu (WMB);
and, conversely, not all offspring of a person's female
'cousins', own or classificatory, are designated as
nganyi or talyu. Whether or not such relatives are so
designated is contingent on whether or not they have
been singled out as prospective in-laws. We may suppose
that certain other designations reported by Radcliffe-
Brown also are contingent in this way. He states:
"Just as my ZD married my talyu, so my ZS marries my
nganyi, and the children of this pair are my marianu
and yagan . . . . This is due to the fact that the
children of a nganyi are in all cases marianu and yagan,
just as the children of a talyu (the brother of a
nganyi) are always kandari" (1913: 182-3). He describes
these terminological relationships as part of "the basis
of the system." At first glance, the implication
appears to be that nganyi (WM) is the appropriate designation of mZSW as such, and yagan (WZ) and marianu
(WB) are the designations for mZSD and mZSS as such.
But this is improbable.

The first thing to note is that there is an apparent inconsistency in the reciprocal relations reported by Radcliffe-Brown. A man is reported to designate his ZSS as marianu 'brother-in-law', but the reciprocal mFMB is reportedly designated as tami 'cross grandfather', although wFMB is reportedly designated as yagan 'husband or brother-in-law'. Now if mZSS may be designated as 'brother-in-law' as the brother of a prospective wife, then why not so designate mFMB as ego's sister's prospective husband? Surely he may be so designated, but only if he is ego's sister's prospective husband, rather than just her potential husband; otherwise he is designated as 'cross grandfather'. Similarly, it must be also that mZSD and mZSS may or may not be designated as yagan (WZ) and marianu (WB), depending on whether or not mZSW is designated as nganyi (WM). If she is not, her children are designated as 'cross grandchild', assuming that tami is normally self-reciprocal, like the other grandparent terms.

All of this suggests that here too, as in the Nyulnyul system, mZSW is designated as 'wife's mother' if and only if she is also mZDHZ, who may be ego's prospective WM in her own right. Of course, her children "are in all cases" marianu (WB) and yagan (WZ), but no doubt they are also ego's tami 'cross grandchildren'.

It suggests also that it is not true that the children of ego's talyu "are always kandari." Almost certainly it is not the case that MMBSC-FFZDC and

MFZSC-FMBDC are always 'mother's mother'-'woman's daughter's child'. The brothers of potential WMs who are not "made" prospective WMs are designated as 'father', and it seems likely that their children are designated as 'brother' and 'sister'. Of course, the children of ego's junior-generation talyu are always ego's kandari because they are ZDHC or ZDC, and ZDC is designated as kandari as the reciprocal of MMB. Also, the child or ego's senior talyu would be ego's kandari if the senior talyu (WMB) were to marry ego's ZD (in which case WMBC would be also ZDC), and if, for the purpose of designating these dual relatives, their relationship to ego through their mother were assigned priority over their relationship to ego through their father. But if ego's WMB marries within his own generation, he normally marries a woman whom ego designates as 'mother' and whose children will be ego's 'siblings', the same as they would be through their father, if their father were not ego's prospective or actual WMB. It makes little or no sense to suppose that WMBC is designated as 'sister's daughter's child' regardless of the kin-class status of WMBW, for we would then have to suppose that WMB as a linking relative is covertly attributed the status of a junior-generation relative, such as the reciprocal ZDH. This in turn would imply that ZDH is covertly attributed the status of WMB, and we would be in a vicious circle. The safest conclusion is that Radcliffe-Brown overgen-

eralized when he said that the children of a talyu are
"always" kandari. Instead, they are kandari if WMBW
happens to be ego's ZD (or classificatory 'sister's
daughter').

As further evidence for this interpretation we may
note that, according to Radcliffe-Brown (1913: 183),
persons of ego's own generation whom ego designates as
'cousin' may be married to persons whom ego designates
as 'sister's daughter's child' or to persons whom ego
designates as 'sibling'. Both arrangements are perfect-
ly "normal" (my phrasing) because ego's MMBSC, for
example, whom ego may designate as 'sister's daughter's
child' or as 'sibling', may be ego's cross cousin's
MMBDC and therefore the two are potential spouses and
siblings-in-law of one another. Their mutual status as
potential (and perhaps prospective) spouses and siblings-in-law is not dependent on how ego classifies
either of them. Therefore, if ego designates his MMBSS
as 'brother' rather than as 'sister's daughter's child'
(because ego's MMBSW is not a woman whom ego designates
as 'sister's daughter'), and if ego's MMBSS marries
properly, this classificatory 'brother' will be married
to one of ego's 'cousins'. It is appropriate to note
that ego himself normally marries a kind of COUSIN (or
a kind of 'sister's son's daughter') - one who has been
made his prospective wife.

Finally, it may be noted that the Mardudhunera mode
of classification of ego's children and siblings' chil-

dren appears to be different from the arrangement in the
Mari'ngar, Murinbata, and many other Kariera-like systems. In the Mardudhunera system male and female ego
use the same terms to denote their sons and daughters;
these terms are extended to the children of ego's same-
sex siblings; and there is a separate term for the
category 'man's sister's child'. Radcliffe-Brown does
not report the term or terms by which a woman designates
her brother's children. Bates's unpublished account
(Ms.) of this system tends to confirm Radcliffe-Brown's
description and suggests that a woman designates her
brother's child as _warri_. This general arrangement is
not inconsistent with description of this as a Kariera-
like system, although it appears inconsistent with the
implications of the parallel-cross neutralization rule
(no. 4, Table 4.4). That rule implies that a man designates his sister's children by the same expression(s)
as she does, but the arrangement here is quite different. One way to resolve this difficulty is to dispense
with the parallel-cross neutralization rule in this
case, that is, to suppose that it is not a feature of
this system. This, however, would leave us without a
means to account for the collateral extensions of the
grandparent terms.

There is a simple way to resolve this difficulty.
It is to suppose that the parallel-cross neutralization
rule _is_ a feature of this system but that its range of
applicability is more restricted than in other Kariera-

like systems. These context restrictions may be expressed in the way the rule is written in kintype
notation. The appropriate expression appears to be:

$$(\ldots m/wFZ. \rightarrow \ldots m/wFB.) = (.wBC\ldots \rightarrow .mBC\ldots),$$
$$(\ldots m/wMB. \rightarrow \ldots m/wMZ.) = (.mZC\ldots \rightarrow .wZC\ldots),$$

that is, let a linking kinsman's father's sister
as a terminus be regarded as structurally equivalent to that linking kinsman's father's brother
as a terminus; conversely, let female ego's
brother's child as a linking relative be regarded
as structurally equivalent to male ego's brother's
child as a linking relative. Similarly for the
case of a linking relative's mother's brother as
a terminus, and so on.

This rule differs from the rule posited for the Mari'ngar system only in that it is not applicable in the case
of ego's FZ or ego's MB as a terminus, nor is it applicable in the case of ego's opposite-sex sibling's child
when regarded as a designated relative. The two rules
have the same structure but different ranges of applicability. Of course, it follows that the superclasses
of this system are somewhat different from those of the
Mari'ngar and other systems dealt with above. In particular, the FATHER'S SISTER class is not a subclass of
the FATHER class, and the MOTHER'S BROTHER class is not
a subclass of the MOTHER class.

SUMMARY

In this chapter we have seen that certain systems described in the literature as Aranda-type systems, and as "based on" a form of second-cousin marriage, are instead structurally quite similar to Kariera-like systems. They feature the same rules of kin-class definition and of kin-class expansion as the Kariera system itself, but they happen to feature also certain superclass-subclass structures which, if not fully understood, may lead the analyst to describe such a system as "very different" from Kariera-like systems and as "based on" four descent lines. From this perspective, however, it is difficult, if not impossible, to account for the many similarities between these systems and Kariera-like systems.

The Nyulnyul and Mardudhunera do not permit cross-cousin marriage, but they do permit marriage between certain types of classificatory 'cousins'. In both systems of kin classification, father's female parallel cousins and mother's female cross cousins are members of the expanded 'father's sister' class, and their daughters are members of the expanded 'cousin' class. Also in both systems, certain kinswomen of the FATHER'S SISTER class are singled out as potential mothers-in-law for a man, and some of these kinswomen become his prospective WMs. The prospective WMs are not designated by the 'father's sister' term but by a special term that may be glossed as 'prospective wife's mother', and their daughters are not designated as 'cousin' but by a special term that may be glossed as 'prospective wife or sister-in-law'. These special terms then designate subclasses of the FATHER'S SISTER and COUSIN super-classes. Other special terms designate prospective WMBs as a special kind of FATHER. Often, these special prospective in-law terms are self-reciprocal. Where the special prospective WM and WMB terms are self-reciprocal, in their expanded senses they designate subclasses not only of the FATHER'S SISTER and FATHER classes but also of their reciprocal class MAN'S CHILD.

The qualifications for being regarded as a poten-tial or prospective WM vary from case to case. Among the Nyulnyul, it seems, virtually any 'distant' classi-ficatory 'father's sister' is eligible to be a man's prospective WM, but his father's female parallel cousins and his mother's female cross cousins are regarded as 'close' relatives and therefore as ineligible. Beyond this range, it is immaterial whether a woman is ego's father's classificatory 'sister' or his mother's clas-sificatory 'cousin' (e.g., his FFFBSD or his MFFBDD), although she may be both at once; she is nonetheless eligible to be his prospective WM. Among the Mardud-hunera, it may be that only ego's parents' 'cousins' are eligible to be his or her prospective in-laws, but it is immaterial whether they are ego's parents' actual or classificatory 'cousins'. As a corollary of this difference between the Nyulnyul and Mardudhunera systems, it may be that some kintypes that are marriageable in Nyulnyul society (e.g., FFFBSDD) are not marriageable in Mardudhunera society. This is simply a result of different rules of subclassification within the FATHER and MOTHER classes.

In both societies it appears that the "orthodox" marriage is between classificatory 'cousins'. That is, men have marital rights in respect to their classificatory 'cousins' (or, rather, their classificatory 'cousins' of a specified subclass). However, men participate in arranging the marriages of their ZDs and may be regarded as 'giving' their ZDs to their ZDHs. In so doing, they establish marital rights in respect of their ZDHZDs as wives for themselves. A man's ZDH's sister may be married to that man's own ZS, in which case his ZDHZD is his own ZSD, and he may regard her as a prospective wife and marry her. If his ZSD is not also his ZDHZD, he has no special marital right in respect of her and he designates her by the usual term for ZSD rather than by the term for prospective wives.

This possibility of intergenerational marriage entails that particular persons may be related in more than one way, and in ways such that alternative classifications are possible. For example, a man's MMBSC is also his MMBSWC, and his MMBSW may be his classificatory 'mother' or his classificatory 'sister's daughter', because ego's MMBS may have married a woman of his own generation or a woman of the second descending genera-

tion in relation to himself. In the one case alter is
only ego's classificatory 'sibling' as the child of
ego's classificatory 'father' (or prospective WMB) and
'mother'; but in the other case he or she is both ego's
classificatory 'sibling' (through alter's father) and
ego's classificatory 'sister's daughter's child'
(through alter's mother). It may be that in such cases
priority is assigned to alter's relationship through
his or her mother. In any event, the result may appear
to be that kintypes of ego's own generation are designated by grandkin terms, for example, that the proper
designation for MMBSC as such is 'sister's daughter's
child' rather than 'sibling'; and this appearance may
further enhance the appearance that these systems are
more Aranda-like than Kariera-like. But the appearance
is misleading. It is not that certain kintypes of ego's
own generation are designated by grandkin terms but that
persons who are related to ego as kintypes of his own
generation are related to him also as kintypes of the
second ascending or second descending generation, and
so on.

In these systems of kin classification there are
no rules of ZSD-FMB marriage or of ZD-exchange that
systematically effect the extensions of kinship terms.
As in all Kariera-like systems, the parallel-cross
status-extension rule determines that MMBSC as such is
classified as 'sibling', and the spouse-equation rule
determines that MMBSW is classified as 'mother'. It

seems, however, that if MMBS is married to a woman whom ego designates as 'sister's daughter', ego may continue to designate her as such and may designate his MMBSC, who is also his classificatory 'sister's daughter's' child, as 'sister's daughter's child'. It is shown in subsequent chapters that there are systems of kin classification in Australia in which certain kintypes in ego's own generation are designated by grandkin terms, but the Nyulnyul and Mardudhunera systems are only superficially similar to them.

Chapter 6

KARADJERI

The focus of this chapter is the system of kin classification of the inland division of the Karadjeri of
Western Australia. This system and several others -
are described by Elkin and Radcliffe-Brown as a special
type, which they designate as the Karadjeri type. They
describe the principal distinctive features of this
type as:

1. Use of three (rather than two or four) terms in
the second ascending and descending generations, distributed in this way, (a) FF, FFZ, (b) MM, MMB, (c) MF,
MFZ, FM, FMB.

2. Terminological differentiation of MBC and FZC.
They attribute these features to a "preference" for
MBD-FZS marriage coupled with a prohibition on FZD-MBS
marriage and a consequent prohibition on the exchange
of sisters in marriage. Radcliffe-Brown (1951: 42)
notes that systems of this type occur in association
with section systems (Karadjeri), subsection systems
(Murngin), patrilineal moieties (Yir Yoront), and appar-

ently, in one society (Larakia) that had no moieties, sections, or subsections.

As noted in Chapter 2, to account for certain differences among the several systems of this type, Radcliffe-Brown introduced the distinction between basic and secondary lines of descent. This permitted him to show, or so he thought, that the apparent seven lines of the Murngin system and the apparent five lines of the Yir Yoront system are artifacts of the diagrammatic devices used to represent these systems; the number of basic lines in both systems is really only three; and so the Murngin and Yir Yoront systems are structurally similar to the Karadjeri system. It was noted also that if Radcliffe-Brown's reasoning is applied to the Karadjeri data, the apparent three lines of this system may be reduced to only two basic lines, and similarly in the Yir Yoront and Murngin cases. This suggests that the basic categories and equivalence rules of Karadjeri-like systems are much the same as the basic categories and equivalence rules of Kariera-like systems. The suggestion is accurate for the Karadjeri system itself except for the presence in that system of an additional equivalence rule, by which a person's father's father is reckoned as structurally equivalent to that person's brother. But this is a relatively low-order rule of the system and its terminological effects are not extensive. The Murngin and Yir Yoront cases are somewhat more complex and are

Table 6.1. Karadjeri (inland) kin classification

Term	Gloss	Primary denotata	Other denotata
1. kaludji	father's father	FF	FFSb; mSC, BSC (also telwel)
2. kami	mother's mother	MM	MMSb; wDC, ZDC (also telwel)
3. djambadu	cross grandparent	MF, FM	MFSb, FMSb; mDC, BDC, wSC, ZSC (also telwel); MBC, FZC, C of (5), (7) (also telwel if younger than ego)
kabali	prospective wife or sister-in-law	distant djambadu of ego's own generation	
yagu	prospective husband or brother-in-law	distant djambadu of ego's own generation, older than ego	
ingalp	prospective husband or brother-in-law	distant djambadu of ego's own generation, younger than ego	
4. djabalu	father	F	FB, FFBS, FMZS, MMBS, MFZS, MZH, SpMB
mugali	prospective WMB	distant djabalu	
5. kamaru	father's sister	FZ	FFBD, FMZD, MMBD, MFZD, MBW, SpM
dalu	prospective WM	distant kamaru	
6. kudang	mother	M	MZ, MFBD, MMZD, FMBD, FFZD, FBW, SpFZ,

7.	kaga	mother's brother	MB	MFBS, MMZS, FMBS, FFZS, FZH, SpF
8.	babalu	brother	B	S of (4), (6)
9.	kabuju	sister	Z	D of (4), (6)
10.	oba	child	C	C of (8), kabali (under 3)
	mugali			reciprocal of mugali under (4)
	dalu			reciprocal of dalu under (5)
11.	djalangga	woman's child	wC	C of (9), mDH, mSW, WBC, HBC, C of yagu or ingalp (under 3)

discussed in Chapter 8.

KARADJERI

Data on the Karadjeri system are available in a
number of sources (Elkin 1928a, 1932, 1964; Piddington
1932, 1937, 1950, 1970; Jolly and Rose 1943-45, 1966).
In addition, I have consulted Elkin's extensive genealogical data. Most of the relevant data on the inland
Karadjeri system extracted from these sources are
presented in Table 6.1.

Super- and subclasses

The classifications of grandparents and their
siblings and the reciprocal kintypes listed in Table
6.1 are somewhat misleading, because incomplete.
Elkin's published accounts of the inland Karadjeri
system give the impression that FF and FFZ (and their
reciprocals) are designated only as kaludji and that
MM and MMB (and their reciprocals) are designated only
as kami. Elkin argues that the terminological distinction between MM and FFZ "follows from" the prohibition
on marriage between FZD and MBS (because this prohibition prevents FFZ from being FWM or MM). In contrast,
Piddington's published (1950: 121-3) and unpublished
(1937) accounts indicate that all these kintypes (FF,
FFZ, MM, MMB) are designated as kami; he says kaludji
is used only by coastal Karadjeri. Elkin's field
report clarifies the situation. It appears that in

practice the tendency is to designate males as kaludji
and females as kami, so that FFZ and MMB may be designated by either term; also mSD and BSD may be designated
as kami (or telwel) because female, and wDS and ZDS may
be designated as kaludji (or telwel) because male. So
the arrangement here is the same as in the Mongkan and
some other Kariera-like systems. This is confirmed by
Jolly and Rose's (1966:98) report that in coastal
Karadjeri FF and his siblings and MM and her siblings
may be designated by either term, kaludji or kami -
probably depending on the sex of alter. If Piddington's
report on the inland Karadjeri is correct (there are no
reasons to suppose it is not), it must be that some
inland Karadjeri do not use kaludji but designate the
parallel-grandparent class as a whole as kami.

The structure of the GRANDKIN superclass of this
system is shown in Figure 6.1. Although, so far as we
know, the grandparent class has no designation, the
reciprocal grandchild class does. Telwel is the common
reciprocal of all three grandparent terms, which may
also be self-reciprocal. Yet another possibility noted
by Piddington is that, in reference to a second descending generation kintype, the appropriate grandparent term
may be coupled with telwel; thus, for example, wDC may
be designated as kami telwel, 'the grandchild kind of
parallel grandmother'. As shown in the figure, the
COUSIN class of this system is a covert subclass of the
CROSS-GRANDKIN class DJAMBADU; that is, it has no spe-

cial designation of its own. It is structurally implied, however, by the presence of the prospective
spouse and sibling-in-law classes (designated as kabali,
yagu, and ingalp), which are subclasses of it. The
evidence for these assertions is discussed below.

According to Elkin and Piddington, FZ may be designated as kamaru 'father's sister' or as djabalu
'father'. Piddington (1937) reports that MB kaga may

Figure 6.1. Karadjeri GRANDKIN class

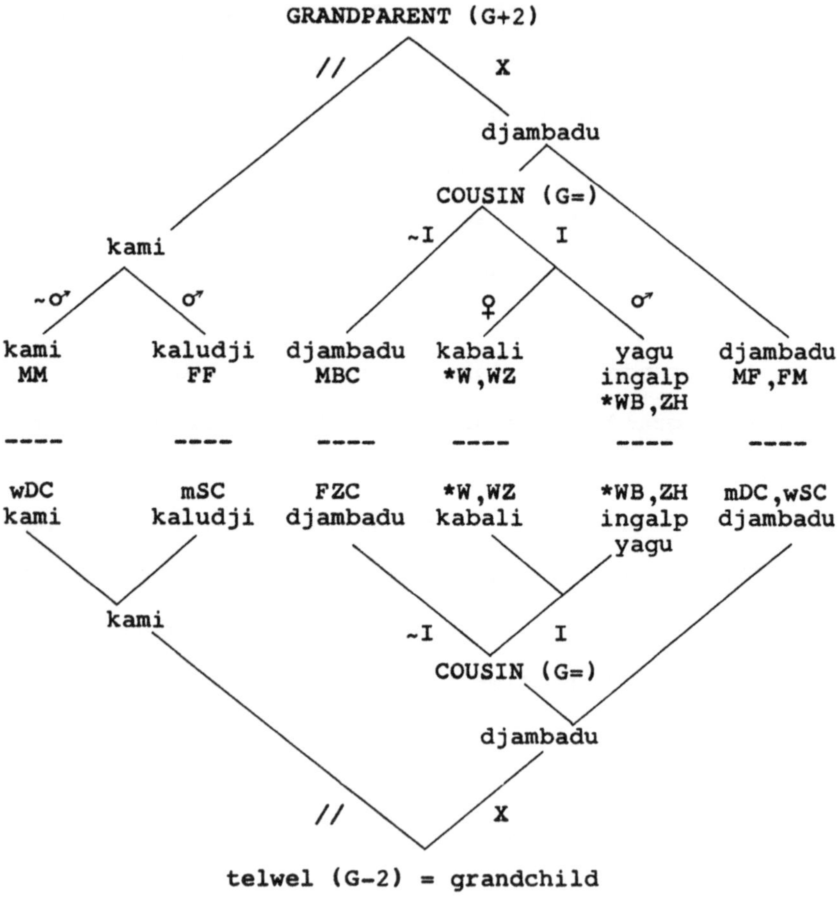

GRANDPARENT (G+2)

//   X

djambadu

COUSIN (G=)

kami   ~I   I

~♂   ♂   ♀   ♂

kami MM	kaludji FF	djambadu MBC	kabali *W,WZ	yagu ingalp *WB,ZH	djambadu MF,FM
----	----	----	----	----	----
wDC kami	mSC kaludji	FZC djambadu	*W,WZ kabali	*WB,ZH ingalp yagu	mDC,wSC djambadu

kami   ~I   I

COUSIN (G=)

djambadu

//   X

telwel (G-2) = grandchild

be designated as <u>kudang</u> 'mother'. It is clear from this and from the relations among the reciprocal terms that here, too, the FATHER'S SISTER and MOTHER'S BROTHER classes are subclasses of the FATHER and MOTHER classes, respectively (see Figure 6.2). The man's child term is also the term for the child class as a whole (cf. Murinbata, Chapter 4). <u>Djalangga</u> 'woman's child' designates a special subclass of the general

Figure 6.2. Karadjeri PARENT-CHILD class

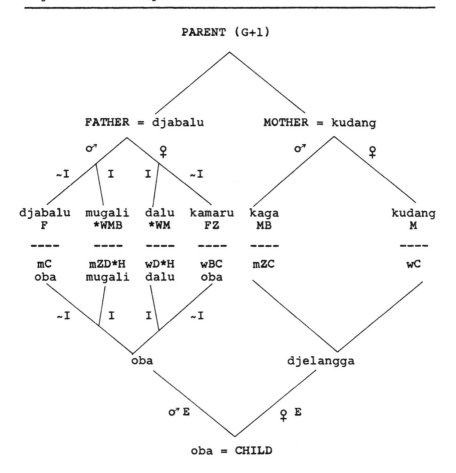

child class.

Jolly and Rose's (1966: 98-9) coastal Karadjeri informants appear to have described members of the class designated as dalu as 'distant' (kajeri, Piddington 1937) 'father's sisters', and members of the class designated as mugali as father's 'distant brothers'. This makes it clear that mugali and dalu designate subclasses of the FATHER and FATHER'S SISTER classes; and because the terms are self-reciprocal they must designate subclasses of the MAN'S CHILD class also. As in the Mardudhunera system (Chapter 5), there are no corresponding specially designated subclasses of the MOTHER and MOTHER'S BROTHER classes.

Not all 'distant' classificatory 'siblings' of one's father are designated as dalu and mugali. According to Piddington (1937, 1970), if a man designates a woman as dalu (he never addresses her as such) this means that he regards her as a prospective WM and must avoid her; he must not go near her camp, he must not mention her name, and he must send her gifts of food through her husband, his 'mother's brother'. Thus dalu signifies not only that the woman so designated is a 'distant' relative of the FATHER'S SISTER class but also that she is regarded as a prospective WM. Similarly, mugali signifies not only that the designated party is a 'distant' relative of the FATHER class but also that he is the brother of a woman who is regarded as a prospective WM or HM.

The criteria for inclusion in these special prospective in-law classes are only partly genealogical.
According to Elkin (1928a), all 'father's sisters' in
ego's own clan are regarded as 'close' (lani) relatives,
but all other classificatory 'father's sisters' are
regarded as 'distant' relatives. Thus, ego's 'distant
father's sisters' include his father's matrilateral
parallel cousins (FMZD, etc.) and his father's wife's
(ego's mother's) cross cousins. In principle, any
genealogically distant 'father's sister' is eligible to
be a man's WM, but should such a kinswoman live in the
vicinity of the camp in which he is growing up, he will
not regard her as a prospective WM (and therefore will
not designate her as dalu, but as 'father's sister').
Piddington (1937) suggests that this is done not because
the duty to avoid one's own WM or potential WMs and
their brothers is excessively burdensome if they live
nearby, but because interpersonal conflict may arise
"from failure to discharge reciprocal obligations
implied in marriage" and the concern is to minimize the
possibility of conflict between genealogically or
socially close relatives.

In certain circumstances an established dalu relationship may be "dissolved." Piddington (1937) notes
that a man and his brother may have the same woman as a
prospective WM. However, if all her daughters become
betrothed or 'promised' to ego's brother, ego will no
longer regard her as a prospective WM - because a man

may not marry his brother's ex-wife or widow, and so
her mother can never become his WM.

If a man designates a woman as dalu, he designates
her daughters as kabali 'prospective wife or sister-in-
law' and her sons as yaqu (if older) or ingalp (if
younger). Among the coastal Karadjeri marriage between
MBD and FZS or FZD and MBS is forbidden because cross
cousins are regarded as 'close' relatives. But among
the inland Karadjeri a man may marry his own MBD if his
MBW is a 'distant father's sister' (or a 'distant daughter') and is regarded as a prospective WM. Thus a man
may designate his MBD as kabali and his MBS as yaqu or
ingalp rather than as djambadu 'cross grandparent' (or
telwel 'grandchild'); but he may not designate his FZD
as kabali or his FZS as yaqu or ingalp (unless this man
is married to or betrothed to his sister). Even so, no
terminological distinction is made between MBC and FZC
as such. Both MBC and FZC may be designated as 'cross
grandparent' (if older than ego) or 'grandchild' (if
younger than ego), but if MBW happens to be ego's
'distant father's sister' and is regarded as a potential
WM, then ego designates his MBC as kabali, yaqu, or
ingalp, as 'prospective wife or sister-in-law' and
'prospective brother-in-law'. So, if we distinguish,
as we must, between the use of kinship terms to denote
particular persons (who may be related to one another
in more than one way) and the use of kinship terms to
designate types or classes of relationships, we must say

that the Karadjeri system of designating relationships, or of classifying types of kin, is not "asymmetrical" but is just as "symmetrical" as the Kariera-like systems discussed in Chapters 4 and 5. The apparent "asymmetry" of the inland Karadjeri system arises from the fact that only 'distant father's sisters' are eligible to be prospective WMs, and MBW may (or may not) be so regarded; therefore, her children may (or may not) be classified as 'prospective wife' and 'sibling-in-law' by male ego. This limited "asymmetry" in the designation of cross cousins is contingent on the kin-class (subclass) status of MBW; it is not a feature of this system that a terminological distinction is made between MBC and FZC as such; the distinction in question is between those COUSINS who are prospective spouses and siblings-in-law and those who are not.

Contrary to Elkin's published report, there is no general prohibition on sister exchange; two men may marry one another's sisters provided that the two men are not cross cousins. Piddington (1937) notes that sister exchange is not common practice but is permissible. In principle, two women may be prospective mothers-in-law to one another's sons (provided they are not also HZ-BW to one another), in which case their sons might, quite properly, marry one another's sisters. Perhaps more commonly, however, a man's ZHM is a woman whom he designates as 'father's sister' rather than as dalu; that is, she is a 'father's sister' who, for one reason or another, has not been made a prospective WM. Her daughter would normally be designated as 'cross grandparent' and would not be regarded as a prospective wife. Even so, Piddington (1937) notes, the man might be permitted to marry the girl, "if no other woman were available" (and, presumably, if no other man had a right to her). A classificatory 'father's sister' who becomes a man's ZHM may subsequently become also his prospective and actual WM.

It is not difficult to see why sister exchange, so-called, is not common, although not prohibited. Among the Karadjeri (as elsewhere in Australia) women are normally betrothed to men much older than themselves, so a man is normally promised a woman whom he would designate as 'grandchild' (rather than 'cross grandparent') if she were not his prospective wife. Therefore, as Piddington (1937) notes, a man's WB is likely to be (though is not necessarily) younger than he is and is designated as <u>ingalp</u> (rather than <u>yagu</u>). It follows that a man's ZHZ is likely to be (though is not necessarily) older than he is and is designated as 'cross grandparent' by him. It may be that ZHZ who is older than ego is so designated even if her mother, ego's ZHM, is designated as 'wife's mother' by ego, for it seems likely that 'prospective wife' is reserved for those daughters of a 'wife's mother' who are younger than ego. Depending on how common it is that ZHZ is older than ego, this might account for the reported rarity of sister exchange.

Thus far we have seen that this system of kin classification is not "asymmetrical" at any generational level. The apparent asymmetry in the second ascending and descending generations disappears once it is realized that the FF and MM categories are subcategories of a parallel-grandparent class. The apparent asymmetry in the first ascending and descending generations results from the fact that the FATHER and FATHER'S SISTER classes feature specially designated prospective in-law subclasses whereas the MOTHER and MOTHER'S BROTHER classes do not; there are special designations for prospective WM (dalu) and prospective WMB, HMB (mugali), but no special designations for prospective WF, HF, WFZ, or HFZ. The apparent asymmetry in ego's own generation results from the fact that FZD is not eligible to be ego's wife and so may not be designated as 'prospective wife' (rather than as 'cross grandparent' or 'grandchild'), whereas MBD may or may not be eligible to be ego's W or WZ and so may or may not be designated as 'prospective wife' (rather than as 'cross grandparent' or 'grandchild'). At further degrees of collaterality, ego's MMBDD and MFZDD may or may not be designated as kabali, and similarly for FFZSD, FMBSD, FMZDD, MMZSD, and MFBSD. Only FFBDD and other daughters of female members of ego's clan are exceptional and like FZD in that they are not eligible to be ego's wives or to be designated as 'prospective wife'.

One implication of all this is that it is misleading to diagram this system as though it consists of or
recognizes three patrilineal descent lines that intermarry asymmetrically or unilaterally (cf. Elkin 1964:
69). Such a diagram implies that FFZSD and FMZDD, like
FZD, are not designatable as <u>kabali</u>, whereas the fact
of the matter is that they are. As Elkin (1928a) notes,
"Marriage with FFZSD would be allowed as long as FFZS
did not marry his MBD, that is my FZ." So "asymmetry"
is not a feature of this system. The equivalence rules
of the system do not place FFZC and FMBC in different
classes; nor do they place MFZC and MMBC in different
classes. FFZC and FMBC are designated as 'mother' or
'mother's brother', and MFZC and MMBC are designated as
'father' or 'father's sister' - or as <u>mugali</u> or <u>dalu</u>
if regarded as prospective in-laws.

Equivalence rules

Clearly, the basic categories and equivalence rules
of this system are much the same as those of the systems
discussed in the preceding two chapters. There is, of
course, one notable difference: Cross cousins are designated as 'cross grandparent' <u>djambadu</u> or as 'grandchild' <u>telwel</u>. This, however, is a relatively minor
difference, for which it is easy to account, simply by
adding another equivalence rule as the lowest-order rule
of the system.

Designation of cross cousins by cross-grandparent-cross-grandchild terms is a common feature of Australian systems of kin classification. Radcliffe-Brown accounted for it by, in effect, positing an underlying rule of structural equivalence of agnatically related kin of alternate generations. This equivalence may be expressed in kintype notation as:

(FF → B) = (mSC → mSb),

that is, let anyone's father's father be regarded as structurally equivalent to that person's brother and, conversely, let any man's son's child be regarded as structurally equivalent to that man's sibling.

This may be described as the "alternate generation agnates equivalence rule" or as the "AGA rule" for short.

The AGA rule written in this way, without context restrictions, implies that a person may designate his or her FF as 'brother' and that a man may designate his SC as 'brother' or 'sister'. Such designations are not uncommon in Australian systems of kin classification, and there is some evidence that they occur also in the Karadjeri system. In Elkin's unpublished genealogical data the term mama appears as a special designation for a much older 'brother' (as an alternative to babalu), and in a few instances it was given as the designation for FF; reciprocally, babalu 'brother' was given as the designation for mSS. It is not reported whether or not mSD may be designated also as 'sister', although it

should follow that she may be. These data indicate some
kind of relationship between the 'father's father' -
'man's son's child' class and the SIBLING class, but in
the absence of more extensive data it is not possible
to be more precise.

We may now return to the cross cousins. Ego's MBC
is, of course, also ego's MFSC and by the AGA rule is
equivalent to MFB or MFZ, both designated as djambadu,
the same as MF. Conversely, ego's FZC is ego's FFDC
and by the AGA rule is equivalent to ego's BDC, also
djambadu. These operations on the cross-cousin kintypes
require that we posit also an auxiliary of the AGA rule,
which is:

  (...wB → ...wFS) = (mZ... → mFD...),

  that is, let a female linking kinsman's brother be
  regarded as structurally equivalent to that woman's
  FS and, conversely, let a man's sister when considered as a linking kinswoman be regarded as structurally equivalent to that man's FD when considered
  as a linking kinswoman.

This rule directs us to regard the kintype MB in the
context MBC as equivalent to MFS and therefore to reckon
MBC's kin-class status as equivalent to that of MFSC;
conversely, it directs us to reckon the kintype FZ in
the context FZC as equivalent to FFD and therefore to
reckon FZC's kin-class status as equivalent to that of
FFDC.

The auxiliary of the AGA rule is a limited "expan-

sion rule" version of the half-sibling-merging rule
(see also Scheffler and Lounsbury 1971: 121-2). It
appears to require operations on kintypes that are
precisely the opposite of operations required by the
half-sibling-merging rule, although in more limited
genealogical contexts. However, the auxiliary of the
AGA rule does not come into conflict with the half-
sibling-merging rule. It is applicable only where no
other rule is applicable, and once it has been applied
the AGA rule itself must then be applied. This implies,
of course, that the AGA rule (together with its auxiliary) is the lowest-order rule of the system. All other
equivalence rules (except the spouse-equation rule)
have priority over it. Because it is the lowest-order
rule of the system, there is no need to place context
restrictions on the rule itself. For example, its
nonapplicability in the case of FFB is determined by
its subordinate status in relation to the same-sex
sibling-merging rule; its nonapplicability in the case
of FFZ is determined by its subordinate status in relation to the parallel-cross neutralization rule; and its
nonapplicability in the case of FFZC is determined by
its subordinate status in relation to the parallel-cross
status-extension rule.

The equivalence rules and relations among them of
the Karadjeri system are listed in Table 6.2.

There is another feature of the Karadjeri system
which, at first glance, may appear to be related to the

AGA rule; but it is doubtful that it is. Elkin (1928a) reports that a man may designate his son as <u>djabalu</u> 'father' and his daughter as 'father' or as <u>kamaru</u> 'father's sister'. Also, a man or woman may designate his or her BS and BD as 'father' and 'father's sister', and the same designations may be applied to any classificatory 'man's child' by male or female ego. Consequently, the terms <u>djabalu</u> and <u>kamaru</u> may be used self-reciprocally. Elkin (1964: 66-7) notes that similar usages occur in many other Australian systems of kin classification. He notes that, typically, they occur when the designated party is "grown up," and not when he or she is still a child, and that, typically, the man's child and woman's child terms are not sex specific.

Table 6.2. Karadjeri equivalence rules

I. (see Table 4.4)

    1. Half-sibling-merging rule

    2. Stepkin-merging rule

    3. Same-sex sibling-merging rule

    4. Parallel-cross neutralization rule

    5. Parallel-cross status-extension rule

    6. Cross-stepkin rule

II. Subordinate to all the above

    7a. The AGA rule: $(FF \rightarrow B) = (mSC \rightarrow mSb)$
    b. auxiliary, $(...wB \rightarrow ...wFS) = (mZ... \rightarrow mFD...)$

III. Subordinate to all the above

    8. Spouse-equation rule (see Table 4.4)

Elkin suggests that as persons mature physically and
socially their sexual identities become more relevant
to social action and are then recognized in the use of
kinship terms that are sex specific, that is, the reciprocals of the terms previously used for them.

Whatever its social motivation, the Karadjeri
practice may be explained (formally) in this way. A
man is permitted to designate his child by the same
term as his child uses to designate him, but not vice
versa - a person may not designate his or her father as
'man's child'. So we may say that the designation (or
one of the designations) for the man's child class is
derived from the designation for the father class. The
simplest way to characterize this derivation is to say
that the designation for the father class is extended,
by the rule of self-reciprocity, to the reciprocal man's
child class. In this way, mS and mD become members of
the FATHER (extended 'father') class. There is a
special term 'father's sister' for the female members
of the FATHER class, so mD, being a female relative and
a member of the FATHER class, may be designated as
'father's sister'. And there is a special term for the
junior-generation members of the FATHER class, that is,
'man's child'. In other words, the junior-generation
members of the FATHER class may be distinguished, by
designating them as 'man's child', from the other members (necessarily senior-generation kin) of the FATHER
class. The members of this residual not-junior (=

senior) subclass of the FATHER class are left with the designation 'father' (or, optionally, 'father's sister' for the females). This means that there are two possible designations for mS, 'father' or 'man's child'; three possible designations for mD, 'father' or 'father's sister' or 'man's child'; two possible designations for FZ, 'father' or 'father's sister'; but only one possible designation for F, 'father'.

If we wish, we may specify a componential definition for the category that includes F and mC, which class results from the application of the rule of self-reciprocity to the father term. But is it not possible to define this class conjunctively. Instead, it has to be defined disjunctively as:

$$[(K.L.G1.+.\sigma') \quad v \quad (K.L.G1.-.\sigma'E)],$$

or, what is the same thing,

$$[(K.L.G1 \ . \ (+.\sigma') \quad v \quad (-.\sigma'E)].$$

This is because the Karadjeri father and man's child classes (unlike the Kariera father and son classes) contrast on more than one dimension of opposition, so that joining them in one superclass is not likely to result in a class that can be conjunctively defined. The father and man's child classes contrast on the seniority dimension. 'Father' has the value senior, and 'man's child' has the value junior. 'Father' has the value male alter, and 'man's child' has the value male ego. Of course, these contrasts are not of the same order. The senior and junior values are (exhaustively) the two values of the seniority dimension, but male alter and male ego are not values of the same dimension. Male alter is a value of the dimension sex of alter, the other value of which is female alter; male ego is a value of the sex of ego dimension, the other value of which is female ego. Even so, the two dimensions are conceptually related to one another, because both are based on the distinction between male and female persons; the difference between them has to do with whether the person in question is ego (or the propositus) or the designated kinsman. Now in general - that is to say, in systems of kin classification throughout the world - the sex of alter opposition is much more commonly used to distinguish and define kin classes (inevitably subclasses of a more general class that includes kintypes of either sex) than is the sex of ego opposition. In and of itself, this distribution suggests that by and large the sex of ego opposition is used in special contexts, that is, to distinguish special or terminologically marked classes (see also Greenberg 1966: 104-5). Where we encounter its use we may expect to find (although we may not be certain that we will find) that any class it is used to define is a specially designated subclass of some other class, and not a structurally independent class. And in general this other class is likely to include and to be based on the reciprocal of the specially designated class. This is the arrangement we have found in the case of

the Karadjeri man's child term; the class designated as
'man's child' (= 'male ego's child') is a specially
designated subclass of the FATHER (expanded 'father')
class. We have seen that the designation for the father
class may be extended, by the rule of self-reciprocity,
to the reciprocal man's child class. This does not
entail neutralization of any of the oppositions that
enter into the definition of the father class, for
although the rule of self-reciproctiy is an extension
rule it is not a rule of neutralization of semantic
oppositions. It is instead a rule of neutralization of
the opposition between one class and its reciprocal
class as wholes. That is, if we posit the class father
we have, by implication, posited also the reciprocal
class man's child; the rule of self-reciprocity specifies that the designation for the posited class is to
be extended to the reciprocal class thereby implied.
Because the designation for the father class is extended
to the reciprocal man's child class, it must be that it
is the father class that is posited (and which has the
unmarked designation) and the man's child class that is
thereby implied (and which has the marked or special
designation).

Of course, the parallel-cross neutralization rule
specifies that female ego's BC is to be regarded as
structurally equivalent to male ego's BC, and the same-
sex sibling-merging rule specifies that a man's BC is
to be regarded as structurally equivalent to his own

child, so that, ultimately, women as well as men may designate their BC as 'father' or 'father's sister'.

The practice of designating BS as 'father' and BD as 'father's sister' is not inconsistent with the practice of designating MBC as 'cross grandparent', that is, as equivalent to mother's 'father' or mother's 'father's sister'. It must be, however, that these practices are based on different rules of kin-class expansion. Note that the extension of 'father' to mS or BS is "asymmetrical"; ego may designate his son or BS as 'father' but alter may not reciprocate with 'man's child'. This "asymmetry" results from the fact that we are dealing here with a special rather than a general rule (cf. Lounsbury 1965: 151). The rule is this case applies only to the father term (and, by implication, to the father's sister term) and results in the extension of that term to the reciprocal man's child category. In contrast, the AGA rule is a general underlying rule of the system; it is not restricted to any particular category, and it is symmetrical. Thus, the extension of 'mother's father' to MBC is paralleled by the extension of 'man's daughter's child' to FZC. Therefore, the two rules cannot be reduced to another more general rule that would cover all the cases covered by the two rules.

Elkin's diagram (1964: 69) of the Karadjeri system indicates that MMBSC and WMBC are designated as __kami__ 'mother's mother' or 'woman's daughter's child', and it

might be thought that this is another example of the
implications of the AGA rule. However, if the parallel-
cross status-extension rule has priority over the AGA
rule, this cannot be, for this priority requires that
MMBSC be classified as 'sibling'. Indeed, this designation is reported by Elkin (1928a), who notes also
that WMBC may be designated as 'sibling'. Here, again,
it seems that we have to deal with the practical consequences of the possibility of ZSD-FMB marriage. That
is, MMBS is classified as 'father' and his children are
classified as 'siblings', provided that his wife is one
of ego's 'mothers'. But MMBSW or WMBW may be ego's own
ZD or classificatory 'sister's daughter', and so ego's
MMBSC or WMBC may be also ego's ZDC or classificatory
'sister's daughter's' child and therefore ego's <u>kami</u>
through his or her mother, but ego's 'sibling' through
his or her father. Thus Piddington (1937) observes
that beyond a fairly narrow range of kin, and beginning
with ego's parents' cousins (parallel or cross), "there
exist merely alternative terms since marriages may be
arranged between people belonging to alternate generations." There is, of course, some truth in this observation, but it should not prevent us from noting that
the alternatives are quite limited and that they are
alternative designations, not for particular kintypes,
but for particular persons who are related to ego in
more than one way. The rules of kin-class definition
and expansion are such that, if all people married

within their own generations, the Karadjeri system would appear altogether Kariera-like, allowing for minor differences of subclass structure. Marriage between kin of alternate generations modifies this appearance, not by introducing the possibility of alternate designations for the same kintypes, but by making it possible for two persons to be related in more than one way and such that they may designate one another by two or more quite different sets of terms.

We may now consider the spouse-equation rule of this system. To all appearances, the rule is that a person's spouse is to be regarded as a member of the COUSIN class (here a covert subclass of CROSS GRANDKIN), that is, the same as the rule posited for Kariera-like systems in Chapter 4 (no. 7, Table 4.4). Certainly, there are no reasons to suppose that the spouse-equation rule of this system is "asymmetrical." Because MBD's status as a prospective wife is dependent on the kin-class status of her mother rather than her father, it would be inappropriate to posit a rule that specifies that WF should be regarded as structurally equivalent to MB in particular. And although part of the rule must be that WM is to be regarded as a member of the FATHER'S SISTER class (because dalu designates a subclass of that class), the classificatory 'father's sisters' who may be WM are not only those to whom a man is related through his mother, that is, his mother's cross cousins; apparently, his father's maternal parallel cousins may also qualify as potential WMs.

Here, too, it seems, the possibility of ZSD-FMB marriage receives no formal expression in the system of kin classification. Elkin and Piddington both report that the spouses of cross cousins and, conversly, the cross cousins of spouses are sometimes designated as 'mother's mother' or 'father's father' (or 'grand - child'), and sometimes as 'brother' or 'sister'. This apparent variation may be accounted for by supposing that, consistent with one of the corollaries of the posited spouse-equation rule (see corollary no. 7g, Table 4.4), the designation for cross cousin's spouse and spouse's cross cousin as such is 'brother' or 'sister'; but sometimes such kinds of relatives are other kinds of relatives as well and are designated according to the other kinds of relatives that they are rather than according to their relationships as cross cousin's spouse or spouse's cross cousin. That is to say, here too we have to distinguish between the deno- tation of persons and the designation of classes or types of relationships.

In brief, the available data offer few if any rea- sons to suppose that the main structural features of the inland Karadjeri system of kin classification differ from those posited in Chapter 4 for Kariera-like systems in general. An appearance to the contrary has been created in published accounts of the Karadjeri system by failure to distinguish between the denotation of

persons, who may be related in more than one way, and
the designation of relationships. The inland Karadjeri
system does differ from the Kariera-like systems discussed in Chapter 4 in that it features the AGA rule,
but this rule is of limited scope and can hardly be
described as a "basic" rule of the system because all
other extension rules (except the spouse-equation rule)
have priority over it.

FORREST RIVER

Without going into the details of the evidence, it
may be noted that the Forrest River systems (Yeidji,
Aranga, Kula, and probably, with some minor differences,
several others) described by Elkin (1928e, 1932, 1962,
1964) are structurally quite similar to the Karadjeri
system. The most notable difference is that in these
systems the only designations available for FF, FFB,
and FFZ, and their reciprocals are 'brother' and 'sister'. There is no distinctive designation for the
potential male parallel-grandparent class. Instead,
this potential class is wholly evacuated and its members
constitute a covert subclass of the SIBLING class.
FFZ's designation as 'sister' may be accounted for by
introducing a corollary of the AGA rule, that is,
$(FFZ \rightarrow Z) = (wBSC \rightarrow wSb)$. As a corollary of the AGA
rule it has the same status as the AGA rule itself in
relation to the other equivalence rules of these systems. That is, it is subordinate to the other rules,

including the parallel-cross status-extension rule, so
that FFZ's designation as 'sister' does not entail
designation of FFZC as 'sister's (or woman's) child'.

Another minor difference between these systems and
the Karadjeri system is that they have no special prospective wife or sister-in-law terms. Female COUSINS
in general are designated as 'father's mother'. But
there is a special designation for male COUSINS, which
distinguishes them from other male members of the CROSS-
GRANDKIN class.

SUMMARY

The preceding discussion has established that the
Karadjeri and Forrest River systems of kin classification are basically Kariera-like systems on which the
AGA rule has been superimposed as a low-order rule of
kin-class expansion. One result is that cross cousins
are terminologically identified with cross grandparents
and cross grandchildren. Another result is that, in
the Forrest River systems, the potential father's
father-man's son's child subclass of the parallel
grandparent-grandchild class is evacuated; FF, FFB, FFZ,
and their reciprocals are designated as 'brother' or
'sister'. In the Karadjeri system, there is a distinctive designation for the father's father-man's son's
child class but some members of this class are designatable also as 'brother' or 'sister'.

The Karadjeri system has been interpreted as "asym-

metrical" and as based on a rule of matrilateral cross-
cousin marriage, but it is not. The appearance that it
is results from the special characteristics of the
Karadjeri criteria of entitlement to the status of
prospective wife, and from failure (on the part of
analysts) to distinguish between the use of kinship
terms to denote persons and the use of kinship terms
to designate kin categories (and to signify the defining
features of those categories).

In some societies with Kariera-like systems of kin
classification, virtually any 'father's sister' is
eligible to be a man's WM. In other societies only
classificatory 'father's sisters' are eligible, and a
classificatory 'father's sister' who is singled out as
a prospective WM is designated by a special expression.
This, basically, is the arrangement in the Karadjeri
case. One implication is that FZ may not be designated
as 'prospective WM' and her daughter, ego's FZD, may
not be designated as 'prospective wife'. MBW, however,
may or may not be a prospective WM. If she is, she is
designated as such, rather than as 'father's sister',
and her daughter is designated as 'prospective wife'.
But if MBW is not regarded as a prospective WM she is
designated as 'father's sister' and her daughter is
designated as 'cross grandparent' or as 'grandchild'.
Thus some men distinguish terminologically between their
FZDs and at least some of their MBDs, but this is not a
matter of distinguishing between the kintypes FZD and

MBD as such. The distinction is between those cross
cousins (or classificatory 'cousins') who are prospective wives and those who are not.

In the Forrest River area, FZD-MBS marriage is not
prohibited and so a man may designate his FZ, as well
as his MBW, as 'prospective WM' or as 'father's sister'.
Moreover, in these systems there is no special term
for prospective wife as distinguished from other 'cousins', and so FZD and MBD are designatable only as
'cousin' (also 'cross grandmother'). It may be, as
Elkin reports, that FZD-MBS marriage is much less common
in this area than is MBD-FZS marriage and that, therefore, FZ is only rarely designated as 'wife's mother' -
whereas she is never so designated in Karadjeri, because
she is not eligible to be ego's WM. There is no need,
however, to suppose (following Elkin) that the Forrest
River systems are (or were) in the process of changing
from Kariera-like to Karadjeri-like systems. There are,
to begin with, no substantial structural differences
between the Forrest River and Karadjeri systems and
Kariera-like systems in general (except for the superimposed AGA rule). Moreover, the differences we have
isolated - in the terminological treatment of prospective in-laws and in the criteria for entitlement to this
status - are merely minor differences in modes of subclassification, and although they are related to differences in marriage rules (in particular to whether or

not a man's FZ is eligible to be his WM), they are <u>not</u> related to rules of cross-cousin marriage that function as rules of kin-class expansion.

Chapter 7

ARABANA

We saw in Chapter 6 that the principal difference
between the Karadjeri and Forrest River systems and the
Kariera-like systems discussed in Chapters 4 and 5 is
that the former have an additional equivalence rule, a
rule of structural equivalence between agnatically
related kin of alternate generations (the AGA rule).
This, however, is a low-order rule superimposed on
systems that are, aside from the consequences of this
rule, quite similar to Kariera-like systems in general.
The most notable consequences of the AGA rule, when
subordinate to the other equivalence rules that characterize Kariera-like systems in general, is that the
potential male parallel-grandparent class becomes a
subclass, either covert or overt, of the SIBLING class;
and the potential cousin (that is, cross-cousin) class
becomes a subclass, again either covert or overt, of
the CROSS-GRANDKIN class. Many Australian systems of
kin classification share these features of the Karadjeri
and Forrest River systems but add to them, again as a

low-order rule, the rule of structural equivalence between <u>uterine</u> kin of alternate generations. One consequence of this is that the potential female parallel-grandparent class also becomes a subclass of the SIBLING class.

The Arabana system has this feature, and so do the Murawari and Wongaibon systems described by Radcliffe-Brown (1923) as "less than fully developed" Aranda-type systems. Because the Arabana data are superior, this chapter focuses on that system and concludes with some brief comments on the Murawari and Wongaibon systems.

ARABANA

The system of kin classification of the Arabana of South Australia is described by Spencer and Gillen (1899: 59-68) and by Elkin (1938a: 438-47). There are no substantial disagreements between these sources, except for Elkin's correction of one or two obvious errors in Spencer and Gillen's report. Elkin's account is by far the more complete of the two, but it too is incomplete in several respects, most notably in the absence of data on the terminological assignments for female ego. Fortunately I was able to interview several Arabana speakers at Port Augusta and Anna Station, South Australia in May, 1972, and to fill in some of the gaps in the published sources.

It is convenient to represent the Arabana system diagrammatically (as in Figure 7.1) as though, in Rad-

242

Figure 7.1. Arabana kin classification (after Elkin 1938a)

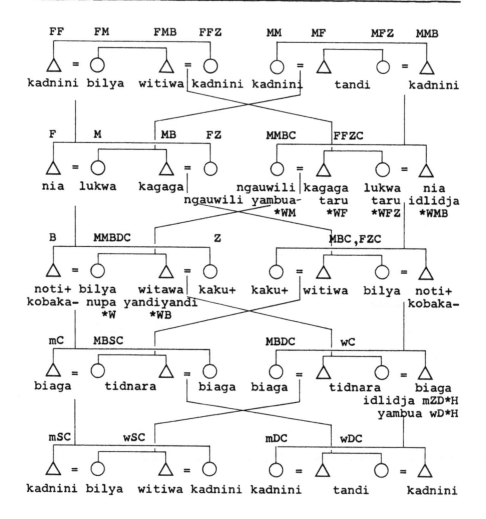

cliffe-Brown's and Elkin's terms, it recognizes four
patrilines and, in this and other ways, expresses the
terminological implications of the prohibition on cross-
cousin marriage and the possibility of marriage between
the offspring of same-sex cross cousins. It should be
fairly obvious, however, that the system is not so
structured. Terminologically, FF's and FM's lines are
not distinguished at any generational level, except for
the prospective in-law subclass distinctions introduced
at the first ascending and first descending generation
levels. Further, although MF and FMB are distinguished
from one another, as _tandi_ and _witiwa_, there are no
corresponding distinctions among their agnatic descendants, again except for the prospective in-law subclass
distinctions introduced at the first ascending generation level and in ego's own generation. This suggests
that the basic categories and equivalence rules of this
system are much the same as those of the Kariera-like
systems discussed in Chapters 4 and 5. But there are
a few differences that require discussion.

First, however, it should be noted that Figure 7.1
differs from Elkin's (1938a: 440) diagram of the Arabana
system in one way that is relevant to the determination
of whether or not this system features the parallel-
cross status-extension rule (no. 5, Table 4.4), which
is the most characteristic rule of Kariera-like systems.
The presence of this rule is indicated by classification
of father's cross cousins as 'mother' and 'mother's

brother' and of mother's cross cousins as 'father' and
'father's sister' (although, as we have seen, these
classifications may be affected also by considerations
of marriageability and, related to this, the possibility
of designation by prospective in-law terms). Conversely, if this rule is present, we expect the children of
ego's female cross cousins to be classified as 'man's
child' and the children of ego's male cross cousins to
be classified as 'woman's child' (or 'sister's child'
for male ego). Elkin's diagram (1938a: 440) indicates
that, for the most part, these classifications are
characteristic of the Arabana system. The apparent,
but not actual, exceptions include the use of idlidja,
rather than nia 'father', as the designation for MMBS,
MFZS, and WMB. But ngauwili 'father's sister' is given
for the sisters of these kintypes, MMBD, MFZD, and WM -
although WM and prospective WM may be designated also
as yambua. Elkin describes idlidja as "the special
term for WMB" (1938a: 442), and it is also the designation of a subclass of the FATHER class. Elkin might
have described it, as he did a similar class in the
Wilyakali system (1938b: 44, 45), as the designation
for a "tribal" or "outside" (that is, classificatory)
'father', who is also a prospective or actual WMB. This
was confirmed by my informants who said "idlidja is a
little bit nia" or "idlidja is close to nia," and who
stated that any senior-generation kinsman designated as
idlidja could be designated also as nia. The subclass

status of the category designated as _idlidja_ is shown
also by the designation of the reciprocal female kintypes MBDD and FZDD as 'man's child' (Elkin 1938a: 440),
which implies that a woman designates her MMBS and MFZS
as _nia_ 'father'. This was confirmed by my informants,
who indicated that there is no special term for HMB,
only 'father'. The corresponding male kintypes MBDS
and FZDS are reportedly designated as _idlidja_ (Elkin
1938a: 440), but surely this is for male ego only and
the designation must signify the relationship actual or
prospective mZDH, the reciprocal of WMB. My informants
indicated that a woman designates her MBDC as 'man's
child' and is designated by them (as their MFZD) as
'father's sister'. _Yambua_ 'wife's mother' is self-
reciprocal and so it is used by a woman to designate
her DH, who may be her MBDS or FZDS.

Another apparent, but not actual, exception to the
implications of the parallel-cross status-extension
rule is that, according to Elkin, FMBD and FFZD are
designated as _tidnara_ 'sister's child', rather than as
lukwa 'mother', although on his diagram (1938a: 449)
the corresponding male kintypes FMBS and FFZS are designated as _kagaga_ 'mother's brother', along with WF also
taru. Yet all the reciprocals are reported as _tidnara_
'sister's child'. The apparent confusion is resolved
when we note Elkin's (1938a: 442) comments, "FMBS and
FMBD are _tidnara_ (or _kagaga_) and _tidnara_ (or _lukwa_),"
and a man may marry the daughter of a "tribal _tidnara_,"

that is, his classificatory 'sister's son's' daughter
(Elkin 1938a: 445). Apparently, the arrangement here
is the same as in other systems analyzed in preceding
chapters. It is not that FMBD as such is 'sister's
child' but that someone related to ego in this way may
be related to him also as the daughter of a classificatory 'sister' - because ego's FMB may marry his classificatory 'sister's son's' daughter who would be ego's
classificatory 'sister'. My informants indicated that
the proper designation for FMBD and FFZD is luka
'mother' or ngamana, a special term for mother's classificatory 'sister', and the proper designation for WFZ
as such is taru, the same as for WF. Taru is self-
reciprocal and so may denote mDH or BDH.

Elkin's diagram shows that MMBSC, MFZSC, FFZDC,
and FMBDC are designated by the sibling terms, and this
is consistent with (and determined by) the parallel-
cross status-extension rule. Elkin's diagram shows
"MMBDD (etc.)" designated as nupa, which he glosses as
'wife', and it shows MMBDS and ZH designated as yandiyandi 'brother-in-law'. But Elkin (1938a: 443) reports
also that MMBDD may be designated also as bilya 'father's mother', the same as FZD or MBD; and, of course,
MMBDD is the daughter of a classificatory 'father's
sister' and of a classificatory 'mother's brother' and
so must be a special kind of COUSIN (see below). My
informants indicated that MMBDD may be designated as
'father's mother', or as nupa if she is considered

marriageable, and that her brother MMBDS may be designated as <u>witiwa</u> 'father's mother's brother' (the same as MBS, FZS), or as <u>yandi-yandi</u> if he is an actual or prospective brother-in-law.

All this demonstrates that the parallel-cross status-extension rule is one of the rules of this system. So is the parallel-cross neutralization rule (no. 4, Table 4.4). My informants stated that FZ <u>ngauwili</u> is "a kind of father," "a female father," and MB <u>kagaga</u> is "a kind of mother." Consistent with this, a man and his sister denote his children by the same term <u>biaga</u> 'man's child'. A woman designates her own child and her ZC as <u>wadu</u> but a man designates his ZC as <u>tidnara</u>, or as <u>wadu</u>. Indeed, a man may designate his own child as <u>wadu</u>, and a woman may so designate her BC. Louise Hercus, a linguist who worked on this language, tells me that <u>wadu</u> means 'child' in the sense of "immature human being." Apparently it is used like English <u>child</u> also as a kinship term, and the kin class it designates has two special subclasses, <u>biaga</u> 'man's child' and <u>tidnara</u> 'man's sister's child'.

According to Spencer and Gillen's data, the collateral extensions of the elder- and younger-sibling terms depend on the relative ages of the linking relatives, not on the relative ages of ego and alter. For example, FB+S is designated as 'elder brother' whether he himself is older or younger than ego, and FB-S is designated as 'younger brother' whether he himself is

older or younger than ego. My data suggest that considerations of the relative ages of linking relatives may
be relevant also to the designations of collateral kin
in G+1 and G-1. For example, although I was told that,
as a general rule, MMBS is designated as 'father', in
one instance I was given 'man's child' as the designation for wMMBS. When I pointed out the apparent contradiction, my informant noted that this MMB was a younger
brother of her MM. So in this case MMB-S is classified
as 'man's child', and reciprocally FZ+DD is classified
as 'father's sister', just the opposite of what we
expect on the basis of the parallel-cross status-
extension rule alone. It will be recalled (see Chapter
4) that Radcliffe-Brown reported that in the Kariera
system reciprocal terms may be reversed, as it were,
on the basis of the relative ages of ego and alter; but
here, it seems, relative age of the linking relatives
is the significant consideration. I was unable, however, to determine whether or not this is a general
rule. Whatever the role of relative age in this context, it must interact to some extent with the effects
of intergenerational marriage - which may result, for
example, in MMBS being also the son of a woman whom ego
designates as bilya 'cousin' (or 'father's mother') or
as nupa 'prospective wife'.

The Arabana parallel-grandparent class is designated as kadnini. The term is used self-reciprocally
to denote FF, FFZ, MM, MMZ, MMB, and mSC, wDC, BSC, ZDC.

Ego's cross cousins may marry intragenerationally or intergenerationally, so their spouses may be 'siblings' or 'parallel grandparents'. Where ego is not related to his cross cousin's spouse in any other way, I was told, the proper term is 'sibling'. So the spouse-equation rule of this system (as in other Kariera-like systems, see no. 7, Table 4.4) specifies that ego's spouse is to be regarded as a COUSIN; conversely, ego's COUSIN'S spouse is to be regarded as a SIBLING.

Although ego's parallel grandparents and their siblings are not designated by the sibling terms, the parallel-grandparent class kadnini is a subclass of the SIBLING class. My informants said of the relatives they designate as kadnini, "they are just like brothers and sisters, although we do not call them that." In other words, the SIBLING statuses of FF, MM, etc., and their reciprocals are covert and not terminologically realized, except for the ways in which they affect the terminological statuses of cross cousins and of kintypes structurally equivalent to cross cousins.

We may account for the (covert) SIBLING status of FF, FFB, FFZ, and their reciprocals by introducing the rule of structural equivalence of agnatic kin of alternate generations and its corollary (posited also for the Forrest River systems, Chapter 6). To account for the (covert) SIBLING statuses of MM, MMZ, MMB, and their reciprocals we must introduce an additional rule, which specifies the limited structural equivalence of uterine kin of alternate generations. This rule (the "AGU rule") is:

$$(MM \rightarrow Z) = (wDC \rightarrow wSb),$$

that is, let a person's mother's mother be regarded as structurally equivalent to that person's sister; conversely, let a woman's daughter's child be regarded as structurally equivalent to her sibling.
This rule also has a corollary:

$$(MMB \rightarrow B) = (mZDC \rightarrow mSb).$$

These rules specify that parallel grandparents and grandchildren are to be regarded as structurally equivalent to siblings. This makes the parallel-grandparent and -grandchild classes subclasses of the SIBLING class, but it does not necessarily imply that parallel grandparents and grandchildren may be designated by sibling terms. The parallel-grandparent class (or subclass) has its own distinctive designation kadnini, and this is the term by which members of that class are designated. That is, they are designated by the expression for the subclass to which they belong, and not by the expression for the superclass to which they also belong.

Although the AGA and AGU rules are here specified without context restrictions, this does not necessarily imply that FFZ and MMB are regarded as structurally equivalent to Z and B when considered as linking kin. If they were, FFZC would be designated as 'sister's (or woman's) child' and MMBC would be designated as 'brother's (or man's) child'. Because these designations are not characteristic of the Arabana system, we may suppose that the AGA and AGU rules as specified above do not affect the statuses of FFZ and MMB <u>as linking kin</u> because they are low-order rules of the system. That is, they are subordinate to all the other rules, including the parallel-cross status-extension rule (although, as shown below, they are so subordinate for different reasons or in different ways).

We may now consider how the terminological statuses of cross cousins are determined. The proper solution to this problem depends on prior understanding of the structure of the CROSS-GRANDKIN class (see Figure 7.2). The relevant data are:

<u>bilya</u> FM, wSC, mZSC, MBD, FZD

<u>witiwa</u> FMB, mZSS, MBS, FZS

Figure 7.2. Arabana CROSS-GRANDPARENT class

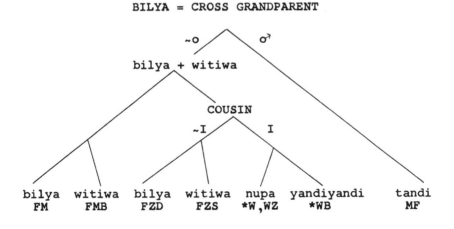

BILYA = CROSS GRANDPARENT

     tandi MF, mDC, MFZ, wBDC

These data differ from Elkin's in that he reports mZSC
as _tandi_ and notes (1938a: 440, note 2) that this seems
odd, because FMB is _witiwa_ and we might expect its reciprocals to be denoted by _bilya_ and _witiwa_. According
to my informants, a man might designate his ZSC as
tandi if their mother, his ZSW, were his classificatory
'daughter', so that ZSC would be also man's 'daughter's'
child. Reciprocally, FMB may be designated as _tandi_;
but only as mother's HMB or mother's classificatory
'father'. Also, according to my informants, a woman
may designate her male cross cousin as _witiwa_ or as
bilya, but a man may designate his male cross cousin
only as _witiwa_.

     In addition, one of my informants said that, as
between _bilya_ 'father's mother' and _tandi_ 'mother's
father', "_bilya_ is the main word, the big word; _tandi_
is like inside _bilya_." As I understand it, this means
that _tandi_ designates a special subclass of the cross-
grandparent class, whose designation is _bilya_; when
this special subclass is removed from the general class
this leaves a residual subclass whose members may be
designated as _bilya_. If this is correct, the distinction made in the opposition between _tandi_ and _bilya_ is
not "male versus female" but "male versus nonmale." Of
course, this is literally true only when the two terms
are used in their primary senses; when the terms are
extended they do not retain this signification.

These data may be accounted for by positing a
cross-grandparent class (see Figure 7.2), which is
expanded collaterally by the same-sex sibling-merging
rule and the parallel-cross neutralization rule, and
to cross grandchildren by the rule of self-reciprocity.
Cross cousins also are included in the expanded cross-
grandparent class (and its reciprocal the cross-grandchild class). In addition, kintypes like MMBDC and
FFZSC are structurally equivalent to cross cousins by
the parallel-cross status-extension rule, and they too
are included in the expanded cross-grandparent class.
It remains only to specify how cross cousins are reckoned as structurally equivalent to cross grandparents
and grandchildren.

This must be by the AGU rule, rather than the AGA
rule, because by the AGA rule and the auxiliary of that
rule posited for the Karadjeri system (no. 7, Table
6.2), MBC would be reckoned as structurally equivalent
to MF and MFZ, and FZC would be reckoned as structurally
equivalent to mDC and wBDC, and therefore designated as
tandi. But they are designated as bilya and witiwa,
that is as equivalent to FM, FMB, and their reciprocals.
The AGU rule accounts for this when supplemented by its
own special auxiliary, which is,

(...mZ → ...mMD) = (wB... → wMS...),

that is, let a linking kinsman's sister be regarded
as structurally equivalent to his mother's daughter; conversely, let a woman's brother, when as a

linking kinsman, be regarded as structurally
equivalent to her mother's son.

In contrast, the AGA rule of this system has no such
special auxiliary; therefore, it is not applicable to
the cross-cousin types. Indeed, the effects of the AGA
rule in this system are so limited that it may be specified with strong context restrictions as,

(.m/wFF. → .m/wB.) = (.mSC. → .mSb.),

(.m/wFFZ. → .m/wZ.) = (.wBSC. → .wSb.),

that is, let ego's father's father as a designated
kinsman be regarded as structurally equivalent to
ego's brother as a designated kinsman; conversely,
let male ego's son's child as a designated kinsman
be regarded as structurally equivalent to male
ego's sibling as a designated kinsman. And similarly for the corollary.

The specification that ego's FF as a designated kinsman
is to be regarded as structurally equivalent to ego's
brother as a designated kinsman does not necessarily
imply that a person designates his or her FF as 'brother'. In the final analysis, how a kinsman of a particular kintype is designated depends not only on rules of
genealogical structural equivalence but also on rules
of subclassification (as well as, perhaps, on other
sociolinguistic considerations). This rule implies
only that ego's FF is to be regarded as a member of the
SIBLING class as well as a member of its parallel-
grandparent subclass, and it is obligatory to designate

one's FF by the expression for this subclass.

Specified in this way, the AGA rule may be regarded as one of the rules of the subset (I in Table 7.1) to which the AGU rule is subordinate.

Now if we suppose that the cross-grandparent class is divided prior to expansion into two subclasses by reference to sex of alter, this implies that FM, FMZ, FMB, and their reciprocals may be designated as bilya 'father's mother', and that MF, MFB, MFZ, and their reciprocals may be designated as tandi 'mother's father'.

Table 7.1. Arabana equivalence rules

I. (unless otherwise noted, see Table 4.4)

    1. Half-sibling-merging rule

    2. Stepkin-merging rule

    3. Same-sex sibling-merging rule

    4. Parallel-cross neutralization rule

    5. Parallel-cross status-extension rule

    6. Cross-stepkin rule

    7. AGA rule,
$$(.m/wFF. \rightarrow .m/wB.) = (.mSC. \rightarrow .mSb.)$$
$$(.m/wFFZ. \rightarrow .m/wZ.) = (.wBSC. \rightarrow .wSb.)$$

II. Subordinate to all the above,

    8. AGU rule
$$(MM \rightarrow Z) = (wDC \rightarrow wSb)$$
$$(MMB \rightarrow B) = (mZDC \rightarrow mSb)$$
    auxiliary
$$(...mZ \rightarrow ...mMD) = (wB... \rightarrow wMS...)$$

III. Subordinate to all the above

    9. Spouse-equation rule

The data depart from this to the extent that FMB and
ZSS are designated as <u>witiwa</u> 'father's mother's brother'
rather than as <u>bilya</u> 'father's mother'. This may be
accounted for by supposing that the male collateral
members of the expanded <u>bilya</u> class are assigned to a
special subclass designated as <u>witiwa</u>. Under the AGU
rule (and its auxiliary) the cross-cousin types reduce
as follows:

$$FZS \rightarrow FMDS \rightarrow FMB \rightarrow FMZ \rightarrow FM$$
$$FZD \rightarrow FMDD \rightarrow FMZ \rightarrow FM$$
$$MBS \rightarrow MMSS \rightarrow ZSS \rightarrow wSS$$
$$MBD \rightarrow MMSD \rightarrow ZSD \rightarrow wSD$$

Assuming that the male collateral members of the expanded <u>bilya</u> class are designated as <u>witiwa</u>, this gives
the desired results, and it accounts for the fact that
a woman may designate her male cross cousins as <u>witiwa</u>.
This is because they are members of the expanded <u>bilya</u>
class but also members of its special subclass <u>witiwa</u>.
If the subclass distinction is neutralized for female
ego in relation to her own-generation kin, she may
designate her male cross cousins as <u>bilya</u>. Apparently
the subclass distinction is not neutralizable for male
ego.

With regard to marriage, Spencer and Gillen (1899:
64) state: "A curious feature in the social organization of the Urabunna tribe is the restriction in accordance with which a man's wife must belong to what we may
call the senior side of the tribe so far as he himself

is concerned. He is only nupa to the female children of the elder brothers of his mother, or what is exactly the same thing, to those of the elder sisters of his father. It follows from this that a woman is only nupa to men on the junior side of the tribe so far as she is concerned." And elsewhere (1899: 61): "A man can only marry women who stand to him in the relationship of nupa, that is, [who] are the children of his mother's elder brothers blood or tribal, or, what is the same thing, of his father's elder sisters. The mother of a man's nupa is nowillie ['father's sister'] to him, and any woman of that relationship is mura [Aranda WM] to him and he to her, and they must not speak to one another."

Thus, according to Spencer and Gillen, a man may marry his MB+D or his FZ+D, or the daughter of any of his mother's classificatory 'elder brothers' or father's classificatory 'elder sisters', but he may not marry his MB-D or his FZ-D or the daughters of any of his mother's classificatory 'younger brothers' or father's classificatory 'younger sisters'. Of course, because designation as elder or younger sibling is contingent on the relative ages of the linking relatives (for collaterals of the second or greater degree), a man's father's 'elder sisters' would (presumably) include the daughters of his mother's mother's elder brothers.

The information given by Elkin's informants is somewhat different. According to Elkin (1938a: 441-2),

they "denied that cross-cousin marriage of either kind
[between a man and his MBD or between a man and his
FZD] was, or had been, permissible." Apparently Elkin's
informants told him that a man could marry his bilya,
and at first he took this to mean that a man could
marry his own cross cousin; but when his informants
realized that he had received the wrong impression,
"they protested in decided terms that such a marriage
was impossible." But, contrary to Elkin's (1938a: 442)
claim, this prohibition is not "reflected" in the classification of ego's own cross cousins as bilya and
witiwa, for, although Elkin's diagram does not show it,
the 'fathers' sisters' daughters and 'mothers' brother's' daughters whom a man may marry also are designated as bilya (Elkin 1938a: 443) and their brothers
are designated as witiwa - though they may be designated also as nupa and yandiyandi as prospective spouses
or siblings-in-law. Terminologically, what demonstrates
that marriage between actual cross cousins is prohibited
is that actual cross cousins may not be designated as
nupa or yandiyandi.

Elkin (1938a: 442) states that, as a rule, a man
may marry the daughter of a 'distant' 'mother's brother'
and 'distant' 'father's sister'; the "parents-in-law
are the cross cousins of ego's parents; in other words
the marriage norm is with the four types of second
cousin which are associated with the Aranda type of
kinship system." But, Elkin adds (1938a: 443), he

"found some hesitation about admitting the correctness of marriage with own second cousins." Some of Elkin's informants agreed. They said that the proper WM for a man is a 'father's sister' "from far away, from another country" (that is, from a different local group and totemic group than his own FZ), and the proper WF for a man is a 'mother's brother' "from far away, from another country" (that is, from a different local group and totemic group than his own MB). Consistent with this, my informants did not volunteer the prospective in-law terms for any kintypes within second-cousin range.

MURAWARI AND WONGAIBON

Radcliffe-Brown (1923) described the Murawari and Wongaibon systems as "less than fully developed" systems of the Aranda type. He acknowledged that they differ from the Aranda system in having only two or three categories in the second ascending generation, and in not distinguishing terminologically between WF and MB, although they do distinguish between FZ and MMBD, WM. There is ample evidence,[1] however, that this is a sub-class distinction. For example, Radcliffe-Brown (1923: 436) reports of the Murawari expression kundikundi, some of the denotata of which are WM, MMBD, and MFZD, that "a man applies this term to some of the women who are ['elder sister'] or ['younger sister'] to his father, but not to his own father's sister." A man's kundikundi "are his mothers-in-law, the women whose

daughters he might marry." This shows that <u>kundikundi</u>
designates a potential or prospective WM class that is
a subclass of the expanded father's sister class, and
the more general implication is that this system features the parallel-cross status-extension rule posited
for Kariera-like systems in general. It could be demonstrated in detail, were it necessary to do so, that
this is true also of the Wongaibon system.

Therefore, the interesting differences between the
Murawari and Wongaibon systems and other systems with
the parallel-cross status-extension rule are in the
terminological assignments of grandkin and cross cousins. In the Murawari system all parallel grandparents
and grandchildren are designated by sibling terms,
although there is an additional special designation for
MM, and FF and MMB may be distinguished from B+ by
reduplication of the elder brother term. There is only
one term for cross grandparents and it is extended to
male cross cousins; female cross cousins are designated
by a different term but this, presumably, is a subclass
distinction. Relatively distant kintypes of ego's
generation reckoned as structurally equivalent to cross
cousins (by the parallel-cross status-extension rule)
are designated by special prospective spouse and
sibling-in-law terms, if their mother's (ego's classificatory 'father's sisters') are regarded as prospective
WMs.

The arrangement in the Wongaibon system is virtually the same; there are some interesting differences in the structure of the cross-grandkin class, but they need not be discussed here. It is worth noting, however, that in a more recent report on this system, J. Beckett (1959: 201) notes that his informant grouped the three terms 'elder sister', 'younger sister' and 'mother's mother' together in a single class, which he glossed in English as 'sisters, grannies'. Similarly, referring to cross grandkin and cross cousins, Beckett's informant spoke of a class of 'grandfathers, cousins'. Clearly, this informant was attempting to describe the SIBLING and CROSS-GRANDPARENT classes of this system. They have no single-term designations and must be expressed, lexically, by the union of the appropriate subclass designations.[2]

Because all parallel grandkin are included in the SIBLING classes of these systems, the AGA and AGU rules must be posited for them. But because in neither system is a distinction made between MF and FM, it is not possible to determine (at least on the basis of the available data) whether the kin-class statuses of cross cousins are governed by the AGA or the AGU rule. It is conceivable that in these systems either rule may be used to reckon the kin-class statuses of cross cousins (and of kintypes structurally equivalent to them). If so, we must posit special auxiliaries for both rules and note that it makes no difference which mode of reckoning is chosen. However, only one of these

rules is <u>needed</u> to reckon the kin-class status of cross
cousins. The principle of parsimony therefore requires
that we posit a special auxiliary for only the AGA or
the AGU rule - but, again, the available data do not
permit a choice.

SUMMARY

This chapter has dealt with three systems of kin
classification that are superficially dissimilar to
Kariera-like systems, but it has been shown that their
principal categories and most of their equivalence rules
are virtually identical to those of the Kariera-like
systems analyzed in preceding chapters. The differences
are attributable to the fact that in these three systems
two additional rules are superimposed on the basic
Kariera-like structure. One is the rule of structural
equivalence of agnatically related kin of alternate
generations, which we encountered also in the Karadjeri
and Forrest River systems; and the other is the rule of
structural equivalence of uterine kin of alternate
generations. These rules determine that parallel grandparents and their siblings are classified as SIBLINGS,
although there may be special parallel-grandparent
categories, which are, however, subclasses of the
SIBLING class. They determine also that cross cousins
and kintypes structurally equivalent to them are classified as CROSS GRANDPARENTS. The data available for
the Murawari and Wongaibon systems are insufficient to

permit determination of whether the kin-class statuses
of cross cousins are determined by the AGA rule or the
AGU rule. In the Arabana system they are determined by
the AGU rule. In all three systems the AGA and AGU
rules do not affect the terminological statuses of
collateral kin in the first ascending and first descending generations, or the terminological statuses of more
distant collateral kin in ego's own generation, except
insofar as they affect the terminological statuses of
those kintypes that are reckoned as structurally equivalent to cross cousins via the parallel-cross status-
extension rule.

Chapter 8

YIR YORONT AND MURNGIN

In this chapter it will be shown that the Yir Yoront
and Murngin systems of kin classification are, at base,
Kariera-like systems on which the AGA rule (no. 7,
Table 7.1) is superimposed. In both systems, however,
the AGA rule has priority in relation to the parallel-
cross status-extension rule, so its effects are more
extensive than in the Karadjeri and Forrest River
systems (Chapter 6). The Murngin system also has the
AGU rule (no. 8, Table 7.1) but it is highly restricted,
like the AGA rule in the Arabana system. The Murngin
system also features more special potential in-law
classes than does the Yir Yoront system.

YIR YORONT

Sharp (1934: 112-3) describes the Yir Yoront vocative system of kin classification by means of a diagram
(see Figure 8.1) with five "patrilineal lines of
descent" linked by matrilateral cross-cousin marriage
into an open-ended series.[1] He describes three of these

Figure 8.1. Yir Yoront kin classification (after Sharp 1934: 112-3, 1937: 78-9)

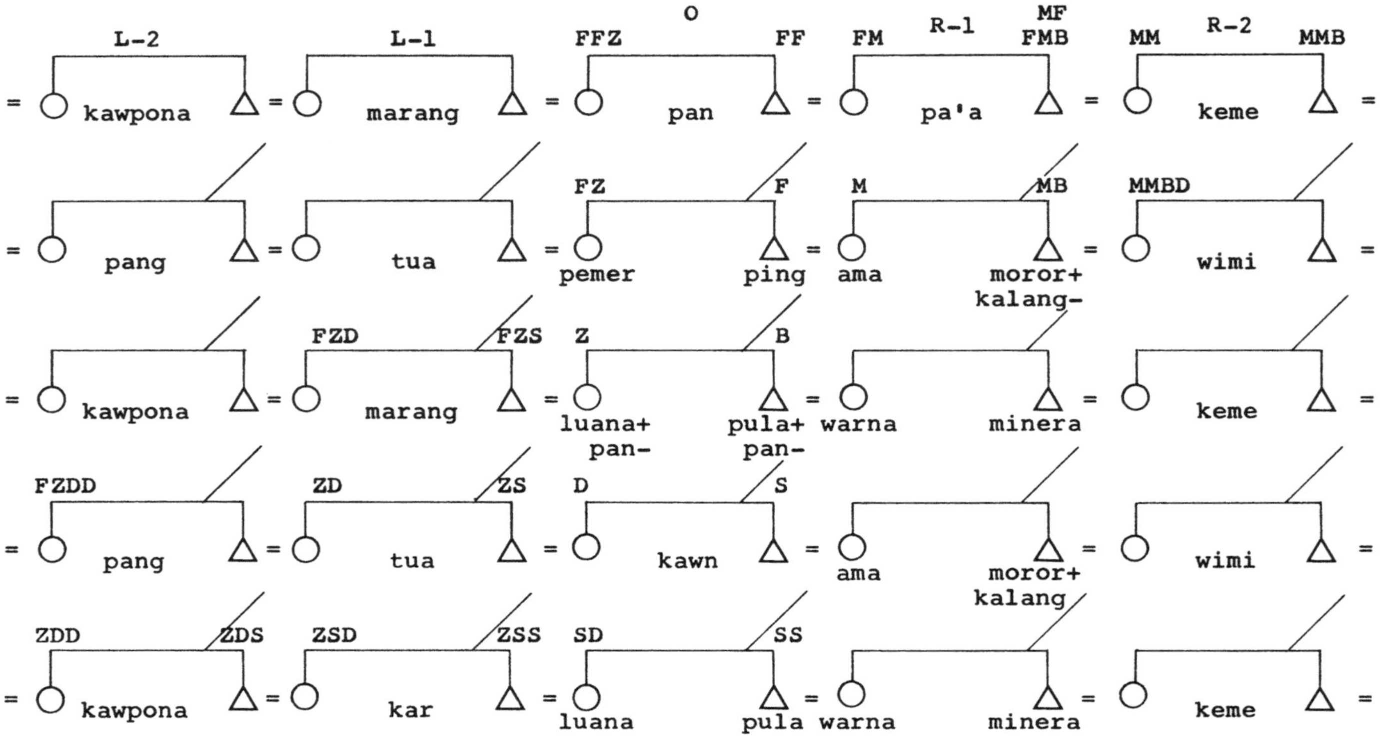

266

lines - FF's, MF's (=FMB's), and MMB's, or O, R-1, R-2
- as "primary." They contain senior terms (see Table
8.1) whose junior reciprocals appear within the same
line, in the case of line O, or in the other two lines -
L-1 in descending generations contains the reciprocals
of the terms of R-1 in ascending generations; L-1 in
ascending generations contains the reciprocals of R-1
in descending generations; and similarly for L-2 and
R-2. That is, L-1 and L-2 are simply the reciprocal
lines implied by R-1 and R-2, whose terms are not self-
reciprocal. Junior terms appear in senior generations

Table 8.1. Yir Yoront reciprocal relations

Senior category	Focal members	Junior reciprocal
1. keme	MM / wDC	kawpona
2. pa'a	FM, MF / wSC, mDC	kar
warna	MBD, etc. /	
	/ FZC	marang
minera	MBS, etc. /	
3. ping	F / mC	kawn
pemer	FZ / wBC	
wimi	MMBC / FZDC	pang
4. ama	M / wC	tua
kalang	MB / mZC	
5. pula	B+ /	
	/ Sb-	pan
6. luana	Z+ /	

in **L-1** and **L-2** because the reciprocal senior terms are extended throughout **R-1** and **R-2** regardless of the generation level of alter in relation to ego - again presumably by the rule of structural equivalence of agnatically related kin of alternate generations (the **AGA** rule). From this perspective the system appears to be "based on" only three patrilines.

This appearance is seriously misleading, for there are good reasons to believe that we might better say the system is based on only <u>two</u> such lines - or, better still, on no such lines at all. D. F. Thomson also recorded information on this system, or on one very closely related to it (see Thomson 1972: 30). In Thomson's data FF and MM are not terminologically distinguished; both are designated as <u>kemer</u> (Sharp's <u>kawpona</u> wDC). In comparative perspective, this apparent disagreement is not strange. Some variant of <u>keme</u> is one of the common parallel-grandparent terms in this area and in some places it covers both the potential male and female subclasses. We know also that in many Australian systems of kin classification FF and FFZ and their reciprocals may be designated either by the same term(s) as MM and MMB or by sibling terms, or they may be designated by a parallel-grandparent term distinct from that for MM and MMB, or by sibling terms. Presumably, this is a case of the first kind. There is a single parallel-grandparent term, applicable to FF and MM and their siblings, but FF and FFZ may be designated

also by the sibling terms. Because we have no reasons
to suppose that MM and MMB also may be denoted by
sibling terms, we cannot assume that the parallel-
grandparent class as a whole is a subclass of the
SIBLING class; instead we must assume that FF and FFZ
have dual statuses, as parallel grandparents and as
SIBLINGS. It does appear strange that FF and FFZ may
be designated by the junior-sibling terms, and mSC and
BSC by the senior-sibling terms - but more of that
below.

There is further evidence that FF's and MMB's lines
(so-called) are not wholly distinct. Sharp (personal
communication 1972) says his informants told him that a
female wimi (e.g., MMBD) is "a kind of pemer [FZ]" and
a male wimi is "a kind of ping [F]."[2] In this system
also the father's sister class is a subclass of the
FATHER class and the mother's brother class is a subclass of the MOTHER class. Thus wimi designates a subclass of the FATHER class and pang (wimi's reciprocal)
designates a subclass of the MAN'S CHILD class. There
are, it seems, no corresponding subclasses of the MOTHER
and WOMAN'S CHILD classes. The arrangement here is much
the same as in the Karadjeri system (Chapter 6) where
dalu (WM) and mugali (WMB) designate subclasses of the
FATHER and MAN'S CHILD classes. The difference is that
dalu and mugali are sex specific (at the senior-generation level) but wimi and pang are not; they are the
simple reciprocals of one another. Like dalu and

mugali, wimi and pang designate potential or prospective
in-law classes. Sharp (1934: 418) reports that the Yir
Yoront say, "I get my wife from that mother's mother's
brother's group." This suggests that here, too, the
orthodox marriage is that of a man with his MMBDD, whose
M and MB he designates as wimi even though they are his
classificatory 'father's sister' and 'father'.

So far, we have seen that some of the apparent
asymmetry of this system, like the apparent asymmetry
of the Karadjeri system, results from the potential
in-laws of the FATHER and MAN'S CHILD classes being
singled out for special designation, whereas the potential in-laws of the MOTHER and WOMAN'S CHILD classes
are not. The system appears asymmetrical also in that
MBC and FZC are designated by different terms. FZC is
designated as marang, MBD as warna, and MBS as minera.
It is evident from the numerous terminological identifications of agnatically related kin of alternate
generations that this system features the AGA rule, and
this suggests that warna and minera designate subclasses
of the CROSS-GRANDPARENT class PA'A. It seems probable
that these subclasses are distinguished as potential
(not prospective) in-law classes for male ego (and as
fraternal sibling-in-law classes for female ego), but
it seems that these classes do not focus on MBD and MBS.
They consist of kin of ego's generation and below who,
by the extension rules of the system, fall into the
CROSS-GRANDPARENT class. So warna may be defined as

(PA'A.~G+2.♀), that is, "CROSS GRANDPARENT, but not of G+2, and female," and <u>minera</u> as (PA'A.~G+2.♂), that is, "CROSS-GRANDPARENT, but not of G+2, and male" (see Figure 8.2). MBD and MBS fall into these classes but they are not the focal members. If, as some of Sharp's observations suggest, a man's rightful wife is his MMBDD, it is not necessary to suppose that MMBDD is the focus of the <u>warna</u> class and that MMBDS is the focus of the <u>minera</u> class. This supposition would not permit us to account for the fact that MBD and MBS also are desig-

Figure 8.2. Yir Yoront CROSS-GRANDPARENT class

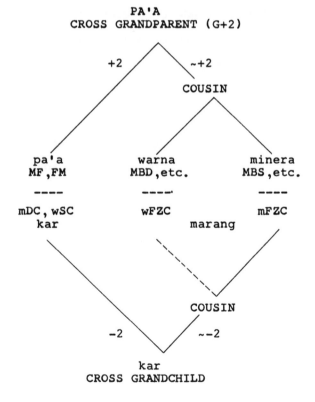

nated by these terms.

Further, the supposition that the warna and minera classes do not focus on MBD and MBS but are, instead, broadly defined classes taken out of the PA'A class, permits us to account for the fact that FMBSC are designated by these terms. The designations of FMBS as 'mother's brother' and of FFZS as 'sister's child' may be accounted for only by supposing that in this system the AGA rule has priority over the parallel-cross status-extension rule. The AGA rule corollary (FZZ → Z) and (wBSC → wSb) applies in the case of FFZC (thus → ZC) but not in the case of FMBC, and this leaves the kin-class statuses of FMBC to be determined by the parallel-cross status-extension rule (no. 5, Table 4.4). By this rule, FMBS's kin-class status is 'mother's brother'. Designation of FMBSC as warna and minera is consistent with this. Note, however, that if the AGA corollary applies in the case of FFZS, it must apply also in the case of FFZSC, who is thereby reckoned as equivalent to ZSC. The reciprocal relationship is FMBSC, and the AGA corollary now applies, so that this kintype is reckoned as structurally equivalent to FMZ and FMB; it is not reckoned as structurally equivalent to MBS and MBD. Even so, FMBSC is designated as warna and minera. The apparent discrepancy is resolved if we suppose (as we have already) that MBD and MBS are not the foci of warna and minera, but that these terms designate broadly defined subclasses of the CROSS-GRANDPARENT class. The

reciprocal marang class must therefore be a subclass of
the CROSS-GRANDCHILD class KAR. Thus kin who fall into
the KAR class but who are in ego's own generation or
above, and not in G-2, are designated as marang.

It may now be seen that, with respect to the designations of cross cousins, the Yir Yoront system is
only superficially asymmetrical. Here, as in the Karadjeri system, there is no distinction between MBC and
FZC as such. Instead, cross cousins are reckoned as
members of the CROSS-GRANDPARENT and CROSS-GRANDCHILD
classes, just as they are in many other Kariera-like
systems (systems with the parallel-cross status-extension rule and the parallel-cross neutralization rule).
But the Yir Yoront system differs from some other
Kariera-like systems in that its cross-grandparent term
is not self-reciprocal. One result of this is that FZC
and MBC are not designated by the same terms, because
FZC are reckoned as equivalent to BDC, and MBC are
reckoned as equivalent to MFSb. This arrangement is
obscured (for the non-Yir Yoront analyst, anyway) by
the singling out of members of the CROSS-GRANDPARENT
class who are not kin of G+2 for special designation by
the terms warna and minera, and of members of the CROSS-
GRANDCHILD class who are not kin of G-2 for special
designation as marang.

The warna and minera subclasses of the CROSS-
GRANDPARENT class and the reciprocal marang subclass of
the CROSS-GRANDCHILD class together constitute a class

that appears to have no special designation of its own, at least within this terminological system. This class is expressed, however, in the Yir Yoront system of signifying classes of kin by gestures (Sharp 1937: 75), both kinds of cross cousins and siblings-in-law being signified by the same body part, the thigh, thus indicating that conceptually (if not terminologically) they are members of one kin class. We may describe this as the Yir Yoront COUSIN class, but it should not be lost sight of that its structure differs somewhat from that of the COUSIN classes of the systems analyzed in preceding chapters; the cross-cousin types are not the foci of this class.

Sharp's (1934: 420) comments on the classification of kin beyond the range shown in his diagram are relevant here. He notes that beyond this range it is less clear how relationships are classified. He says his informants were "unable to say what their WMM or WMBW should be called." He adds, "For them the ideal structure hangs in mid-air at its lateral terminations, a fact which causes them some mental discomfort when it is pointed out, and leads to attempts to make good the deficiency." When asked in the abstract for the designation for male keme's (WMF's, MMB's) wife, there was "a slight tendency" to say pa'a (FM, MFZ); and when asked about male wimi's (MMBS's, WMB's) wife, the "tendency" was to say ama (M).[3] These "tendencies" are readily understandable if this is basically a Kariera-

like system. It would then follow that Sharp's informants, replying in the abstract, reasoned that a male
keme (WMF, MMB) is like one's FF (or FFB) and so his
wife is like one's FM pa'a; and male wimi (MMBS, WMB)
is like one's F (or FB) and so his wife is like one's
M ama. Sharp (1934: 420) adds: "In the actual genealogies this tendency does not appear. It is found,
rather, that EGO may call an individual appearing in
lines R-3 and L-3 by any one of the several terms proper
to that person's generation and sex except those terms
included in the immediately adjoining lines, the relationships being traced through nearer relatives so that
these kin are brought back irregularly into the structure."[4] In other words, he found that ego's male kemes'
(MMB's, WMF's) wives were designated variously as pa'a
(FM, MFZ), warna, or marang, or even as pan (FFZ) or
kawpona (ZZD); and the male wimis' (MMBS's, WMB's)
wives were designated variously as ama (M) or tua (ZD),
or even as pemer (FZ) or pang (FZDD). And these designations depended on how ego was related to these women
in other, often closer, ways. Presumably some of this
variation is attributable to "wrong" marriages (Sharp
1934: 428), but most of it surely is not.

Consider the case of wimi's wife who is ego's ama
'mother'. This appears to imply that ego's MMBS has
married one of his own marang, presumably not his own
FZD but a classificatory 'father's sister's' daughter,
such as his MFZDD who was not also his own FZD. Sharp's

diagrammatic representation of the system suggests that this is not permitted, but it may be that such marriages are permitted. Among the Ngantjara to the north,[5] FZ, FFBD, FMZD, MFZD, and MMBD are designated as ngatinye 'father's sister', but MMBD is singled out as a special case and is designated as ngatinye ngaintj 'tabu father's sister' or, as they say, "poison auntie." She is regarded as male ego's rightful WM and her daughter is designated as tumam, the same as MBD. The daughters of the other 'father's sisters' are designated as muuya. A man may marry his MB-D, and his MB-W may be his MMBD, but he may not marry his FZD and so MBW cannot be MFZD. However, this does not preclude the possibility of marrying MFZDD, even though she (MFZDD) is designated as muuya 'father's sister's daughter'. The prohibition is on marriage to a FZD, and not on marriage to all women designated as muuya. In the event of such a marriage, ego changes his designation for her from muuya to tumam. My informant had made such a marriage and insisted there was nothing irregular about it. But he said it would be wrong to marry his FFBDD or his FMZDD, regardless of his relationship to their fathers, because their mothers are 'close sisters' of his father.

If the same arrangement is possible in the Yir Yoront case, MMBS's wife may be ego's 'mother' without at the same time being either ego's own mother or the actual FZD of ego's MMBS. This requires only that a man may marry a woman whom he designates as marang,

though he then changes her designation from _marang_ to
warna; but there is no designation for _wimi_'s (WMB's)
wife as such (or it is 'mother') and so ego continues
to designate this woman as 'mother'.

Another arrangement that occurred in Sharp's genealogical data is that some _wimis_' (WMB's) wives were
ego's 'sister's daughters'. The implication is that a
man may marry his FZDDD or his ZDHZD. Neither of these
relationships (which coincide in the second descending
generation of L-3 on Sharp's diagram) is given a designation on Sharp's diagram, but his observations on lines
R-3 and L-3 indicate that it could be _warna_ or _marang_.
This suggests the possibility of sister's daughter
exchange, in which arrangement one man marries a woman
whom he designates as _warna_ and the other marries a
woman whom he designates as _marang_. Sharp does not
mention this possibility. He writes instead of men
"giving" their sisters and daughters to other men (1934:
415). But he writes also of men getting their wives
from their MMB's "groups" (1934: 418), and he says, "a
woman usually selects a husband for her daughter by
arrangement with one of the sisters of her own husband,
the choice, of course, being definitely limited by the
marriage rule" (1934: 427). The clear implication is
that here, as elsewhere in Australia, women have a
significant voice in arranging their daughters' betrothals. A woman may do this by agreeing to give her
daughter to her FZD's son, in which case he marries his

MMBDD. Surely, a woman arranges her daughter's marriage in cooperation with her brother and her husband, so that MMBS may be seen as giving his ZD to ego as a wife, and MMBDH (who may be ego's own MB) may be seen as giving his daughter to ego. If ego is obliged to give his ZD to his WMB, it follows that ego's _wimi_'s wife may be ego's 'sister's daughter'. Ego's 'sister's daughter' may be his WMB's (or MMBS's) _warna_ or _marang_, but it must be that marriage to a woman of the latter category is not in itself "wrong."

It seems clear enough from the preceding observations that the Yir Yoront system of kin classification is not at base "asymmetrical" and that, probably, it is structurally quite similar to the Karadjeri system. It differs from the Karadjeri system, however, in classifying FFZC and FZH as 'sister's child', rather than 'mother' or 'mother's brother'. There are no reasons in this case to suppose that FFZC's status as 'sister's child' is dependent on the possibility of marriage between kin of alternate generations. Instead it must be attributed to the AGA rule. This may be done by assuming that the AGA rule is assigned priority over the parallel-cross status-extension rule, so that the parallel-cross status-extension rule governs the kin-class statuses of all parents' cross cousins except FFZC. Conversely, it governs the kin-class statuses of all cross cousins' children except MBSC.

From Sharp's diagram, it appears that the conven-

tional designation for FZH is 'sister's child' and,
conversely, that the conventional designation for WBC
is 'mother' or 'mother's brother'. This is the reverse
of the usual arrangement in Kariera-like systems. It
may be, of course, that the designations indicated in
Sharp's diagram reflect only what would be the case if
a man were married, in orthodox fashion, to one of his
MMBDDs, such that his WBC would be his MMBDSC (or MBSC)
and therefore would be classified as 'mother' and 'mother's brother' both prior to and after ego's marriage.
We do not know that these same designations are assigned
to FZH and WBC in the event of an irregular or wrong
marriage. Thomson (1972: 30) recorded some instances
in which FZH was designated as 'mother's brother' and
WBC as 'sister's child', and some instances in which
they were designated in the ways indicated in Sharp's
diagram. Thomson's genealogical data are not extensive,
but they suggest that WBC are designated as 'mother' or
'mother's brother' when they happen to be close kin such
as MBSC; and they are designated as 'sister's child'
when they happen to be close kin who were so designated
prior to ego's marriage (as happens when a man marries
his MFZDD or the daughter of a distant classificatory
'father's sister'); or when WBC are not otherwise closely related to ego. This, according to my informants,
is the practice in the Ngantjara and Thayorre systems
just to the north of the Yir Yoront. My informants
indicated that either arrangement is satisfactory, be-

cause the daughters of distant classificatory 'father's sisters' are acceptable wives, and when such marriages occur it sometimes happens that FZH is a kinsman to whom one is otherwise related as a fairly close classificatory 'sister's child'. In such cases, it is neither necessary nor proper to change one's designation for FZH from 'mother's brother' to 'sister's child'.

It is possible of course that the Yir Yoront practice is different. If, in the event of an irregular or wrong marriage ego may designate his or her FZH and WBC by the terms appropriate to an orthodox marriage, we must suppose that the cross-stepkin rule of this system is a weaker (less generally applicable) version of the rule that is more typical of Kariera-like systems in general (see Table 4.4, no. 6). In particular we must suppose that the rule

$$(...mZH \rightarrow ...mWB) = (mWB... \rightarrow mZH...)$$

is not part of this system. Thus FZH is not structurally equivalent to FWB or MB and WBC is not structurally equivalent to ZHC or ZC. But the complement to this rule, that is,

$$(...wBW \rightarrow ...wHZ) = (wHZ... \rightarrow wBW...)$$

is part of this system. Thus MBW is structurally equivalent to MHZ or FZ, and HZC is structurally equivalent to BWC or BC. Because FZH and WBC are not affected by the cross-stepkin rule, if need be, as in the case of an irregular or wrong marriage, their kin-class statuses may be determined by the AGA rule. Thus:

FZH → FFDH → BDH → mDH,

WBC → WFSC → WFSb, that is, WF, WFZ.

This brings us to the problem of specifying the spouse-equation rule of this system. This is difficult to do in the absence of information, which Sharp does not provide, on how in-laws are classified in the event of an irregular or wrong marriage. All we know from Sharp's account is how in-laws are classified in the event of an orthodox marriage, that is, the marriage of a man to his MMBDD (who may be also his MBD). In this event, the pre- and post-marital designations are the same; and there is no need to posit a spouse-equation rule to account for these designations, because in this case a man's WF, for example, is his 'mother's brother' independently of the in-law relationship between them. Thomson (1972: 30) recorded some instances in which wBSW was classified as 'sister's child' (not 'mother') and, conversely, HFZ was classified as 'mother' (not 'sister's child'). Again, this could result from the marriage of a man to his MFZDD or the daughter of a distant classificatory 'father's sister'. In Thomson's genealogical data, no other relationships between ego and alter are indicated in these cases, and there are instances of the same woman classifying some of her BSWs as 'mother' - they were also her MBSDs - and others as 'sister's child'. Again, according to my Ngantjara and Thayorre informants, either arrangement may occur and neither is in any sense "wrong."

We can only speculate that in Yir Yoront, in the
event of an irregular or wrong marriage, the designations given to in-laws are those appropriate to an
orthodox marriage. If so, the spouse-equation rule of
this system is somewhat different from the rule posited
for the systems analyzed in preceding chapters. For
this system it is not sufficient to specify only that
male ego's wife as a designated relative is to be regarded as a member of the COUSIN class, for this class
has a somewhat different structure than the COUSIN
classes of simpler Kariera-like systems. Instead we
must specify that male ego's wife as a designated relative is to be regarded as a member of the specially
designated _warna_ subclass of the CROSS-GRANDPARENT
class; conversely, female ego's husband as a designated
relative is to be regarded as a member of the reciprocal
specially designated _marang_ subclass of the CROSS-GRAND-
CHILD class. The corollaries of this rule also are
correspondingly different. For example, instead of "let
spouse's father as a designated relative be regarded as
a member of the MOTHER'S BROTHER class," we must, again,
differentiate between the two sexes of ego and specify
"let male ego's WF as a designated relative be regarded
as a member of the FATHER'S SISTER'S HUSBAND class."
But the latter class has no distinctive designation of
its own. As noted above, FZH's status may be the same
as that of mDH, via the AGA rule, and in orthodox marriage mDH is 'sister's son'. Therefore, HF is to be

regarded as a member of the SISTER'S (or WOMAN'S) CHILD
class.

We can now see, from another perspective, why
Sharp's informants exhibited a "tendency" to specify
MMBW and WMM as pa'a 'cross grandparent'. In the case
of MMBW this follows from the cross-stepkin rule. Just
as MBW is to be regarded as a member of the FATHER'S
SISTER class, so is MMBW to be regarded as a member of
the MOTHER'S FATHER'S SISTER class, but this is the
same as the FATHER'S MOTHER class. In the case of WMM
it follows from the spouse-equation rule, which specifies that WM is to be regarded as a member of the same
class as MBW - the FATHER'S SISTER class - and that WMM
is, therefore, to be regarded as a member of the FATH-
ER'S MOTHER class. Some of Sharp's informants appear
to have applied these rules without difficulty, at least
within this range. Beyond this range, however, it may
be somewhat more difficult to do so. Ego's MMBW or his
WMM may not be a woman whom he otherwise designates as
'cross grandparent' - she may be a kinswoman whom he
otherwise designates as warna or marang. Because of
his relationship to her through his MMB or his WM he
may change his designation for her to pa'a, but it
appears that he is not obliged, by the rules of the
system of kin classification, to change his designation
for her or his designations for other persons to whom
he is related not only through her but in other ways as
well. Thus the terminological statuses of relatives

who would appear diagrammatically in Sharp's line R-3, may or may not be determined by the classificatory statuses of MMBW or WMM as such, but they may be - and quite probably usually are - determined by the terminological statuses of MMBWB and WMMB and their wives <u>as</u> <u>other</u> <u>kinds</u> <u>of</u> <u>kin</u>. From this perspective, it is doubtful that, for the Yir Yoront, "the ideal structure hangs in mid-air at its lateral terminations" (Sharp 1934: 420). Instead, it has no lateral terminations; it may be extended indefinitely, but there is no practical need to do this in a relatively small and endogamous community within which most relationships can be traced, one way or another, within a relatively narrow range.

We must now account for the designation of FF and FFZ by junior-sibling terms and, conversely, for the designation of mSC by senior-sibling terms.

It is easy to account for this, formally anyway, by supposing that in this system the AGA rule has the form (FF → B-) and (mSC → mSb+), with the corollary (FFZ → Z-) and (wBSC → wSb+). But, we may ask, why should the AGA rule have this particular form in this system? Conceivably, the answer has to do with the possibility that "ego may marry his son's son's cross cousin, who is classified with mother's brother's daughter" (Sharp 1934: 417). That is, a man classifies his WBSD as <u>warna</u>, and this woman is his son's son's FMBSD <u>warna</u> - or perhaps even his son's son's MBD, if ego's son married his MBD. Thus, as Sharp puts it, a man's FF may "compete" with him for women in his own generation, but not vice versa. We may suppose, however, that this competition is restricted by certain rules of priority of right in relation to particular kinswomen. As we have seen, some of Sharp's observations suggest that a man has a rightful claim to his MMBDD as a wife. Presumably, although ego's FF might be related to this woman in such a way that he, too, would classify her as warna, ego rather than his FF would have priority of right to her. It is probable that in the case of two uterine brothers who share the same MMBDDs, the older brother has priority of right in relation to the younger. So it may be that, if a man were to designate his SS as 'younger brother', this might be taken to imply that he assumes not only a right but also a prior right to their common warna who are more closely related to the SS. To avoid this possible unwanted implication, SS is designated as 'elder brother' and, reciprocally, FF is designated as 'younger brother'.

John von Sturmer (personal communication) has suggested another interpretation. He notes that among the neighboring Ngantjara most kin terms signify that the designated relative is either senior or junior in relation to the speaker or propositus, and in general senior kin have the right to initiate interaction with their junior kin. In the case of grandparents and grandchildren, however, it is the grandchild who has the right to initiate interaction. Therefore it is regarded as appropriate for a man's son's son to address him as 'younger brother', rather than as 'elder brother', and so on. In other words, vocative use of a junior-kin term signifies indexically the speaker's right to initiate interaction.

Finally, we may consider Sharp's (1934: 421) observation that the terms of this system "may be extended to seven or nine generations if need be by applying to persons in the additional generations the corresponding terms of the alternate preceding generation." This implies that G+3 kin are designated by G+1 terms and G-3 kin are designated by G-1 terms. Unlike McConnell (see Mongkan, Chapter 4), Sharp does not mention the possibility of designating G+3 kin also by G-1 terms. However, in Thomson's genealogical data, a few such relationships are recorded, and in those cases G+3 kin are designated by G-1 terms and G-3 kin are designated by G+1 terms, for example, mSSC are given as 'father' and 'father's sister', and mDDC also are given as 'father' and 'father's sister'; reciprocally FFF and MMF are given as 'man's child'. On this evidence it seems probable that the Yir Yoront arrangement is the same as the Mongkan.

The equivalence rules of the Yir Yoront system are listed in Table 8.2. Note that the rule for reckoning the kin-class statuses of G±3 kintypes (no. 7) must be assigned priority over the AGA rule (no. 8). If this were not done the AGA rule would apply in reckoning the kin-class statuses of G+3 kintypes and this would give incorrect results. Similarly, the AGA rule (no. 8) must be subordinate to all other rules, except rule 5, the parallel-cross status-extension rule, in relation to which it has priority. An alternative way to express this fairly complex relationship between the AGA rule and the parallel-cross status-extension rule on the one hand and the other rules of subset I on the other hand

Table 8.2. Yir Yoront equivalence rules

I. 1. Half-sibling-merging rule

2. Stepkin-merging rule

3. Same-sex sibling-merging rule

4. Parallel-cross neutralization rule

5. Parallel-cross status-extension rule

6. Cross-stepkin rule:
   $(...wBW \rightarrow ...wHZ) = (wHZ... \rightarrow wBW...)$

7. G+3 rule (see Ompela and Mongkan, Chapter 4)

II. 8. AGA rule, subordinate to all the above except rule 5, in relation to which the AGA rule (but not its auxiliary C) has priority,
   a. $(FF \rightarrow B) = (mSC \rightarrow mSb)$,
   b. corollary, $(FFZ \rightarrow Z) = (wBSC \rightarrow wSb)$,
   c. auxiliary, $(...wB \rightarrow ...wFS) = (mZ... \rightarrow mFD...)$.

III. 9. Spouse-equation rule, subordinate to all the above,
   a. $(W \rightarrow CR.GR.PAR..\sim G+2.\female) = (H \rightarrow CR.GR.CH..\sim G-2.\male)$,
   b. $(WF \rightarrow MO.BRO.) = (mDH \rightarrow man's SIS.SON)$,
   c. $(HF \rightarrow woman's FA.SIS.HUS.) = (mSW \rightarrow WF.BRO.DAU.)$
   d. etc.

---

would be to treat the AGA rule and the parallel-cross status-extension rule as an internally ordered subset (II), which in turn would be subordinate to subset I (the same as in Table 8.2 but minus the parallel-cross status-extension rule). This alternative would not be entirely satisfactory, however, because the parallel-cross status-extension rule in no way conflicts with any of the other rules of subset I. That is, there are no contexts in which rule 5 and one of the other rules of subset I are both applicable, and so there is no need to specify an order between rule 5 and the other rules of subset I. Needless to say, I have been unable to find a simpler way of specifying the relations among the AGA rule and the parallel-cross status-extension rule and the other rules of subset I.

MURNGIN

The following discussion is based primarily on Warner's (1930, 1931, 1933, 1958 [1937]) accounts of this system, supplemented where necessary by information drawn from more recent accounts and by information provided, in personal communications, by Nicolas Peterson, Warren Shapiro, and Bernard Schebeck.[6]

Warner presented his data on the Murngin system of kin classification in the form of a diagram in which seven patrilineal descent lines are linked systematically by matrilateral cross-cousin marriage (see Figure 8.3 and Table 8.3). According to Warner, the system is

Figure 8.3. Murngin kin classification (Warner 1958: 58)

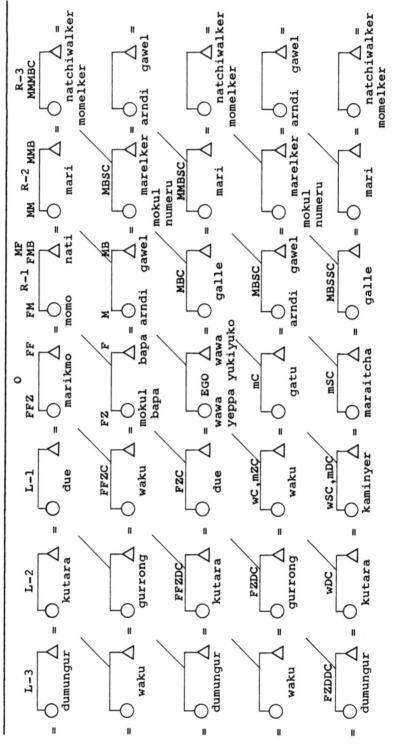

"closed," but not, as might be expected, by designating the wives of men of line R-3 by the same terms as the women of the corresponding generations of L-3. Instead, the wives of men of R-3 are designated by the same terms as the women of the corresponding generations of R-2, and the husbands of women of L-3 are designated by the same terms as the women of corresponding generations of L-2 (Warner 1930: 210-1). This seems strange in view of Warner's claims that the system is "based on" matri-

Table 8.3. Murngin reciprocal relations

1.	marikmo	FF / mSC	maraitcha
2.	mari	MM / wDC	kutara
3.	nati	MF / mDC	kaminyer
4.	natiwalker	MMMBS / mFZDDC	dumungur
5.	galle	MBS / mFZC	due
6.	momo	FM / wSC	
7.	momelker	MMMBD / wFZDDC	dumungur
	galle	MBD / FZC	due
8.	bapa	F / mC	gatu
9.	mokul	FZ / wBC	
10.	marelker	MMBS / mFZDC	gurrong
11.	mokul numeru	MMBD / wFZDC	
12.	arndi	M / wC	waku
13.	gawel	MB / mZC	
14.	wawa	wB, mB+ /	
		/ mB-, wZ-	yukiyuko
15.	yeppa	mZ, wZ+ /	

lateral cross-cousin marriage, that patrilateral cross-cousin marriage is prohibited, and that a man may not marry his sister's husband's sister. His remarks on how the system is "closed" appear to imply that, to the contrary, a man may marry his FZD (or a woman who stands in the same kin-class relationship to him as his FZD does). That is, it appears to imply some kind of "symmetry" rather than, or in addition to, "asymmetry" in the system (see also White 1963: 121-2). Moreover, the apparent symmetry is not confined to the relationships between lines R-2 and (potential) R-4 and lines L-2 and (potential) L-4. Note that lines R-1 and R-3 share some terms (<u>arndi</u> 'mother' and <u>gawel</u> 'mother's brother'), and so do lines L-1 and L-3 (<u>waku</u> 'woman's child'). Further still, some of the terms of R-3 are morphologically derived from terms of R-1, and some of the terms of line R-2 are morphologically related to terms of line 0 (cf. Warner 1930: 210). This suggests that some of the terms of line R-3 designate subclasses of classes that originate, as it were, in line R-1, and similarly for L-3 and L-1, and for lines R-2 and L-2 in relation to line 0.

Super- and subclasses

From these observations, and on the basis of the preceding analyses, it is clear that the Murngin system features a number of super- and subclasses similar to those of other Australian systems of kin classification. Warner himself was aware of some of these superclasses

(1930: 210) and perhaps even others (1933: 69), but he
did not analyze them thoroughly, and his seven-line
diagram has the effect of emphasizing the subclasses at
the expense of the superclasses. It thereby emphasizes
the relatively superficial differences between the
Murngin system and Kariera-like systems of kin classification. If the superclass relations are not taken
sufficiently into account, it is easy to get the impression that the system is radically different from the
Kariera-like systems discussed in preceding chapters.

As already noted in Chapter 2 (Karadjeri-type
systems), Radcliffe-Brown (1951: 46-9) subsequently
observed that the seven lines of Warner's diagram are
reducible to five or even three "basic" lines. The
three lines to the left, he argued, contain the reciprocal relationships and terms of the three lines to the
right and, being logically implied by them, cannot be
regarded as "basic." Further, R-3 must be a "secondary"
line in relation to R-1, and similarly for L-3 and L-1.
He suggested that the "special terms" momelker (MMMBD),
ngatiwalker (MMMBS), and dumungur (FZDDC) are used to
"distinguish certain relatives by marriage . . . within
a more general category or class" (1951: 49). Thus, for
example, ngatiwalker designates "a special kind of
ngati," 'mother's father', to whom ego is related by
marriage (as well as genealogically, perhaps).

As noted also in Chapter 2, Radcliffe-Brown might
have pressed the analysis even further and argued that

there are only two "basic" patrilines (see also White
1963: 121-2). The designation for MM and MMB _mari_ is
morphologically related to the designation for FF and
FFZ _marikmo_ (or _mari'mo_). This suggests a PARALLEL-
GRANDPARENT class, the designation for which must be
mari; but a special subclass is recognized and designated as _marikmo_, thus leaving the term for the class
as a whole to be also the designation for the residual
subclass. If so, it follows that the reciprocal GRAND-
CHILD class has the designation _kutara_; the special
subclass is designated as _maraitcha_ (mSC) and the
residual subclass as _kutara_ (wDC). This interpretation
is supported by the designation of MMBD as _mokul_, the
same as FZ; MMBD may be distinguished from FZ, and
presumably, certain other classificatory 'father's
sisters', by designating her as _mokul numeru_ 'tabu
father's sister' (Warner 1930: 211). MMBS, however, is
designated as _marelker_ and not as _bapa_ 'father'. But,
of course, it may be that the _marelker_ class is a subclass of the FATHER class. Indeed, it must be. _Mokul
numeru_ and _marelker_ have a common reciprocal _gurrong_
(FZDC, wDH), and must therefore belong to the same
superclass. If MMBD is a special kind of FATHER'S SIS-
TER it follows that MMBS is a special kind of FATHER
(similar to father's BROTHER). Ultimately, both of
these subclasses belong to the FATHER class (for here,
too, a woman designates her BC by the same term as he
does, _gatu_ 'man's child'; and a man designates his ZC

by the same term as she does, <u>waku</u> 'woman's child').
Reciprocally, the class designated as <u>gurrong</u> must be
a subclass of the MAN'S CHILD class.

One implication of all this is that, to put it in
Radcliffe-Brown's terms, as between the FF and MMB lines
(and, conversely, the mSC and ZDC or wDC lines) one is
"primary" and the other "secondary." Although FF is
designated by a marked form of <u>mari</u>, it must be that
the FF line is, in Radcliffe-Brown's sense, "primary."
The categories of the MMB line in the first ascending
generation are subclasses of the categories <u>derived by</u>
<u>expansion</u> from the categories of the same generation
in the FF line. But it should be clear by now that
questions about descent lines and their number are mis-
directed. Descent lines are not basic underlying
structural features of Australian systems of kin classification, Radcliffe-Brown's claims to the contrary
notwithstanding.

The various superclass-subclass relations so far
noted are readily deducible from Warner's data, and they
are strongly confirmed by subsequent reports. Shapiro
(1967, 1969: 39-40, 130-3) and Peterson (1971) note that
there is some variation among the so-called Murngin, by
dialect and region, in how they use certain terms.
Among the inland or southern Murngin some of the special
(probably potential in-law) terms are not used at all
and other terms take their place; and in some areas
these special terms are only seldom used where they

might be and, again, other terms take their place. The
reported substitutions are:

1. <u>mari</u> 'mother's mother' or 'parallel grandparent'
   for <u>marikmo</u> 'father's father';

2. <u>kutara</u> 'woman's daughter's child' or 'parallel
   grandchild' for <u>maraitcha</u> 'man's son's child';

3. <u>momo</u> 'father's mother' for <u>momelker</u> MMMBD, etc.,
   or 'potential WMM';

4. <u>ngati</u> 'mother's father' for <u>ngatiwalker</u> MMMBS,
   etc., or 'potential WMMB';

5. <u>kaminyer</u> 'cross grandchild' for <u>dumungur</u> FZDDC or
   'potential DDH and DDHZ' for a woman;

6. <u>bapa</u> 'father' for <u>marelker</u> MMBS, etc., or 'potential WMB';

7. <u>mokul</u> 'father's sister' for <u>mokul</u> <u>numeru</u> MMBS,
   etc., or 'potential WM';

8. <u>gatu</u> 'man's child' for <u>gurrong</u> FZDC or 'potential
   DH and DHZ' for a woman.

Shapiro (1967: 355) indicates that the special subclass
labels are used mostly by coastal (northern) and insular
peoples "who take the marriage regulations of the section system" less seriously than do the inland Murngin.

Further still, <u>due</u> (FZC, etc.) designates a subclass of the KAMINYER or CROSS-GRANDCHILD class and,
conversely, <u>galle</u> (MBC, etc.) designates a subclass of
the MOMO and NGATI classes or the CROSS-GRANDPARENT
class. This, too, can be deduced from Warner's data,
but it is indicated most clearly by Peterson's report

(personal communication) that some inland Murngin use the expressions momo (MF) and ngati (FM) to denote MBC, MMBDC, etc. For the same reasons that it must be supposed that, in the Yir Yoront system warna and minera do not focus on MBD and MBS but designate broadly defined subclasses of the PA'A or CROSS-GRANDPARENT class, we must suppose that galle likewise designates a broadly defined subclass of the MOMO + NGATI or CROSS-GRANDPARENT class. It does not focus on MBD and MBS. Instead its primary designatum is the subclass of those CROSS GRANDPARENTS who are kin of ego's generation or the second descending generation. Conversely, due designates a subclass of the CROSS-GRANDCHILD class and its primary designatum is the subclass of those CROSS GRANDCHILDREN who are kin of ego's generation or the second ascending generation.

We may now consider how the due (FZC, etc.) and dumungur (FZDDC) classes are related. Warner (1930: 210) says "Due has been altered to dumungur," and Radcliffe-Brown (1951: 48) inferred that dumungur "belongs to the general category of due . . . but is distinguished by a term that is derivative from due." There are, however, several reasons why this is improbable. Note that if the dumungur class were a subclass of the DUE class, it would follow that the reciprocal momelker (MMMBD) and ngatiwalker (MMMBS) classes are subclasses of the GALLE class, the reciprocal of DUE. But we know, from Shapiro's and Peterson's studies, that

when dumungur are designated by any other term it is
kaminyer 'cross grandchild' (not due), and when momelker
and ngatiwalker are designated by any other terms they
are momo and ngati 'cross grandmother' and 'cross grandfather' (not galle). Therefore, it must be that the
dumungur class belongs to the same superclass as the
due class (i.e., KAMINYER or CROSS GRANDCHILD) but is
not itself a subclass of the DUE class. Reciprocally,
it must be that the momelker and ngatiwalker classes
belong to the same superclass as the galle class (i.e.,
MOMO + NGATI, CROSS GRANDPARENT) but are not themselves
subclasses of the GALLE class. Furthermore, it may be
that the expression dumungur is not morphologically
derived from the expression due. Bernard Schebeck, a
linguist who has worked on this language, says that
this derivation is morphologically quite unlikely (personal communication, 1972).

Because here, too, cross cousins and structurally
equivalent kintypes are classified as CROSS GRANDKIN,
it must be assumed that the AGA rule is one of the
equivalence rules of this system. Therefore, it may be
that the FF and mSC classes are subclasses of the
SIBLING class, or that FF and mSC have dual kin-class
statuses, as 'parallel grandkin' and as classificatory
'siblings'. Peterson reports (personal communication)
that his informants described mari in general, that is,
MM, MMB, wDC, and ZDC as well as FF, FFZ, mSC, and BSC,
as 'partial' or 'half' (marganga) 'siblings', and he was

told that persons who designate one another as 'father's father'-'man's son's child' or as 'mother's mother'-'woman's daughter's child' may also call one another 'brother' or 'sister'.[7] Apparently, this is not done between 'real' and 'close' kin of these classes, but only between relatively distant ones. These data indicate that the several parallel-grandkin classes of this system are subclasses of the SIBLING class. Again, the Murngin system is similar to the Yir Yoront, but differs from it in treating the MM and wDC classes (as well as the FF and mSC classes) as subclasses of the SIBLING class. One implication of this is that the Murngin system features the AGU rule as well as the AGA rule. It is shown below that the ranges of the two rules are, however, somewhat different.

These super- and subclass relations are illustrated in Figures 8.4, 8.5, 8.6, and 8.7. We must now ask how the potential in-law subclasses are distinguished within their superclasses. This necessitates consideration of some aspects of Murngin "marriage rules."

"Marriage rules" and subclassification

Warner repeatedly asserts that Murngin kin classification, and for that matter much of Murngin social structure, is "based on" a rule of matrilateral cross-cousin marriage. "Preferably," he says, a man marries a MBD, but if none is available he may marry another kinswoman whom he designates as galle, for example, his

MMBDD (Warner 1930: 142; 1931: 194-5). Warner also repeatedly asserts that "ideally" a man "gives" his daughter as a wife to his ZS; but if he has no ZS or if his ZSs already have wives he may give his daughter to a distant or far-off 'sister's son' (1930: 237). Warner's phrasing often suggests that the alleged "preference" for MBD-FZS marriage could better be described as a matter of men having rights to their MBDs as wives. Warner (1930: 231-2; 1958: 99) notes that affairs and marriages between men and kinswomen of

---

Figure 8.4. Murngin FATHER-MAN'S CHILD class

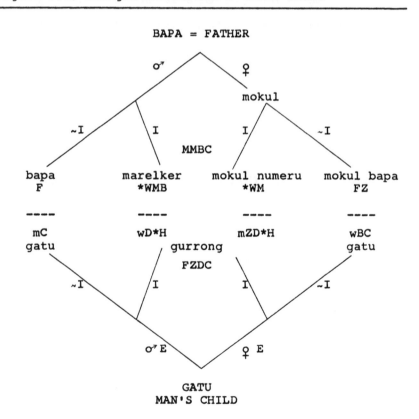

Figure 8.5. Murngin MOTHER-WOMAN'S CHILD class

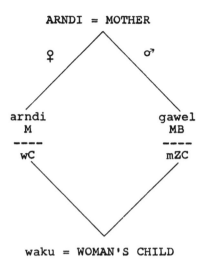

Figure 8.6. Murngin SIBLING class

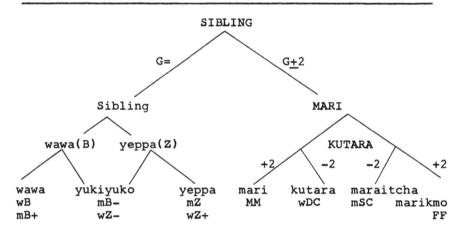

classes other than <u>galle</u> are not uncommon, and it is even "expected" that classificatory 'cross grandparents' and 'cross grandchildren' will be lovers or sweethearts. Marriages between such kin are not strongly condemned. Even so, Warner describes such marriages as "illegal." It may be, however, that they are "illegal" only insofar as, in marrying a classificatory 'cross granddaughter', for example, a man marries a woman to whom he has no prior right. That is, it may be that such marriages are not so much "illegal" as they are "extra-legal" - an

Figure 8.7. Murngin CROSS-GRANDKIN class

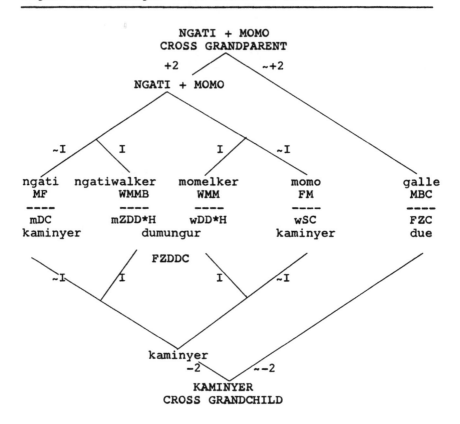

expression Warner (1930: 231) also uses to characterize them.

Shapiro (1967: 354) reports that the right to bestow a woman as a wife belongs to her M and MB, and not to her F; ideally, the woman whom a man claims as his wife is his MBD and MMBDD at once (MMBD being also MBW), but if not he has a stronger right to a MMBDD than he does to an MBD. That is, it is a man's right to demand that his MMBD give him her daughter as his wife. Shapiro (1969) notes that in some areas there is a growing tendency to permit men to betroth their daughters, and in the past men occasionally did this without the consent of their wives and wives' brothers. But his informants insisted that the proper procedure is for a woman to be 'promised' as a wife by her M and MB. Although a man may not rightfully bestow his daughter on his ZS, he may urge his wife to give their daughter to his ZS, and his ZS may rightfully seek his support in getting his (the ZS's) MMBD to meet her obligations.

Clearly, then, kinswomen designated as mokul numeru 'tabu father's sister' are so designated, and thereby distinguished from "ordinary" classificatory 'father's sisters', as potential WMs (or BWMs). Similarly, kinsmen designated as marelker are so designated, and thereby distinguished from "ordinary" classificatory 'fathers', as potential WMBs. The reciprocal category is designated as gurrong, and this category must consist

of a woman's daughters' potential husbands and their
sisters (also a man's potential ZDHs and ZDHZs). These
potential in-law statuses are determined by genealogical
criteria. According to Shapiro (1969), a man's principal potential WM's are his FATHER'S SISTERS of his MM's
"sib" (lineage); a man has no special right to his MMBD
as such. Therefore, a man may have several potential
WMs and WMBs, and several men (uterine brothers or
maternal parallel cousins) may share their potential
WMs and WMBs. The potential for competition and conflict among these men is perhaps mitigated to some
extent by the practice of bestowing women not only as
wives but also as wives' mothers (Shapiro 1969, 1970).
The right to bestow a woman as a WM also belongs to her
M and MBs. Therefore, to secure a particular woman as
a prospective (not merely potential) WM, a young man
must look to one of the mothers of his potential WMs
and to her brothers to allocate one of her daughters to
him. These arrangements are made not by the young man
(prospective husband) himself but by his F or MB, or
perhaps both. Therefore, it is probable that the
special significance of the expressions momelker and
ngatiwalker is that they designate one's potential WMMs
and WMMBs. A man's principal momelker and ngatiwalker
are those of his momo and ngati who belong to his MMM's
lineage. The daughters of a man's momelker are his
mokul numeru, potential WMs, but they may or may not
belong to his MM's lineage. If they do he has a special

right to them as prospective WMs because of the double
consanguineal-cum-in-law relationship; if they do not
his right to them is contingent on the rights of men to
whom they are women of the MM's lineage.

Moreover, it is advantageous for a man if the kinswoman assigned to him as a prospective WM is married to
his MB or to a close classificatory 'mother's brother'.
Although the right to allocate a woman as a prospective
wife or WM rests with her M and MB, her father has a
recognized interest in the matter, and he may urge his
wife to allocate her daughter to a particular man,
expecially his ZS. Therefore, where a man's MMB and
his MB have married their own MMBDDs, so that his MBW
is also his MMBD and his MMMBDD, he has the strongest
possible claim against this woman as his prospective
WM, because he has claim to her as a potential WM and
claim to her as the daughter of a potential WMM. If
'promised' to him as prospective WM and WMM, she and
her mother are then committed to give him their D and
DD to be his wife. Moreover, it gives him the additional advantage of having his own MB to assist him in
finally securing the daughter of his prospective WM as
a wife, because his MB may intercede as the husband of
the prospective WM and as the father of the prospective
wife. Although a man may not allocate his own daughter
to his sister's son, he may put pressure on his wife to
fulfill her obligations and, at the same time, fulfill
his duties to his ZS to help him get a wife. He can

304

most easily fulfill this duty if his wife is his own
(or close) MBD and MMBDD.

Of course, in practice it may be only seldom that
such a situation presents itself. More often, a man
has available to him only one of these bases on which
to claim a woman as a prospective W or prospective WM.
It often happens, too, that several men have claims
against the same woman on the same basis or on different
bases. Among uterine brothers and uterine parallel
cousins, priority of right is assigned to elder 'brothers' in relation to younger 'brothers', and men who have
already secured wives or prospective wives for themselves should defer to their 'brothers' who have not
(Warner 1930: 237-8; Shapiro 1969). Presumably, a man
who can claim a woman as a prospective WM because she
is a member of his MM's lineage has priority of right
in relation to a man who may rightfully claim her as
prospective WM because she is the daughter of a woman
of his MMM's lineage. If this were not the arrangement,
it would be difficult if not impossible to decide among
rival claimants to a woman as prospective WM, at least
on any genealogical basis - because it would mean that
more distant kinsmen of a woman have priority of right
in relation to close kinsmen, and it is probable that
there will be more of the former than of the latter.

Equivalence rules

Having clarified the super- and subclass structure

of this system and the "marriage rules" on which some
of these structures are based, we may now consider the
rules of kin-class expansion and the probable relations
among them. It should be clear that these are the same
as in the Yir Yoront system (see Table 8.2), except for
the addition of the AGU rule. This latter rule has
relatively few effects on the classification of kin.
Indeed, it is relevant only to determination of the
subclass statuses of MM, MMZ, MMB, and their reciprocals
(and structurally equivalent kintypes) as kinds of
SIBLINGS. Because it has no effects (except indirectly)
on the kin-class statuses of more distant kintypes, it
may be specified with strong context restrictions as,

$$(.m/wMM. \rightarrow .m/wZ.) = (.wDC. \rightarrow .wSb.),$$

$$(.m/wMMB. \rightarrow .m/wB.) = (.mZDC. \rightarrow .mSb.),$$

That is, let ego's mother's mother as a designated
kinswoman be regarded as structurally equivalent
to ego's sister as a designated kinswoman;
conversely, let female ego's daughter's child as
a designated kinsman be regarded as structurally
equivalent to female ego's sibling as a designated
kinsman. (And similarly for the corollary.)
Specified in this way the AGU rule may be regarded as a
member of the subset of rules (I in Table 8.4) to which
(with the exception of the parallel-cross status-extension rule) the AGA rule is subordinate.

It should be noted how these rules account for some
of the terminological assignments implied by Warner's

Table 8.4. Murngin equivalence rules

I. (unless otherwise noted, see Table 4.4)

   1. Half-sibling-merging rule

   2. Stepkin-merging rule

   3. Same-sex sibling-merging rule

   4. Parallel-cross neutralization rule

   5. Parallel-cross status-extension rule

   6. Cross-stepkin rule (see Table 8.2)

   7. G±3 rule (see Ompela and Mongkan, Chapter 4)

   8. AGU rule

      a. (.m/wMM. → .m/wZ.) = (.wDC. → .wSb.)

      b. (.m/wMMB. → .m/wB.) = (.mZDC. → .mSb.)

II. 9. AGA rule, subordinate to all the above except
      rule 5, in relation to which the AGA rule (but
      not its auxiliary c) has priority,

      a. (FF → B) = (mSC → mSb)

      b. corollary, (FFZ → Z) = (wBSC → wSb)

      c. auxiliary, (...wB → ...wFS) =

                              (mZ... → mFD...)

III.10. Spouse-equation rule, subordinate to all the
      above.

---

diagram, especially a few that are apparently contra-
dicted in other accounts of Murngin kin classification.

We may begin by noting a difference between Warner's diagram and one provided by Berndt (1955: 86;
1971: 199). On Warner's diagram the position occupied
by FFFZS and FFZH is given the designation due (FZS,
ZH), but on Berndt's diagram it is given the designation kaminyer (mDC, wSC). The rules posited above imply
that FFZH is structurally equivalent to ZH and that
FFFZS is structurally equivalent to FZS; thus if the
rules are correct we expect FFFZS and FFZH to be designated as due. Conversely, we expect the reciprocals
MBSSS and WBSS to be designated as galle, and this is
the designation given on Warner's and Berndt's diagram
to the position occupied by MBSSS and WBSS. That is,
in Berndt's diagram there is an apparent inconsistency
of reciprocals. The inconsistency is only apparent,
however, because due designates a subclass of the
kaminyer class. Among some Murngin, kintypes designatable as galle are designated also or instead as momo
(FM) and ngati (MF); conversely, it must be that kintypes designatable as due are designated also or instead
as kaminyer. So, it seems that Warner's and Berndt's
diagrams each show only one of the designations that
may be assigned to certain kintypes.

Shapiro (1969: 42-3) implies that the proper designation for FFZDS is kutara (ZDS), and this is accounted for by the rules posited above. But Shapiro says,

according to his informants, the son of male _kutara_ is _bapa_ 'father', not _gurrong_ as in Warner's diagram. Only if this _kutara_ is married to ego's _due_ FZD is his son classified as _gurrong_ (FZDC), and then as FZDS rather than as FFZDSS. The implication appears to be that FFZDSS as such is classified as _bapa_ 'father' and FZDS as such is classified as _gurrong_, and if the two relationships should happen to coincide the relationship through alter's mother (also the closer genealogical relationship) takes precedence. But perhaps not. _Gurrong_ is the reciprocal of _marelker_; _gurrong_ designates a subclass of the GATU or MAN'S CHILD class and _marelker_ designates a subclass of the BAPA or FATHER class. To appreciate the possible implications of the relationships among these terms we must note also that the designation of a particular relative depends not only on the rules of structural equivalence and on super- and subclass relationships, but _may_ depend also on considerations of relative age. Shapiro (1969: 38-9) and Peterson (1971: 173) report that the elder-sibling and grandparent terms (and the special terms morphologically based on the grandparent terms) always denote the older member of a dyad and may not be used for the younger member. Conversely, the junior-sibling terms and the grandchild terms (and _dumungur_ FZDDC) always denote the younger member of a dyad and may not be used for the older member. All other terms, according to Shapiro, "may be used for either member of the dyad."

Presumably, this means that their use may or may not depend on relative age; relative age is an optional, rather than mandatory, consideration in deciding on the designation of, for example, FZDS. Should ego's FZDS be older than ego, ego might refer to him as gurrong (FZDS) or as marelker (MMBS) - or as bapa (F), that is, by the designation for the higher-order class FATHER of which the marelker class is a subclass. The case of FFZDSS must be similar. The equivalence rules of the system place this kintype in the FATHER class, but a person so related to ego may be younger than ego and so may be designated as 'man's child', rather than as 'father'. But if this person happens to be also the son of ego's due (FZD) or the husband of ego's waku (ZD) he may be designated as gurrong - and the gurrong class is merely a subclass of the GATU or MAN'S CHILD class.

Shapiro agrees with Warner that the son of ngati-walker (MMMBS, WMMB) is 'mother's brother' (not 'sis-ter's child' as reported by Webb in Lawrence and Murdock 1949), but he says this designation is dependent on the kin-class status of ngatiwalker's wife, "ideally" mari 'mother's mother'. It must be, however, that either term, 'mother's brother' or 'sister's child', is appro-priate, depending on alter's age relative to ego, and whether the relationship is reckoned through alter's father or mother. Shapiro (1969: 43) says that ngati-walker's SS is not ngatiwalker but galle (MBS, etc.), because the son of ngatiwalker is 'mother's brother', whose son is galle. But, again, it must be that either designation ngatiwalker or galle is appropriate. Presumably, the SS of ngatiwalker is designatable as ngatiwalker also, by the AGA rule; but the class designated as ngatiwalker is a subclass of the NGATI or MOTHER'S FATHER class, and the galle class also is a subclass of the NGATI (and MOMO) class. As a MOTHER'S FATHER of ego's own generation, the SS of ngatiwalker therefore may be designated as galle. It may be that relative age is a consideration here, too. Perhaps galle is reserved for men more or less the same age (or younger) than ego, in which case the SS of a ngatiwalker who happens to be older than ego would appropriately be designated as ngatiwalker himself.

Knowing that the designations actually given to particular kinsmen may depend on relative age, as well as on underlying structural equivalences of kintypes, helps to account for the apparently contradictory reports about how kin of the third and fourth ascending and descending generations are classified. Warner (1930: 211) states: "Marikmo's [FF's] father is old gatu [S] and the latter's father is wawa [B+]. Ngati's [MF's] father is gawel [MB] and his sister arndi [M]; their first ascending generation is gawel [sic., galle?]. The other lines continue alternating as they do in the five generations. The son of maraitcha [mSS] is gatu." This last statement seems inconsistent with the first, because if FFF is designated as 'man's child' we expect the reciprocal term 'father' for the reciprocal relationship mSSS. But we know that the designation actually assigned to a particular kinsman may be the reciprocal of the term assigned by the under- lying rules of structural equivalence, because of con- siderations of relative age. Thus the kintype SSS may be structurally equivalent to the kintype F, but a man may designate his SSS as 'man's child', because his SSS is necessarily younger than he is.

Lawrence and Murdock's (1949) account of Webb's data for the third and fourth ascending and descending generations differs from Warner's account only in a few details. Webb states specifically that the designation for MMM and MMMB is waku, which is the reciprocal of the terms that Warner's statement might lead us to expect. And Webb reports mSSS as bapa (F) and mSSD as mokul (FZ).

The information provided by Shapiro (1968: 348; 1969: 42-3) differs radically from that provided by Warner and Webb, except that Shapiro agrees with Warner that FFF is 'man's child' and with Webb that mSSS is 'father' and mSSD is 'father's sister'. Like Webb, he reports MMM and MMMB as 'sister's child'. Shapiro claims that kin in G+3 in Warner's line R-1 are desig- nated as 'sister's child' (not 'mother' and 'mother's brother', as per Warner) and that kin in G+3 in line R-2 are designated as 'man's child' (not 'father's

sister' and 'potential WMB', as per Warner); for the
reciprocals in G-3 in L-1 Shapiro has 'mother' and
'mother's brother', and for G-3 in L-2 he has 'father'
and 'father's sister'. All these designations are
precisely the reciprocals of the terms implied by
Warner's general statement.

The apparent contradictions immediately disappear
if we suppose that the Murngin mode of assigning designations to kin of the third and fourth generations
is the same as the Mongkan (see Ompela and Mongkan,
Chapter 4). A few differences arise, however, from the
fact that in Murngin some G-1 kin are designated normally by G+1 terms; mSW, for example, is designated as
'mother'. If it is reasoned that FFF is similar to the
F of maraitcha (mSC), the junior reciprocal of marikmo
(FF), and is therefore like a man's son or gatu, it
must be reasoned also that FFM is like the mother of
maraitcha (mSC), and the mother of maraitcha is mSW,
whose designation is 'mother'. In this way, a G+1 term,
rather than a G-1 term, is assigned to a G+3 kintype.
In all other cases, reckoning the kin-class statuses of
G+3 kintypes in this way results in the assignment of
G-1 terms to them. But, inevitably, kin of these types
are older than ego and, thus, as Warner reports, they
may be designated by the reciprocal G+1 terms.

It may be worth noting that the kin-class statuses
of the descendants of G+3 cross-collateral kin are not
affected by the kin-class statuses of these linking G+3

kin. Thus, for example, MMMB is classified (or may be classified) as <u>waku</u> (ZS), but MMMBS is classified as <u>ngatiwalker</u>, a special kind of NGATI (MF) and not as <u>kaminyer</u> (ZSS). MMMBS's classification as a kind of NGATI is accounted for by the parallel-cross status-extension rule, whereby MMMBS is reckoned as structurally equivalent to MFB.

Some further comments on Murngin marriage rules and their relationship to the system of kin classification are in order. We have seen that the Murngin system of kin classification is not based on a rule of matrilateral cross-cousin marriage. The claim that it is implies the hypothesis (heretofore only loosely formalized and casually tested) that the collateral extensions of the terms of the system are governed by a rule of structural equivalence between a man's MBD and his wife (cf. Warner 1930: 210). Warner's diagrammatic representation of the system is based on this assumption. But the assumption is false. There is no preference for MBD-MZS marriage, and although MBD is classified as <u>galle</u>, any particular MBD may or may not be among a man's most rightful potential wives (depending on whether or not her mother is ego's 'father's sister' of his MM's lineage). The closest kinswoman among a man's rightful potential wives is his MMBDD, but it seems that he has no more right to kinswomen of this type than he has to any other <u>galle</u> whose mothers belong to his MM's lineage. And even if a man's MMBDD

were his most rightful potential wife, there is no way
in which the Murngin system of kin classification may
be analyzed as based on a rule of MMBDD marriage, if
this implies the hypothesis that the collateral extensions of the categories and terms of the system are
governed by a rule of structural equivalence between a
man's MMBDD and his wife. This hypothesis cannot
account for the fact that MMBDD and MBD are members of
the same kin class, nor can it account for the fact that
MMBD is regarded as a special kind of classificatory
'father's sister' - to give but two examples from among
the many ways in which this hypothesis is inadequate.

Although men have preferential rights to certain
kinswomen of the galle class, all female members of
which are regarded as potential wives, this does not
mean that a man may not "legally" marry kinswomen of
some other class or classes. But it does of course
imply that a man has no "legal" (that is, rightful)
marital claims to all kinswomen of the galle class!
This may or may not affect his ability to acquire women
of other kin classes as wives, because they may or may
not have other kinsmen who do have the right to claim
them. After all, a man's right to certain kinswomen
must be a right as against other men who are not related
to those women in the same way. This does not automat-
ically disbar the other men from taking these women as
wives, provided there are no men who have rights to
them, or if there are such men provided they are willing

to waive their rights (cf. Shapiro 1969: 78; Hiatt
1965).

Warner (1930: 231-2) describes several "varieties
of extra-legal sexual relations" (pre- and postmarital).
He says all are "considered illegal and condemned by
the people," but he goes on to note "each frequently
leads to a permanent union that has full tribal recognition." Presumably some such unions are considered
"illegal" only insofar as the men concerned acquire as
wives women to whom they have no rights. Warner says
that certain categories of kin are typically involved
in such affairs. These include, for male ego, galle
(MBD, etc.), momo ('father's mother'), small arndi
(mSW, etc.), waku ('sister's daughter'), and rarely
momelker (MMMBD, WMM). Elsewhere (1958: 99) Warner
mentions 'mother's father' and 'man's daughter's daughter' as possible lovers or sweethearts who may marry.
He says that of these several kinds of tolerated, if
"illegal," unions, those between 'father's mother' and
'woman's son's son', and between 'mother's father' and
'man's daughter's daughter' are the least condemned.
Consistent with this, Elkin (1953: 414) says that marriage to the daughter of a classificatory 'sister's
son' is an "alternative marriage which is in favor
today." This would be a marriage between a man and one
of his kaminyer, who could be his classificatory
'daughter's daughter' as well as his classificatory
'sister's son's daugher'.

In the light of the analysis presented above of the Murngin system of kin classification, it is hardly surprising that classificatory 'father's mothers' and 'mother's father's sisters' (both <u>momo</u>) are among the other classes of kinswomen whom a man may marry with little or no social condemnation. They are, after all, members of the same superclass CROSS GRANDPARENT as the <u>galle</u> class and they are, in effect, the classificatory 'sisters' of his <u>galle</u>. They are, to be precise, his <u>galle</u>'s classificatory 'elder sisters' (FF's 'sisters'), but they are not necessarily older than his <u>galle</u> or himself; and so if they are younger they are satisfactory sexual and marital partners - maybe no less so than ego's younger <u>galle</u> with respect to many of whom ego's marital rights are weak and highly contingent.

The possibility of marriage to a classificatory 'father's sister's daughter' <u>due</u> is not mentioned by Warner (or by Berndt or Shapiro). Yet according to Peterson (1971: 179) "a classificatory <u>due</u> is within the marriageable category." Peterson explains that there is no prohibition against marriage to women of the <u>due</u> category in general, although there is a prohibition against marriage with FZD and the daughters of 'father's 'close' classificatory 'sisters' (specifically FFBD and FMZD). MFZD is likewise classified as 'sister' by ego's father, but she is not regarded as ego's 'full' or 'close' classificatory 'father's sister', and her daughter is not prohibited as a wife except in special circumstances. A man's MFZD may be his own FZ or his FFBD or FMZD, and in this event of course he may not marry her daughter. Also, as among the Karadjeri, there is no general prohibition on two men being married to one another's sisters. This is prohibited only when one man X marries his own MBD, MFBSD, or MMZSD, so that if her brother Y were to marry X's sister, Y would have to marry his FZD, FFBDD, or FMZDD. It is unclear from the accounts I have seen whether or not a man might marry his MFZDD if his sister is married to her brother. The probability of being able to arrange such a marriage would be slight in any event, even if X had a sister who could be promised to Y. Y would have no right to X's sister through her mother, Y's MFZD. And the fact that Y's sister had been given to X would give Y no right to X's sister - although in some Australian societies it would give Y's MB a claim to X's ZD.

Apparently the Murngin do not practice sister's daughter exchange. If one man gives his ZD to another this does not entitle him to claim the other's ZD as a wife. It may be that such a right is not recognized because it would conflict with the rights of other men to the same women as MMBDDs, and this avoids creating additional situations in which several men may rightfully claim the same woman. Of course, in those societies in which both kinds of right are recognized, it is probable that one is assigned priority in relation to the other; and it seems probable that priority is assigned to the right to a woman who is one's MMBDD.

Matri-determination and matrilines

Finally, we should consider Maddock's (1970) claim
that "Murngin terms are normally matri-determined,"
that is, in general, the term a person applies to a
kinsman is determined by how he or she is related to
that kinsman's mother. According to Maddock: "Warner's
evidence is that Murngin terms are matri-determined
unless the child's father and mother have married incorrectly, in which case certain persons will call the
child by the term they would have used had its father
married correctly. A woman who marries incorrectly may
be referred to as if she were in the correct relationship to her husband, that is, the term is viri-deter-
mined. Murngin terms appear to be patri- or viri-
determined only as an adjustment to wrong marriages"
(1970: 80). Maddock does not explicate the implications, as he sees them, for a systematic analysis of
Murngin kin classification, but he does conclude that
"galle should be specified as MMBDD," rather than as
MBD (1970: 81). It is not clear from Maddock's phrasing
whether he intends to argue that the galle category is
somehow defined by reference to MMBDD (we have seen that
it cannot be so defined) or, more plausibly, that it is
misleading to focus on MBD as the "preferred" wife when,
instead, a man's marital rights to a woman depend on how
he is related to her mother rather than on how he is

related to her father. If the latter is his argument, of course his point is well taken.

Warner's evidence on what is done in cases of irregular and wrong marriages is not as clear-cut as Maddock makes out. In the passages cited by Maddock, Warner says nothing to indicate that ego's designation for alter is "normally" determined by ego's designation for alter's mother. He stresses, instead, the terminological unity of clans vis-à-vis nonmembers who are related to them genealogically or through marriage, and he appears to believe that "normally" ego's designation for alter is the same whether their relationship is reckoned through alter's father or mother - "normally" being when alter's father and mother have made a "proper" marriage. His diagram of the system is constructed on this assumption. But, he adds, "wrong" marriages do occur and "the Murngin have arranged a rather simple scheme to make the new marriage fit into the general kinship system and adjust the relationships of the offspring to other members of the society" (1958: 105). He goes on to describe the "scheme": "If one of ego's own family marries wrong the relationship term of the woman this person married is changed so that it fits what she would be called were she the proper galle of ego's relative. Thus, when ego's own galle is taken by bapa, the former no longer calls her galle, but arndi. This is true throughout. When ngatiwalker marries ego's mokul instead of ego's momelker [sic. mari], ego no

longer calls the woman mokul, but momelker [sic. mari];
yet the correct term for the woman is known, as will be
seen presently" (1958: 105).

Unfortunately, Warner does not explain what he
means by "ego's own family" in this context, but it is
clear he is referring to more than ego's clan (or patrilineage) - and it seems safe to suppose that he means
ego's close kin, regardless of clan affiliation. That
is, if a close kinsman marries "wrongly," ego does not
change his designation for this close kinsman but
instead changes his designation for the wife to the
conventional designation for the wife of such a kinsman.
Warner then goes on to explain what he means by "yet
the correct term for the woman is known, as will be seen
presently." He makes it clear that although ego may
change his designation for the wife of a close kinsman
who has married wrongly or irregularly, ego's former
relationship to the woman is not thereby "lost," as it
were, for any and all social purposes (see also Shapiro
1969: 40). Warner states: "When children are born
from a wrong marriage, the father calls them by the
terms all other parents give their children (gatu).
But everyone else calls the children by the term used
if the mother had married normally. To use the native
expression, 'The father is thrown away'. There is only
one exception: if a father marries ego's galle [MBD,
etc.], ego would not call the children gatu [mC], but
wawa [B] and yeppa [Z]. In all other cases the rule

would hold; if ego's father married ego's momelker
[WMM], ego would call the offspring marelker [WMB] and
mokul [FZ], and not brother and sister. There are
recorded cases of this. When ngatiwalker [WMMB] marries mokul [FZ] and the mokul is ego's mother's mother's
blood-brother's daughter, he looks to this mokul to
give him a daughter for his wife. Any offspring from
this union would be called galle [MBD, etc.] by ego"
(1958: 105-6). It should be noted that some of these
examples have to do with what is probably a very special
case, the case of a man's potential WM married improper-
ly or irregularly, such that, if ego were to reckon his
relationship to her and to her offspring through her
husband, he would ipso facto deprive himself of a potential wife. So, it is hardly surprising that in these
cases the father is "thrown away." The examples show
also that although ego may change his designation for a
particular kinswoman as the wife of a particular kind
of kinsman, his former relationship to her does not
thereby become wholly irrelevant, and it may be made
relevant to his classification of her offspring. That
is to say, ego may alter someone's status as a designated kinsman but not his or her status as a linking
kinsman.

Although Warner here asserts that, with only one
or two exceptions, "everyone" designates the offspring
of an irregular marriage as though their mother had
married "normally," he elsewhere asserts something quite

different. He says (1958: 27) the general rule in
cases of wrong marriage is that members of the clan
(patrilineage) of the husband reckon their kin-class
relationship to the offspring of the marriage through
the husband and disregard their relationship to the
offspring through the wife - although they do not for-
get it, because it may be important for other purposes.
However, for a person who does not belong to the clan
of the husband, the general rule is to disregard one's
relationship to the husband and to designate the offspring of the marriage according to how one is related
to them through their mother. By way of example, Warner
cites the case of a man (ego) whose father took as a
second wife a woman whom ego designated as momo (FM).
Ego would call their son 'brother', "for he would trace
the relationship through his father, because of the
influence of the clan and family." But, says Warner,
ego's ngati (MF) would call the child waku (ZS),
instead of kaminyer (DC) as he calls ego, because,
"being outside the clan, [ngati] would trace his relationship to the child of the wrong marriage through
the mother," his sister or classificatory 'sister'.
Warner's phrasing tends to suggest that there is a
general rule, "throw away the father," which is suspended for members of the clan of the father. But his data
may also be interpreted as indicating that the general
rule (a rule-of-thumb, not an iron-clad prescription)
is to reckon one's relationship to the offspring of an

irregular or wrong marriage through one's clan mate
(male or female) or one's closer kinsman who is a partner to that marriage. In real situations, however, it
must be that a number of considerations interact to
determine how anyone classifies the partners to and the
offspring of an irregular or "wrong" marriage. These
considerations include the relative clan affiliations
of ego and alter, the relative distance of one relationship as opposed to the other, the possibility that
one relationship may be consanguineal and the other by
marriage, and - perhaps most important of all - what
the reckoner stands to gain or lose by way of potential
WMs and wives by choosing to emphasize one relationship
rather than another.

None of this is to deny that the Murngin may say
that, as a general rule, in cases of irregular or
"wrong" marriage "the father is thrown away" or ignored
for the purpose of designating the offspring of the
marriage. But if they do say this, it must be with the
unstated provision "other things being equal," and without intending to imply that the father's kin-class
status (re ego) is wholly irrelevant for all social
purposes. It is understandable that they should choose
to emphasize this general rule-of-thumb instead of
others they must know. For, in general, "throwing away
the father" probably has the effect of maximizing
placement in the potential in-law categories.

Whatever the Murngin practice in cases of irregular

or wrong marriage, it is unnecessary and misleading to
suggest that "Murngin terms are normally matri-determined." Certainly, it is possible to diagram the distribution of terms in relation to kintypes on a series
of "matrilineal descent lines," as well as on a series
"patrilineal descent lines," or on both at once as
Radcliffe-Brown did (1951: 46). But this sort of exer-
cise is a poor substitute for systematic structural
semantic analysis, if intended as such, and the use of
such diagrams has tended to deflect attention from more
productive, because more realistic, questions. The
realistic questions have nothing to do with "descent
lines," patrilineal or matrilineal, but are: How are
the basic classes and subclasses of the system defined
and what are the rules whereby they and their designations are extended to more distant kintypes? Plainly,
the basic classes are not defined by patrilineal or
matrilineal criteria. The AGA rule has the effect of
extending terms along "patrilineal lines," but the AGU
does not; it determines only that the MM and wDC classes
are subclasses of the expanded 'sibling' class. Yet,
as Shapiro (1938: 348) has noted, the Murngin can and
do readily recite certain "matri-sequences." They say,
for example, that mari's (MM's) daughter is arndi (M),
whose daughter is yeppa (Z), whose daughter is waku
(ZD), whose daughter is kutara (wDD), whose daughter
again is arndi (M). Thus, these terms are ordered, as
it were, in a four-generation "matrilineal cycle." Of

course, this conforms neatly to the situation in the
Murngin subsection system, wherein the DDDD and MMMM of
any woman are members of the same subsection as that
woman. But to account for this terminological repeti-
tion or "cycling" it is not necessary to suppose that
a four-generation "matrilineal cycle" is a basic structural feature of the system of kin classification. This
apparent "matrilineal cycle" is nothing more than a
by-product of the means by which the terms of the system
are extended to kin of the third and fourth ascending
and descending generations.

SUMMARY

The Yir Yoront and Murngin systems of kin classification are not based on matrilateral cross-cousin marriage and they do not differ from Kariera-like systems
in recognizing three rather than only two "lines of
descent." They have been represented diagrammatically
as though they were so structured, but only at the expense of ignoring the superclass and subclass relationships that reveal their underlying similarities with
Kariera-like systems.

The Yir Yoront and Murngin systems are similar to
Kariera-like systems in general in that they feature
the parallel-cross status-extension rule, and the
parallel-cross neutralization rule; and like some
Kariera-like systems - such as the Karadjeri system -
they feature also the AGA rule. But they differ from

these other systems in the ways in which they combine these rules. In the Karadjeri system the parallel-cross status-extension rule is assigned priority in relation to the AGA rule, so the AGA rule has only a relatively limited range of effectiveness. In the Yir Yoront system the AGA rule has priority in relation to the parallel-cross status-extension rule and so it has a somewhat greater range of effectiveness than it does in the Karadjeri system. The arrangement in the Murngin system is the same, but the Murngin system adds the AGU rule in a highly restricted form, so that it has a relatively limited range of effectiveness. Apparently as an adjustment to the greater range of effectiveness permitted to the AGA rule, the Yir Yoront and Murngin systems feature modified versions of the cross-stepkin and spouse-equation rules that are more typical of Kariera-like systems in general.

Chapter 9

WALBIRI AND DIERI

We have seen that in many Australian systems of kin
classification the parallel-cross status-extension rule
(no. 5, Table 4.4) is supplemented by a rule of structural equivalence of agnatically related kin of alternate generations (AGA rule), and it may be supplemented
also by a rule of structural equivalence of uterine kin
of alternate generations (AGU rule). The effects of
these rules are highly variable and depend on how they
are ordered in relation to the parallel-cross status-
extension rule and in relation to one another. So far
we have dealt with systems in which the AGA and AGU
rules are subordinate to the parallel-cross status-
extension rule, or with systems in which the AGA rule
has priority in relation to the parallel-cross status-
extension rule but the AGU rule (if present) does not.
In this chapter we consider some systems with both the
AGA and AGU rules but without the parallel-cross status-
extension rule. In most such systems, for example,
Walbiri and Aranda, the AGA rule has priority in rela-

tion to the AGU rule and is supplemented by an auxiliary
that permits MBC to be reckoned as equivalent to MFSC
(MFSb) and, reciprocally, FZC to be reckoned as equivalent to FFDC (BDC). But in a few systems, Dieri, for
example, the AGU rule is assigned priority in relation
to the AGA rule and is assigned an auxiliary, which
permits MBC to be reckoned as equivalent to MMSC (ZSC)
and, reciprocally, FZC to be reckoned as equivalent to
FMDC (FMSb). Thus, in the Walbiri and Aranda systems
(AGA prior), the 'cousin' class is a subclass of the MF
and mDC classes; but in the Dieri system (AGU prior),
the 'cousin' class is a subclass of the FM and wSC
classes. A few other superficial differences between
the Walbiri and Aranda systems and the Dieri system are
attributable also to this rather minor but distinctive
structural difference.

Of the many Australian systems of kin classification that are apparently so structured, the Walbiri,
Aranda, and Dieri systems are by far the best described.
The following discussion begins with the Walbiri system
because a few critical items of information are available for it that are not available for the Aranda
system.

WALBIRI

The principal source on the Walbiri system of kin
classification is M. Meggitt's Desert People (1962).
Kenneth Hale provided some additional unpublished data,

and Dr. and Mrs. Nicolas Peterson made it possible for me to spend several days at Yuendumu in June 1972 and to make some limited inquiries of my own.

Meggitt's diagrams (1962: 84 ff.; 1972: 68) of the Walbiri system of kin classification are reproduced here as Figure 9.1 (see also Table 9.1). As in other cases, the diagram does not reveal the super- and subclasses of the system, which must be made clear before we can determine the equivalence rules.

Table 9.1. Walbiri reciprocal relations

1. jabala	FM / wSC	jabala
ngumbana	*WB / *ZH	ngumbana
gandia	*BW / *ZH	gandia
2. djamiri	MF / mDC	djamiri
wangili	MBC / FZC	wangili
3. wabira	F / mC	ngalabi
bimari	FZ / wBC	
maliri MBDC, FZDC / MFZC, MMBC		maliri
4. ngadi	M / wC	guru
ngamini	MB / mZC	
wandiri MBSC, FZSC / FFZC, FMBC		wandiri
5. babali	B+ / B-	gogono
waringi	FF / mSC	waringi
6. gabidi	Z+ / Z-	ngauwuru
djadja	MM / wDC	mindiri

330

Figure 9.1. Walbiri kin classification
(after Meggitt 1962: 84; 1972: 68)

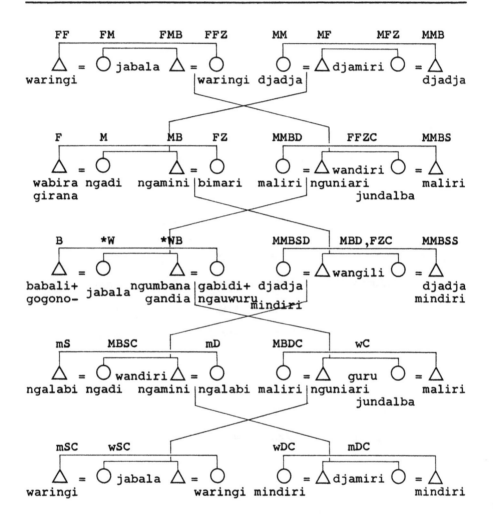

Super- and subclasses

We may begin by noting that in this system, too, the FZ class is a subclass of the FATHER class and the MB class is a subclass of the MOTHER class (see Figure 9.2). There is a special expression <u>ngamadi</u> for the female collateral or MZ subclass of the MOTHER class but not, it seems, for the male collateral or FB sub-

---

Figure 9.2. Walbiri PARENT-CHILD class

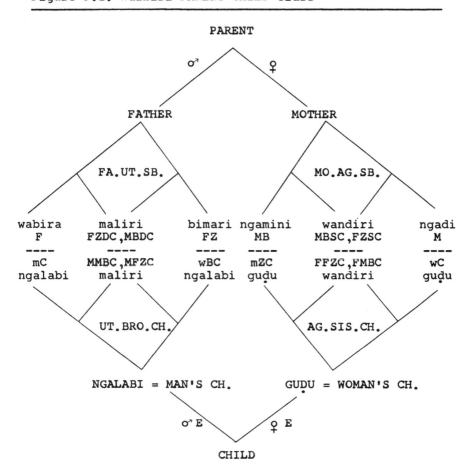

class of the FATHER class. Meggitt does not report that
FZ may be designated as 'father' or as 'female father'
or that MB may be designated as 'mother' or as 'male
mother', but he does say, "the mother's brother so
resembles the mother that his daughter [MBD] is like a
half-sibling" (1962: 193). Further, it seems probable
that the MB term ngamini is morphologically derived from
the mother term ngadi, as is the special MZ term
ngamadi. The CHILD class of this system is divided into
two subclasses 'man's child' ngalabi, and 'woman's
child' guru (or gudu). As in many other Australian
systems, the man's child term is extended to mBC and
wBC, and the woman's child term is extended to wZC and
mZC. An alternative designation for mD or wD and kintypes structurally equivalent to them is jundalba, and
an alternative expression for wS and kintypes structurally equivalent to it is nguniara. Apparently there is
no simple expression for the CHILD class as such, but
there is a simple expression jundalba 'daughter' for
the female subclass of the CHILD class.

The FATHER (including FATHER'S SISTER) and MAN'S
CHILD classes include a special subclass designated as
maliri (self-reciprocal), and the MOTHER (including
MOTHER'S BROTHER) and WOMAN'S CHILD classes contain a
special subclass designated as wandiri (also self-
reciprocal). The subclass status of the wandiri category is evident from the alternative designation of
senior wandiri (FFZC, FMBC) as 'woman's child' and,

conversely, the alternative designation of junior wandiri (MBSC, FZSC) as 'mother' or 'mother's brother' (Meggitt 1962: 157, 199). My informants indicated these relationships by saying the senior wandiri are 'half-guduna' (that is, classificatory 'woman's child'), and the junior wandiri are 'half-ngamadi' and 'half-ngamini' (that is, classificatory 'mother's sister' and 'mother's brother'). Similarly, they said the senior maliri (MMBC, MFZC) are 'half-ngalabi' and the junior maliri (FZDC, MBDC) are 'half-kida' (Meggitt's wabira) or 'half-bimari'; that is, MMBC and MFZC are classificatory 'man's child' and, reciprocally, FZDC and MBDC are classificatory 'father' or 'father's sister'. So, in this system, parents' cross cousins and cross cousins' children are in the categories reciprocal to those in which they are found in simple Kariera-like systems; they are also the foci of subclasses of the FATHER and MAN'S CHILD and the MOTHER and WOMAN'S CHILD classes.

The parallel-grandparent classes of this system are subclasses of the SIBLING class (Figure 9.3). FF and FFZ "are thought to be like senior siblings, and [man's] son's children like junior siblings" (Meggitt 1962: 141, 199). Apparently, distant waringi 'father's fathers' are commonly designated by sibling terms: "For everyday purposes, few people bother to specify the generation-levels of their distant waringi, especially those of other communities or settlements; they

are all 'siblings'." Also, a man's FF and SS are described as his 'half-brothers' (Meggitt 1962: 145). It
seems that the terminological opposition between
'siblings' and 'parallel grandfathers' is neutralizable
in everyday discourse, especially when the reference is
to classificatory 'parallel grandfather' (male or
female).

Further evidence that <u>waringi</u> designates a subclass of the SIBLING class is found in <u>tjiliwiri</u> or 'up-
side-down Walbiri' (Hale 1971). In this mode of discourse, used in certain ritual contexts, the ordinary
expressions for categories are replaced by their anto-

Figure 9.3. Walbiri SIBLING class

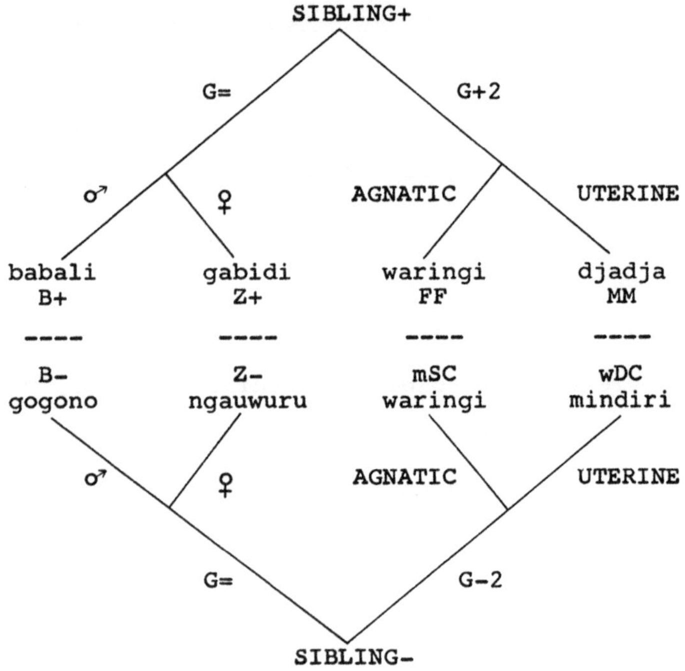

nyms. In the case of a kinship term, its antonym is its reciprocal or, if the expression is self-reciprocal, the expression most closely related to it by class inclusion. For example, in speaking tjiliwiri, if a man wishes to refer to his father, instead of saying 'my father' he says 'your son'; similarly, if he wishes to refer to his son, instead of saying 'my son' he says 'your father'. If he wishes to refer to his FF waringi, however, he would not say 'your waringi' (intending the interpretation 'your SS, that is, my FF') but he would say 'your elder brother', because 'elder brother' is the class most closely related to 'father's father' by class inclusion. The expression 'your elder brother' is ambiguous in tjiliwiri because 'elder brother' is the reciprocal of 'younger brother', which might also be understood as the referent.

The parallel-grandmother term djadja and its reciprocal mindiri also designate subclasses of the SIBLING class. A person's MM and MMB are "regarded as senior 'half-siblings'," and a man's MMB "is the same as an elder brother" (Meggitt 1962: 147, 163, 193). Meggitt does not indicate whether or not distant djadja (MM) and mindiri (wDC) may address one another by sibling terms, but it must be that they sometimes speak of one another as 'siblings'. Because djadja (MM), unlike waringi (FF), is not self-reciprocal, its tjiliwiri substitute is not 'elder brother' or 'elder sister' but is its own reciprocal mindiri (wDC); conversely, the tjiliwiri substitute for mindiri is djadja.

It will be convenient to distinguish these two subclasses of the SIBLING class by describing the waringi (FF) class as the AGNATIC SIBLING class and the djadja (MM) class as the UTERINE SIBLING class.

The cross-cousin class wangili is a subclass of the MOTHER'S FATHER or DJAMIRI class. Meggitt does not report that cross cousins may be designated as djamiri, nor does he report that a cross cousin is regarded as like a sibling of MF and MFZ or, conversely, of mDC and BDC. However, in the Walbiri system of designating kin classes by means of signs made with the hands (Meggitt 1954), the sign for cross cousins is the same as that for MF and MFZ. And in tjiliwiri the substitute for djamiri is wangili and, vice versa, the substitute for wangili is djamiri. This indicates that these two classes are related by class inclusion, and the only plausible supposition is that wangili 'cousin' designates a subclass of the DJAMIRI or MOTHER'S FATHER class (and the reciprocal MAN'S DAUGHTER'S CHILD class).

It was noted above that the maliri class is a subclass of the FATHER and MAN'S CHILD classes; and the wandiri class is a subclass of the MOTHER and WOMAN'S CHILD classes. It may now be added that these subclasses are definable in this way. The senior wandiri (FFZC, FMBC) are the children of ego's AGNATIC SISTERS and, reciprocally, the junior wandiri (MBSC, FZSC) are ego's mother's AGNATIC SIBLINGS. The senior maliri

(MMBC, MFZC) are the children of ego's UTERINE BROTHERS
and, reciprocally, the junior maliri (FZDC, MBDC) are
ego's father's UTERINE SIBLINGS.

The wandiri class is a potential in-law class
(Meggitt 1962: 157). The senior wandiri are potential
fathers-in-law and their sisters, and the junior wandiri
are potential daughter's husbands and son's wives (for
male ego). A man's potential WF is the son of his
AGNATIC SISTER. Similarly, the maliri class is a
potential in-law class (Meggitt 1962: 150). The senior
maliri are potential mothers-in-law and their brothers;
the junior maliri are potential daughter's husbands and
son's wives (for female ego). A man's potential WM is
the daughter of his UTERINE BROTHER.

Equivalence rules

We may now consider the equivalence rules that must
be posited for this system. It should be more or less
obvious that these are much the same rules as were
posited for the systems analyzed in the preceding
chapters, although of course the relations among them
must be somewhat different. The principal difference
is that this system does not feature the parallel-cross
status-extension rule (no. 5, Table 4.4). In this
system cross cousins' children are not classified as
'man's child' or 'woman's child' but, just the opposite,
as 'father' and 'father's sister' (MBDC and FZDC) or as
'mother' and 'mother's brother' (MBSC and FZSC) -

although they may be classified also as <u>maliri</u> and
<u>wandiri</u>. Also, MMBDC, FFZSC, etc., are not classified
as 'cousins' but as <u>jabala</u> 'father's mother' or 'woman's
son's child', or they are designated by special potential spouse and sibling-in-law terms.

These differences, and others, between this system
and the systems analyzed in the preceding chapters may
be accounted for by supposing that in this system the
AGA and AGU rules have greater scope than they do in
other systems, where they are subordinate to the
parallel-cross status-extension rule or where only one
of them occurs. Thus the principal analytic problem is
to specify the relations between the AGA and AGU rules,
and between them and the other equivalence rules of the
system.

We may begin by noting that the kin-class statuses
of cross cousins must be governed by the AGA rule.
Because the 'cousin' class is a subclass of the MOTHER'S
FATHER class, it must be that, as in many other Australian systems, MBC are structurally equated with their
senior AGNATIC SIBLINGS, MF and MFZ; and, conversely,
ego equates his or her senior AGNATIC SIBLING, FF, with
his or her brother and thus equates FZC with BDC. This
requires the AGA rule and its special auxiliary
$(\ldots wB \rightarrow \ldots wFS) = (mZ\ldots \rightarrow mFD\ldots)$. If the AGU rule
had a similar auxiliary and was thereby permitted to
apply in reckoning the kin-class statuses of cross
cousins, MBC would be structurally equated with ZSC and

FZC with FMSb; cross cousins would be designated as 'father's mother' and the 'cousin' class would be a subclass of the FATHER'S MOTHER class rather than, as it is, a subclass of the MOTHER'S FATHER class.

It must be that the inclusions of FFZC in the 'woman's child' class and, reciprocally, of MBSC in the 'mother' and 'mother's brother' classes are determined by the AGA rule corollary (FFZ → Z) = (wBSC → wSb). Inclusion of MMBC in the 'man's child' class and, reciprocally, of FZDC in the 'father' and 'father's sister' classes is determined by the AGU rule corollary (MMB → B) = (mZDC → mSb). Of course, the corollary of the AGA rule does not apply in the cases of FMBC and the reciprocal FZSC, nor does the corollary of the AGU rule apply directly in the cases of MFZC and the reciprocal MBDC. The AGA and AGU rules may be made applicable to these kintypes only by first reckoning them as structurally equivalent to other kintypes to which these rules are applicable.

The simplest way to do this is to posit additional auxiliaries of the AGA and AGU rules. These are,

1. AGA auxiliary:

(FM... → FFW...) = (...wSC → wHSC),

2. AGU auxiliary:

(MF... → MMH...) = (...mDC → ...mWDC).

These, of course, are limited "expansion rule" versions of the stepkin-merging rule, but they do not conflict with the stepkin rule. As auxiliaries, they apply only when no other rules are applicable, and once one of its auxiliaries has been applied the AGA or the AGU rule (as the case may be) must then be applied. Thus, the AGA and AGU rules must be subordinate to the stepkin-merging rule - and, as we shall see, they are subordinate to all other equivalence rules (except the spouse-equation rule). For reasons that will become apparent in due course, this auxiliary of the AGA rule must be assigned priority in relation to the other (opposite-sex sibling-expansion) auxiliary of that rule (the AGU rule has no other auxiliary). That is, where it is possible to apply either auxiliary of the AGA rule, the stepkin-expansion auxiliary must be applied in prefer-ence to the opposite-sex sibling-expansion auxiliary.

The AGA rule is now applicable to the kintypes FMBC and FZSC as follows.

FMBC → FFWBC, AGA auxiliary,

FFWBC → BWBC, AGA rule,

mBWBC → mWBC, same-sex sibling-merging rule,

mWBC → mZC, cross-stepkin rule,

mZC → wC, parallel-cross neutralization rule.

wBWBC → wZHC, corollary of cross-stepkin rule,

$$(wBWB \rightarrow wZH) = (mZHZ \rightarrow mWZ),$$

wZHC → wHC, same-sex sibling-merging rule,

wHC → wC, stepkin-merging rule.

Reciprocally, FZSC is reckoned as structurally equiva-lent to M or MB.

Similarly, the AGU rule is now applicable to the

kintypes MFZC and MBDC as follows.

MFZC → MMHZC, AGU auxiliary,

MMHZC → ZHZC, AGU rule,

wZHZC → wHZC, same-sex sibling-merging rule,

wHZC → wBWC, cross-stepkin rule,

wBWC → wBC, stepkin-merging rule,

wBC → mC, parallel-cross neutralization rule.

mZHZC → mWZC, corollary of cross-stepkin rule,

mWZC → mWC, same-sex sibling-merging rule,

mWC → mC, stepkin-merging rule.

Reciprocally, MBDC are reckoned as structurally equivalent to F or FZ.

We may now consider how the classifications of
second cousins are governed by these rules. Many of
these reductions are fairly straightforward, provided
we assume that the AGA rule (and its corollary) has
priority over the AGU rule (and its corollary). This
order between the AGA and AGU rules (see Table 9.2) is
implied by the classification of cross cousins (see
above) and by the classification of the kintype MMBSC,
to which both rules are potentially applicable, as
'mother's mother' or 'woman's daughter's daughter'
(djadja or mindiri). Under the AGA corollary,
MMBSC → MMSb, djamiri; but under the AGU corollary
MMBSC → BSC, waringi, which is incorrect. The cases
of FFZDC, FFZSC, and FMBSC are similar. In each case
the corollary of the AGA rule is directly applicable
and produces the appropriate results (ZDC, FSC, FMSb,

342

respectively).

In the cases of FMBDC and its reciprocal MFZSC, the AGA and AGU corollaries are not directly applicable, nor are any other rules, so we must invoke the special half-sibling expansion auxiliary of the AGA rule. FMBDC's kin-class status may then be reckoned as follows.

FMBDC → FFWBDC, AGA auxiliary,

FFWBDC → BWBDC, AGA rule,

mBWBDC → mWBDC, same-sex sibling-merging rule,

Table 9.2. Walbiri equivalence rules

I. 1. Half-sibling-merging rule

   2. Stepkin-merging rule

   3. Same-sex sibling-merging rule

   4. Parallel-cross neutralization rule

   5. Cross-stepkin rule

II. 6. AGA rule, subordinate to all the above,
       a. (FF → B) = (mSC → mSb),
       b. corollary, (FFZ → Z) = (wBSC → wSb),
       c. auxiliary, has priority in relation to (d),
          (FM... → FFW...) = (...wSC → ...wHSC),
       d. auxiliary, (...wB → wFS) = (mZ... → mFD...)

III. 7. AGU rule, subordinate to all the above,
        a. (MM → Z) = (wDC → wSb),
        b. corollary, (MMB → B) = (mZDC → mSb),
        c. auxiliary, (MF... → MMH...) =
                      (...mDC → ...mWDC).

IV. 8. Spouse-equation rule, subordinate to all the above,
       a. (Sp → FA.MO..G=), self-reciprocal,
       b. (SpF → AG.SIS.CH.) = (mCSp → MO.AG.SB.),
       c. (SpM → UT.BRO.CH.) = (wCSp → FA.UT.SB.),
       d. etc.

mWBDC → mZHDC, cross-stepkin rule,

mZHDC → mZDC, stepkin-merging rule.

wBWBDC → wZHDC, corollary of cross-stepkin rule,

wZHDC → wHDC, same-sex sibling-merging rule,

wHDC → wDC, stepkin-merging rule.

Reciprocally, MFZSC's kin-class status is reckoned as structurally equivalent to MMSb, etc.

The cases of MMBDC and its reciprocal MFZDC are somewhat different. The AGU corollary applies directly in both cases, so that MMBDC → BDC, MFZDC → MFSb, djamiri, 'mother's father'. These results seem anomalous, for MMBDC and MFZDC are classified as jabala, equivalent to FM and FMB or their reciprocals. Even so, there are reasons to suppose that this reckoning is correct. As noted above, ultimately, designation by a particular kinship term may depend not only on the rules of genealogical structural equivalence but also on the rules of subclassification.

Meggitt nowhere reports that MMBDC and MFZDC may be designated as 'mother's father' djamiri rather than as 'father's mother' jabala; yet he does say (1962: 163) a man's female cross cousin wangili (a special kind of MOTHER'S FATHER) "is like a sister" to his jabala 'father's mothers' of his own generation, and (1962: 193) "MMB is the same as an elder brother, so that his daughter's daughter is like one's own daughter's daughter." Of course, a person's female cross cousins and his female second cousins such as his

MMBDD and MFZDD do not normally designate one another
as 'sister'. They are related as FMBDC-MFZSC or as
FFZDC-MMBSC, and are classificatory 'mother's mother'
and 'woman's daughter's child' of one another. That
is, they are UTERINE SIBLINGS. Note, then, that ego's
MFZDC is a UTERINE SIBLING of ego's MF and, reciprocally, ego's MMBDC is the DC of ego's UTERINE BROTHER.
Note also that ego's MF's AGNATIC SIBLING (ego's MBC)
is in the same category as FM and, reciprocally, ego's
AGNATIC BROTHER'S DC (ego's FZC) is in the same category as ego's DC (or BDC in the case of female ego).
But, in contrast, ego's MF's UTERINE SIBLING (MFZDC) is
in the other subclass of the same superclass as MF, and
ego's UTERINE BROTHER'S DC is in the other subclass of
the same superclass as ego's DC.

This arrangement may be interpreted as follows.
As between the two cross-grandparent classes (see
Figure 9.4) 'father's mother' and 'mother's father',
the MF class is a specially designated subclass of the
cross-grandparent class, and the FM class is the residual subclass. This implies that the designation for the
cross-grandparent class as such, if ever verbally
realized, is jabala. It implies also that the kintypes
MF and FM are both JABALA, although normally, it seems,
only FM is designated as jabala. Reciprocally, mDC and
wSC are both JABALA, although normally only wSC is
designated as jabala. In other words, there is a
covert JABALA class, the foci of which are FM and MF.

On this analysis, the rules of structural equivalence
equate MMBDC and MFZDC with the foci (or their reciprocals) of the JABALA class. This makes it appropriate
to designate them as jabala. Similarly, however, the
rules of structural equivalence place MBC and FZC in
the JABALA class, but they are designated as djamiri,
'mother's father' or 'man's daughter's child'. Djamiri,
however, is the designation for the special subclass of
the JABALA class, which subclass we may now define as
consisting of MF and his AGNATIC SIBLINGS. These
JABALA may be designated as djamiri; and the remaining
members of the JABALA class, including MMBDC and MFZDC,
are designated as jabala. The special wangili 'cousin'
class of the DJAMIRI class consists of MF's AGNATIC
SIBLINGS and, reciprocally, the DC of ego's AGNATIC
BROTHERS.

Figure 9.4. Walbiri CROSS-GRANDPARENT class

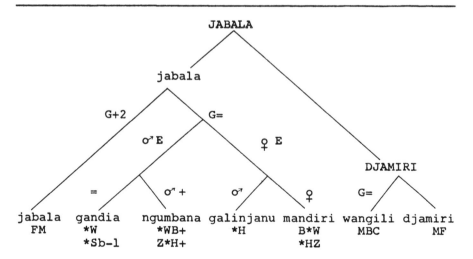

Other kintypes of the <u>djamiri</u> class include MBSSC,
FZSSC, and their reciprocals, and other kintypes of the
<u>jabala</u> class include FZDDC, MBDDC, and their recipro-
cals. By the equivalence rules posited above, all
these kintypes reduce to MFSb and BDC and, therefore,
are members of the JABALA class. MBSSC and FZSSC,
however, are ego's MF's AGNATIC BROTHER'S AGNATIC
SIBLINGS, and the reciprocals FFFZC and FFMBC are ego's
AGNATIC BROTHER'S AGNATIC BROTHER'S DC. This inclusion
of the AGNATIC SIBLINGS of AGNATIC BROTHERS of MF in
the same class as MF is, of course, a simple extension
of the inclusion of the AGNATIC SIBLINGS of MF in the
same class as MF. Again, all other G+2 and G-2 kintypes
that fall into the JABALA class are designated as
<u>jabala</u>. The <u>jabala</u> class itself is further divided
into a number of subclasses, but these may be discussed
in connection with Walbiri marriage rules. Before
taking up that subject we should note how the equivalence rules posited above determine the composition of
the several SIBLING subclasses.

In ego's own generation the kintypes MMBSC-FFZDC
and MFZSC-FMBDC are classified as UTERINE SIBLINGS,
<u>djadja</u> or <u>mindiri</u>. The corollary of the AGA rule
applies directly to the first two kintypes, but it
applies to the latter two types only after application
of the stepkin-expansion auxiliary of the AGA rule and
the AGA rule itself. Kinsmen of these types are the
AGNATIC SIBLINGS of ego's MM or, conversely, the DC of

ego's AGNATIC SISTER. Other kintypes classified as
UTERINE SIBLING include FZDSC and MBDSC and their
reciprocals. The rules posited above equate these
kintypes with Sb. It would be stretching the point to
argue that kinsmen of these types are or may be regarded
as MM's AGNATIC SIBLINGS. Note, however, that ego's
FZDSC is the child of ego's father's UTERINE BROTHER
and therefore may be regarded as like a UTERINE SIBLING
to ego. Therefore, the extended MM class includes not
only MM's AGNATIC SIBLINGS but also the children of
ego's parents' same-sex UTERINE SIBLINGS. Note that
inclusion of FZDSC in the UTERINE SIBLING class is
similar to inclusion of MMSb in that class, for MMSb
is ego's mother's UTERINE SISTER'S child.

No kintypes of ego's own generation are included
in the AGNATIC SIBLING class, but in G-2 the kintypes
MBSDC and FZSDC are designated as _waringi_, and so are
their reciprocals in G+2. Again, it would be stretching
the point to argue that kin of these types are, or may
be regarded as, AGNATIC SIBLINGS of ego's FF. They are,
however, the children of ego's parents' same-sex AGNATIC
SIBLINGS, and therefore may be regarded as like ego's
AGNATIC SIBLINGS - in the same way that ego's father's
same-sex AGNATIC SIBLING'S child (i.e., FFFC) is ego's
AGNATIC SIBLING.

We may now consider the special subclasses of the
JABALA 'father's mother' class, all of which consist of
ego's _jabala_ of his or her own generation. From Meg-

gitt's account it appears that <u>jabala</u> of ego's own
generation are divided into two classes by reference
to sex of ego (see Figure 9.4). For female ego this
class is again divided into two subclasses by the
criterion of sex of alter - the males are designated
as <u>galinjanu</u> and the females as <u>mandiri</u>. Male ego's
own-generation <u>jabala</u> may be designated as <u>gandia</u>,
but certain males of this class are designatable also
as <u>ngumbana</u>. These are potential-in-law classes. A
woman's <u>galinjanu</u> are her potential husbands and her
<u>mandiri</u> are her potential sisters-in-law. A man's
female <u>gandia</u> are his potential wives, and his <u>ngumbana</u>
and his male <u>gandia</u> are his potential brothers-in-law
(Meggitt 1962: 85, 144, 160-3). Meggitt glosses
<u>ngumbana</u> as 'senior brother-in-law' but says (1962:
161) that "relative age does not determine the alloca-
tion of the terms. Instead, <u>ngumbana</u> is the man whose
own father was the 'mother's father' of the actual
father of the other, who is <u>gandia</u>." This suggests
that FMBSS is <u>ngumbana</u> and, reciprocally, FFZSS is
<u>gandia</u>, for FMBS is classificatory 'mother's father'
to F. Of course, this rule cannot be applied in the
case of MMBDS-MFZDS and Meggitt does not report which
of these two kintypes is designated as <u>ngumbana</u>. There
are several possibilities that may be noted. First,
the designation may be allowed to depend on how ego's
father is related to ego's MMBDH or MFZDH. Second,
because MMBDS is structurally equivalent to BDD, a

junior kintype, it may be that this kintype is designated as <u>qandia</u> and its reciprocal MFZDS, structurally
equivalent to MFB, is designated as <u>ngumbana</u>. Third,
because MMBD is ego's mother's classificatory 'mother's
father's sister', but MFZD is mother's classificatory
'brother's daughter's daughter', it could be reasoned
that MMBDS is <u>ngumbana</u> because his mother is the
'mother's father's sister' of ego's mother. On the
basis of the available data we have no way to choose
between these alternatives - but this, it seems, is a
relatively minor matter.

We may ask why it is that the potential spouse and
sibling-in-law classes are subclasses of the JABALA
class. Why is it that certain kintypes within this
class (and not some other class) are singled out for
special designation as potential spouses? It is easy
to see that this may follow from regarding FF as like
a brother and even as a special kind of BROTHER. FF's
status as a kind of BROTHER does not affect the kin-
class status of his wife with respect to ego, because
she is ego's FM as well as FFW; but it may be reasoned
that, even so, if ego's AGNATIC BROTHER'S wife is designated as <u>jabala</u>, it would be appropriate for ego and
his own brothers to marry kinswomen whom they designate
as <u>jabala</u>, especially kinswomen of their own generation.
If the class of potential wives were not restricted to
kinswomen of a man's own generation, he would be permitted to compete with his AGNATIC BROTHERS (his FF and

SS, etc.) for the same women as wives. And, of course, both logically and temporally, ego's 'father's mothers' are potential wives of his FF before they are potential wives, or even like potential wives, to ego. Meggitt (1962: 144) notes there is a special designation jabala djamu for a man's classificatory 'father's mothers' and 'sister's son's daughters' of his second ascending and second descending generations, and "although they are possible sexual partners for [ego], ideally they are the proper wives for his father's father and son's son respectively."

Although it may be that only jabala of ego's own generation are designated as gandia 'potential wife', it seems probable that many Walbiri men marry jabala of their second ascending and descending generations and that such marriages are not regarded as significantly different from marriages between jabala of the same generation. Meggitt (1965: 163) describes marriage with "classificatory 'MMBDD'" as the "ideal" and marriage with "classificatory 'MBD/FZD'" as "a less favored alternative." He reports (1962: 68; 1965: 164) that 91.6 percent of all marriages (in his sample of 617) are of the former type and another 4.2 percent are of the latter type. The remaining 4.2 percent are distributed among six other terminological categories, and the categories "FM" and "ZSC" (i.e., classificatory jabala of ego's second ascending and descending genera-tions) are not included among them. Although Meggitt does not say so, it seems a fair assumption that these classificatory _jabala_ are included in his "classificatory 'MMBDD'" category; that is, by "classificatory 'MMBDD'" Meggitt intends "classificatory _jabala_" in general and not just _jabala_ of ego's own generation.

It seems, moreover, that even those JABALA who are not designated as _jabala_ 'father's mother', but as _djamiri_ 'man's daughter's daughter' or _wangili_ 'cousin', also are regarded as potential spouses. Meggitt (1962: 163) describes classificatory 'cousins' as "possible alternative spouses," and there is no stigma attached to such marriages. Meggitt's informants pointed out that a man's classificatory 'cousin' is a potential wife for his MMB (if she is his MMB's classificatory 'sister's son's daughter', and not his own ZSD), and a man's MMB is the same as an elder brother; thus, "as a man may, through the levirate, marry the ex-wife of a 'brother', he may with some propriety marry the potential wife of the MMB" (1962: 162). Further, they noted, a man may marry the 'sister' of his wife; his classificatory 'cousin' is like a sister to his classificatory _jabala_, so he may marry the former as well as the latter.

Meggitt also recorded one instance each of marriage between a man and his classificatory 'mother's father's sister' and classificatory 'daughter's daughter', both _djamiri_ (1962: 86). He says (1962: 145) the people regarded these marriages as "in poor taste" but also

"as jural marriages and not merely as long-term lia-
sons." According to Meggitt the people "justified"
these marriages by saying that there were no 'father's
mothers' or 'cousins' available, and also by pointing
out that such women are appropriate alternative wives
for a man's FF and SS, who are "the same as" his brothers. "As he may wed the ex-wife or widow of his
'brother', so his marriage with the potential wife of
a kind of 'half-brother' is also allowable" (1962: 145).

Clearly, the Walbiri regard classificatory 'cousins' as potential spouses and siblings-in-law - but not,
as it were, in their own right. They are ego's potential in-laws because they are, first of all, the potential in-laws of ego's UTERINE SIBLINGS or, conversely,
because they are the UTERINE SIBLINGS of ego's own
specially designated potential in-laws. That is, they
are ego's potential in-laws only by analogy and by
extension.

The situation with regard to ego's _jabala_ of his
own generation is somewhat different. They are regarded
as _like_ siblings-in-law because they belong to the same
kin class as the spouses and siblings-in-law of ego's
AGNATIC SIBLINGS, but they are especially like siblings-
in-law to ego himself or herself because they are kin
of ego's own generation. This does not prevent a person
from regarding all classificatory _jabala_, regardless of
generation level, as potential spouses and siblings-in-
law, and he may marry _jabala_ of other generations if

they become available. Thus ego and his AGNATIC BROTH-
ERS share their potential spouses, but not their designated potential spouses. Similarly, ego's principal
UTERINE BROTHER, ego's MMB, and the latter's AGNATIC
BROTHERS (e.g., ego's MMBSS) share their potential
spouses, but not their designated potential spouses.
The potential wives of ego's UTERINE BROTHERS are ego's
female MOTHER'S FATHERS (including ego's classificatory
'cousins') and, conversely, ego's potential wives are
the female MOTHER'S FATHERS of his UTERINE SIBLINGS.
These two sets of AGNATIC BROTHERS, who are related to
one another as UTERINE BROTHERS, regard one another's
potential wives as additional potential wives, but
again not as designated potential wives.

There is, however, one exception to the rule that
jabala of ego's own generation are potential spouses
and siblings-in-law. According to Meggitt (1962: 85),
"People who are actual MMBDD and MMBDS (jabala and
ngumbana) may not marry . . . for they are thought to
be too closely related and too much like siblings." A
person is regarded as closely related to his or her
MMBDC because he "shares the maternal spirit of his or
her own MMB, a man who in many respects is regarded as
a kind of elder brother" (Meggitt 1962: 85, 163).
Therefore, a person's MMB's daughter's children are
regarded as like his or her own brother's daughter's
children (Meggitt 1962: 193). So, in this society,
although MMBDC and MFZDC are classificatory jabala

354

(FM, mZSC), they are not regarded as potential spouses and siblings-in-law. Even so, Meggitt (1962: 85, 160) indicates that MMBDS and MFZDS designate one another as ngumbana and gandia. This is not inconsistent with description of the classes designated by these terms as potential in-law classes, because the classes are not defined by the potential in-law statuses of their members. The ngumbana and gandia classes are subclasses of the G= jabala class and as such they must be genealogically (or consanguineally) defined. The status of being a potential in-law is an implication (but not a universal implication) of membership in these classes. Similarly, designation of MMBD and MFZD as maliri is not inconsistent with the fact that a man may not marry his MMBDD or MFZDD, because the maliri class is not defined by the potential in-law statuses of its members; it is defined genealogically as "the child of a UTERINE BROTHER and, reciprocally, father's UTERINE SIBLING." Again, the status of being a potential mother-in-law is an implication (but not a universal implication) of membership in this class.

Finally, on the matter of equivalence rules, we may consider whether or not this system features a spouse-equation rule and, if it does, the specifications of that rule. It is difficult to come to definite conclusions on these questions because Meggitt does not report on the terminological implications of marriage between "prohibited" categories of kin (4.2 percent of

his sample). Some nine men in his sample married
women whom they designated as 'mother', but we are not
told how they then designated their in-laws. For
example, did they continue to designate their wives'
fathers only as 'mother's father', or did they designate
them also as gudu 'sister's child' or as wandiri, or by
either of these terms? Without information of this
kind we cannot say whether or not this system features
a spouse-equation rule by which a person designates his
or her spouse's kin (at least close kin) as though his
or her spouse were a kinsman of the jabala class, in
the event that his or her spouse is not a kinsman of
that class.

Meggitt does report on the terminological implications of marriage between classificatory 'cousins'. In
general, a person continues to designate his in-laws by
the terms by which he designated them prior to his
marriage. A man continues to designate as 'father's
sister' the kinswoman who happens to be his WM but who
is also his classificatory 'father's sister'; and he
continues to designate as 'mother's brother' the kinsman
who happens to be his WF but who is also his classificatory 'mother's brother' (Meggitt 1962: 150, 157). It
may be, however, that these in-laws are designatable
also by the terms that a person would use for his in-
laws if he were married to one of his G= jabala. This
is indicated by Meggitt's (1962: 150) observation that,
in the case of a marriage between 'cousins', wDH and

mZDH are kinsmen whom ego designated as ngalabi 'man's
son' prior to the marriage, and ego may continue to
designate them as such after the marriage, or "both may
be called wabira [F]." Of course, in the case of marriage between jabala of the same generation, wDH and
mZDH would be kinsmen of the maliri subclass of the
FATHER or WABIRA class. Meggitt does not indicate that
the corresponding designations may be used for other
in-laws, in the case of marriage between classificatory
'cross cousins', or, as already noted, in the case of
an "irregular" or "wrong" marriage – but presumably
they may be and Meggitt has not bothered to mention the
possibility for the several other in-law types.

We may therefore suppose – or, at least, treat it
as a hypothesis – that, in the event ego's spouse is
not ego's own-generation jabala, ego may designate his
in-laws as though his spouse were his own-generation
jabala. This would not necessarily imply alteration of
the kin-class statuses of the persons concerned as
linking kin, but only that as in-laws they are designatable by certain terms and that to signify the kinds
of in-laws they are these are the appropriate terms.

On this hypothesis, the spouse-equation rule of
this system specifies that ego's spouse is to be regarded as a FATHER'S MOTHER (or, reciprocally, as a WOMAN'S
SON'S CHILD) of ego's own generation. One corollary of
this rule is that SpF is to be regarded as a member of
the WOMAN'S CHILD class and, more specifically, as a

member of the <u>wandiri</u> or AGNATIC SISTER'S CHILD subclass of that class; conversely, mDH and mSW are to be
regarded as members of the MOTHER class and, more specifically, as members of the <u>wandiri</u> or MOTHER'S AGNATIC
SIBLING subclass of that class. Another corollary is
that SpM is to be regarded as a member of the MAN'S
CHILD class and, more specifically, as a member of the
<u>maliri</u> or UTERINE BROTHER'S CHILD subclass of that
class; conversely, wDH and wSW are to be regarded as
members of the FATHER class and, more specifically, as
members of the <u>maliri</u> or FATHER'S UTERINE SIBLING subclass of that class.

Of course, yet other corollaries may be adduced,
but there is no need to do so here. The important
point is that, apparently, the spouse-equation rule of
this system is quite different from the spouse-equation
rule typical of Kariera-like systems. The difference
is, of course, a necessary adjustment to other differences between the two kinds of systems, and especially
to the absence of the parallel-cross status-extension
rule in the Walbiri system and others like it. It bears
repeating that such structural differences as there are
between the two kinds of systems are not attributable to
differences in their respective spouse-equation rules
but, just the opposite, the differences in their spouse-
equation rules are attributable to differences in other
extension rules.

Walbiri marriage

We have seen that in some Australian societies a
man has a special rightful claim to his MMBD as his
prospective WM and to her daughter as his prospective
W; other women of the same kin class as his MMBD are
regarded only as potential WMs, and other women of the
same kin class as his MMBDD are regarded only as potential wives. Clearly this is not the arrangement in
Walbiri society, for a man is prohibited from marrying
his MMBDD or MFZDD, though he may marry other relatively
distant kinswomen of the same kin class, jabala; and,
with the exception of MMBDD and MFZDD, jabala of ego's
own generation are potential spouses and siblings-in-
law. Also, most women of the kin class that includes
MMBD and MFZD are regarded as potential WMs, and all
men of the kin class that includes FFZS and FMBS are
regarded as potential WFs. A man's prospective WF is
chosen from among his kin of this class, by his F, FF,
and MB. Their choice is made public at the time he is
ritually circumcised (Meggitt 1962: 139; Peterson 1969:
32-3). They choose one or more of his father's 'cousins' (actual or classificatory) to be his circumcisers,
and to give one or more of his daughters to the young
man. The set of a man's circumcisers may include
AGNATIC BROTHERS of his father's 'cousins' such as his
father's MBSSS or FZSSS, and these men also may become
his WF in the event that the principal circumciser is
unable to fulfill his obligations; or one of them may

be chosen as the principal prospective WF to begin with.
These and other 'brothers' of the prospective WF are
known as the boy's jualbiri; it is their duty to ensure
that the prospective WF "hands over the bride."

The right of betrothal of a woman rests, however,
with her MB and MMB, and not with her father (Meggitt
1962: 121). A girl's MB and MMB consult with her father
about her betrothal, but they are not obliged to follow
his advice or wishes. It may be, however, that they
commit themselves on the matter of the disposition of
their ZD and ZDD, at least in a preliminary way, at the
time of the circumcision. This is suggested by
Meggitt's report (1962: 155) that the WB of the boy's
circumciser may actually participate in the ceremony
and is often chosen to be one of jualbiri, in this case
"that member of the boy's marriage line who ensures
that the circumciser later gives the lad a wife." This
seems a curious way to put it, for the circumciser's WB
is the MB of the circumciser's daughter and the man who
has the right to dispose of her in marriage to begin
with. In any event, the duty to provide a young man
with a wife arises out of a ritual contract and is not,
in this society, prescribed to particular types of kin.
But the right to enter into such a contract is delimited
by kinship criteria; only a man's father's 'cousins' or
their AGNATIC SIBLINGS may be chosen as his prospective
WFs, and only his mother's classificatory 'cousins'
(and presumably also their AGNATIC SIBLINGS) may be

chosen as his prospective WMs and WMBs. In this way,
the range of potential in-laws is narrowed down to a
smaller class of kin who become prospective in-laws
as well.

ARANDA

The Aranda system of kin classification, as described by C. Strehlow (1913) and by Spencer and Gillen
(1899, 1927), appears to be virtually identical to the
Walbiri system, and there is no need to discuss it in
detail.[1] It may be noted, however, that the complete
data provided by Strehlow and Spencer and Gillen on the
classification of kin in the third and fourth generations may be accounted for in the same way as the corresponding classifications in the Mongkan and Murngin
systems (see Chapters 4 and 8). Of course, because the
equivalence rules and subclasses of the Aranda system
are somewhat different from those of the Murngin and
Mongkan systems, some of the classifications of G-3 and
G-4 kin are different also.

For example, in the Murngin system FMM is designated as _gatu_ 'man's child' and the reciprocals wDSC
are designated as _bapa_ 'father' and _mokul_ _bapa_ 'father's
sister', but in the Aranda system the designation for
FMM is _murra_ and the reciprocals wDSC also are designated as _murra_. Aranda _murra_ corresponds to Walbiri
maliri. It designates the class of father's UTERINE
SIBLINGS, a subclass of the FATHER class, and, recip-

rocally, the class of UTERINE BROTHER'S children, a
subclass of the MAN'S CHILD class. Because no alternative designations are reported for FMM and wDSC in the
Aranda system, it is not immediately clear whether FMM
is classified as a member of the FATHER class or as
member of the MAN'S CHILD class, and similarly for the
reciprocals. It must be, however, that FMM is classified as a member of the FATHER class and wDSC is classified as a member of the MAN'S CHILD class, just the
opposite of the arrangement in the Murngin system.
This may be explained as follows. In both systems, FMM
must be classified as though she were wSW or mZSW -
because, to use the Aranda terms, FMM is the mother of
ego's senior apulla (FM); the mother of ego's junior
apulla (wSC, ZSC) is wSW or ZSW, and the designation
for this relationship is murra. In the Aranda system
wSW and ZSW belong to the special FATHER'S UTERINE
SIBLING subclass of the FATHER class; reciprocally,
HM and HMB belong to the special UTERINE BROTHER'S
CHILD subclass of the MAN'S CHILD class. Of course,
in the Aranda system FMM is father's UTERINE SISTER
and in this respect is similar to FZDD, who is classified as murra and who is a potential wife for female
ego's son. (In Aranda society a man may marry his own
MMBDD or MFZDD.) In contrast, in the Murngin system
there is no special FATHER'S UTERINE SIBLING subclass
of the FATHER class; FZDC are designated as gurrong,
but the gurrong class is a subclass of the MAN'S CHILD

class, because in this system the parallel-cross status-extension rule has priority in relation to the AGU rule. Moreover, in Murngin society FZDD is not a potential wife for female ego's son; female ego classifies the potential wives of her sons as 'brother's child' gatu. In short, although FMM may be regarded as similar to wSW or ZSW in both systems, the classificatory results of this posited similarity in the Aranda system are just the opposite of what they are in the Murngin system, principally because the Aranda system does not feature the parallel-cross status-extension rule and thus permits the AGU rule a greater range of applicability than it has in the Murngin system.

DIERI

The Dieri system of kin classification is, in most respects, quite similar to the Walbiri and Aranda systems, but in some ways it is strikingly different. The most notable difference is that in the Dieri system cross cousins are terminologically identified with FM and wSC as kami, rather than with MF and mDC as in the Walbiri and Aranda systems; and the second cousins MMBDC, FFZSC, etc., are terminologically identified with MF and mDC as nadada, rather than with FM and wSC as in the Walbiri and Aranda systems. This difference has led at least one analyst (Korn 1971) to argue that the Dieri system is based on matrilineal rather than patrilineal descent and that it is a "four-line matri-

lineal symmetric alliance system of social classification." This analysis is founded on the assumption that
the terms of this system are monosemic "social category"
labels rather than polysemic kinship terms, and this
assumption is tenable only if one ignores, as Korn does,
the abundant evidence for polysemy and for terminological extension (see Gason 1879: 156; Howitt 1891: 44,
58). It is not difficult to show that the Dieri system
is no more based on matrilineal descent lines than
other Australian systems of kin classification are
based on patrilineal descent lines. The basic categories and equivalence rules of the Dieri system are
precisely the same as those of the Walbiri and Aranda
systems; the differences noted above result from assigning priority and an opposite-sex sibling-expansion
auxiliary to the AGU rule rather than to the AGA rule.
This system, like the Walbiri and Aranda systems, does
not feature the parallel-cross status-extension rule.

The main sources on the Dieri system are Gason
(1874, 1879 in Taplin 1879); Howitt (1878, 1891, 1904),
who acquired much of his information from missionaries,
but some of it from his own fieldwork with Dieri informants; and Elkin (1931a, 1938b). Additional information
has been taken from Howitt's notes and replies from his
correspondents (on deposit in the Royal Victorian
Museum, Melbourne), from the notes and manuscripts of
Ruether (on deposit in the South Australian Museum,
Adelaide), and from the field notes of Fry (also on

deposit in the South Australian Museum). Also, I had
the good fortune to be able to interview one of the few
remaining Dieri speakers, Ben Murray, at Farina, South
Australia, in May 1972. Mr. Murray provided important
information that helps to resolve some of the apparent
contradictions and to clarify some of the analytical
problems posed in the published literature.

Some aspects of the Dieri system of kin classification are illustrated in Figure 9.5. Again, the diagram does not reveal subclass relationships that must
be identified before the equivalence rules of the system can be specified.

Super- and subclasses

We may begin with the CROSS-GRANDPARENT class (see
Figure 9.6). According to Ben Murray, as between kami
(FM) and nadada (MF), "kami is the main word," that is,
the designation for the CROSS-GRANDPARENT class as a
whole is kami; nadada designates a special restricted
subclass of the general kami class. Consistent with
this, my informant indicated that virtually all kintypes
designatable as nadada are designatable also as kami.
The only ones not so designatable are MF, MFB, and their
male reciprocals; that is, use of the special subclass
designation is mandatory in relation to these kintypes
but optional in relation to all other members of this
subclass. The special 'mother's father' subclass is
further subdivided into a number of special subclasses.

Figure 9.5. Dieri kin classification

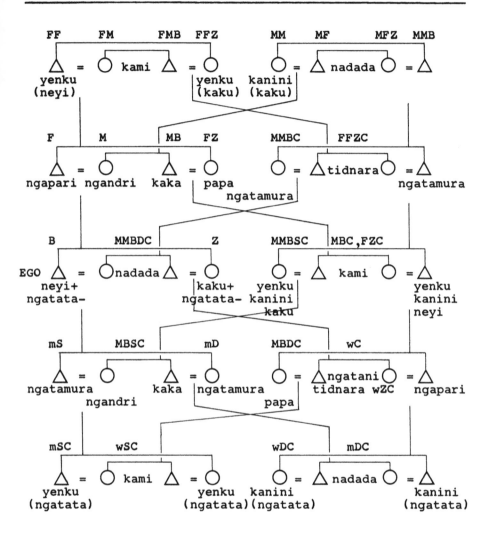

Ego's <u>nadada</u> of the second ascending and descending generations are distinguishable from ego's <u>nadada</u> of his own generation by designating the latter as <u>nadada ngatata</u>, that is, by postposing the junior-sibling term. Howitt (1904: 163) noted this possibility and assumed (wrongly) that the same designation could be given to <u>nadada</u> of the second descending generation and, thus, that a man regards his DC as his 'younger brother' and

---

Figure 9.6. Dieri CROSS-GRANDKIN class

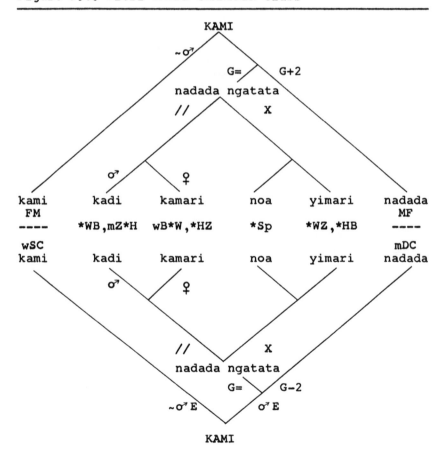

'younger sister' (Howitt 1904: 164). These false
assumptions led to much confusion in Howitt's analysis
of the Dieri system (cf. Radcliffe-Brown 1914), for
example, to the suggestion that a woman regards her DC
and a man regards his ZDC as nadada-noa (1904: 164).
This, however, is plainly false, for Howitt had already
stated that such relatives are kanini 'mother's mother'
and are regarded also as 'siblings'. In fact, nadada
of ego's own generation are not considered to be classificatory 'siblings' (i.e., members of the SIBLING
class), but they are regarded as like siblings because
they are kin of ego's own generation. To indicate this
they may be designated as nadada plus the junior-sibling
term.

Nadada of ego's own generation are classifiable
also as prospective spouses and siblings-in-law. Those
of the opposite sex are designated as noa 'prospective
spouse'. Prospective siblings-in-law of the same sex
as ego are designated as kadi (male) and kamari (female), and prospective siblings-in-law of the opposite
sex are designated as yimari.

The PARALLEL-GRANDPARENT class of this system is a
subclass of the SIBLING class (see Figure 9.7). This
is evident from the following. Howitt (1904: 162)
states, "the children of a woman are considered as
being the younger brothers and sisters (ngatata) of her
father." (See also Elkin 1938: 71.) No doubt the
"father" in this statement is a misprint for "mother"

368

(cf. Radcliffe-Brown 1914: 53), because on the next page (1904: 163) Howitt adds: "The kanini term is reciprocal between a woman and her daughter's children. She is sometimes called kanini-kaku, the kanini elder sister, apparently because her grandchildren are regarded as being on the same level as herself, being her younger brothers and sisters. Thus they are on the same level with her as her brother's children are to him. The brother of the kanini is called also the kanini-neyi, that is the kanini elder brother, her

Figure 9.7. Dieri SIBLING class

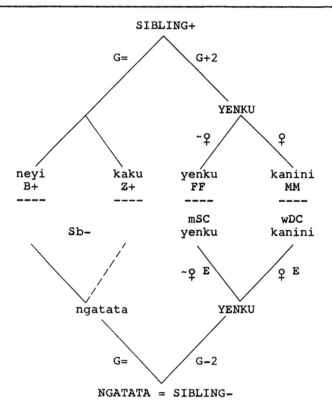

daughter's children being his younger brothers and
sisters." Thus we have the following terminological
possibilities:

MM <u>kanini</u>, also <u>kanini-kaku</u> ('elder sister')

MMB <u>kanini</u>, also <u>kanini-neyi</u> ('elder brother')

wDC⎫
    ⎬<u>kanini</u>, also <u>kanini-ngatata</u> ('younger sibling')
ZDC⎭

Howitt does not report a similar set of alternative
designations for FF and mSC, but he does say that,
because a woman's grandchildren are regarded as her
younger brothers and sisters, "they are on the same
level with her as her brother's children are to him."
Doubtless this is another of Howitt's many typographical
errors and "brother's children" should be "brother's
son's children." This implies that FF <u>yenku</u> is designatable as <u>yenku-neyi</u>, FFZ is designatable as <u>yenku-</u>
<u>kaku</u>, and mSC, BSC are designatable as <u>yenku-ngatata</u>.
Again, these possibilities are not reported by Howitt
but they were readily volunteered (with no prompting on
my part) by my informant, who gave the following:

FFZ+ <u>yenku</u>, <u>yenku-kaku</u>, or simply <u>kaku</u>

FFZ- <u>yenku</u>, <u>yenku-ngatata</u>, or simply <u>ngatata</u>

FFB+ <u>yenku</u>, <u>yenku-neyi</u>, or simply <u>neyi</u>

FFB- <u>yenku</u>, <u>yenku-ngatata</u>, or simply <u>ngatata</u>

Therefore, according to my informant, a sibling of ego's
parallel grandparent may be designated by a parallel-
grandparent term, or by a sibling term appropriate to
his or her sex and age relative to the linking parallel

grandparent, or by the two terms combined. Conversely, a parallel grandchild of ego's sibling may be designated by the appropriate self-reciprocal parallel-grandparent term, or by the sibling term that ego uses for his or her linking sibling. My informant insisted that although parallel grandparents and parallel grandchildren themselves are "like brother and sister," they may not be addressed by the appropriate sibling terms; in direct address their SIBLING status must remain covert, but it may be indicated in third-person reference to them by postfixing the appropriate sibling term to the appropriate parallel-grandparent term.

In the Dieri system, MMB and mZDC may be designated by the FF term, and FFZ and wBDC may be designated by the MM term. This was reported to Howitt by his correspondent O. Siebert (in Howitt's papers, Box 8, Folder 2), but it was not subsequently reported by Howitt (as far as I know). Structurally the arrangement here is slightly different from the Mongkan arrangement (see Chapter 4, Ompela and Mongkan). My informant stated that, as between kanini (MM) and yenku (FF), "kanini is the main word." Therefore, the term for the parallel-grandparent category as such is kanini. Not all parallel grandparents are so designated, however, because the male members of the class are singled out as a special subclass and designated as yenku, thus leaving the nonmale (= female) members of the class to be designated as kanini. Again, this may be done before

or after widening of the class. If it is done before
widening of the class, any kintype reckoned as structurally equivalent to FF is designated as yenku, even
if that type is female - thus FFZ may be designated as
yenku. If it is done after widening of the class, FFZ
is designated as kanini, because not-male. And similarly for MMB (cf. Elkin 1938b: 68-9 on the similar
Yantruwanta system).

It should follow also that FF and mSC are designatable as kanini, because if the male versus nonmale
opposition is suspended the designation for the class
as a whole then applies to its male as well as to its
not-male members. Unfortunately, I did not think to ask
my informant about this possibility, but there is some
evidence for it in Elkin's account. Elkin (1938b: 56)
says, the "double term" yenku kanini "may be used for
[a man's] son's children who are strictly speaking
yenku; the explanation given is that they are half
kanini." The English expression half is often used by
Australians to talk about dual class membership, either
when the two classes in question are related as superclass-subclass or when the kinsman in question is a
member of two structurally unrelated kin classes. Thus,
in describing his son's child as yenku kanini or 'half
kanini', Elkin's informant may have been stating in
effect that the class designated as yenku is a subclass
of the class designated as kanini.

In the Walbiri and Aranda systems the kintypes

MMBSC and FMBDC and their reciprocals are classified as
UTERINE SIBLINGS, that is, as structurally equivalent
to MM, MMB, and their reciprocals. The kin-class
statuses of these kintypes in the Dieri system are
somewhat obscure in the published literature. Although
it is clear that they are classified as SIBLINGS and
may be denoted by the sibling terms, it is not clear to
which of the two special subclasses they belong, yenku
or kanini. The early sources, including Howitt, do not
report on the designations of these kintypes, but
Radcliffe-Brown (1914), in his commentary on Howitt's
reports, hypothesized that they are designated as
kanini. According to Elkin, the situation is more
complex. In Elkin's diagrams (1931a: 497; 1938b: 53)
the designation for MMBSD is given as kaku 'elder sister' but the designation for MMBSS is given as yenku
kanini "neyi". In the text Elkin explains: "[MMBS's
children] are classified with brother and sister, or,
what is much the same thing, with father's father and
his sister, yenku; in the latter case, they are described as being 'light' yenku. But occasionally,
there is an approximation to the Aranda terminology,
for these same children are sometimes called yenku
kanini, and so too, the same double term may be applied
to son's children who are really yenku; the explanation
given is that they are half yenku, meaning that they
belong to the same moiety as the kanini. This use of
kanini, however, was, in the cases which came to my

notice, purely secondary. If a man were not sure of the term to be applied to a certain relation, he would think of the latter's moiety and generation level, and give any or all of the terms that could apply. Thus, the term for MMBSS might be given as <u>yenku</u> <u>kanini</u> <u>ngatata</u> (that is, FF, MMB, B-)" (1931a: 495). Here the emphasis is on the classification of MMBSC as <u>yenku</u> 'father's father' or as 'sibling', and their possible designation as <u>kanini</u> 'mother's mother' is said to be, somehow, "secondary." Later (1938b: 56-7) Elkin repeated these observations, but added "MMB and his SS . . . are <u>kanini</u> and <u>niyi</u> (brother) respectively"; and "the son's children of MMB (<u>kanini</u>) are <u>niyi</u> (brother) and <u>kaku</u> (sister)." In these statements the emphasis is on the classification of MMBSC as 'brother' and 'sister'.

My informant indicated the same three possibilities with regard to the designation of MMBSC. He first volunteered the sibling terms but, on cross examination, said that these kintypes could also be designated as <u>yenku</u> or <u>kanini</u>, but he thought 'brother' and 'sister' were the more appropriate terms because they are relatives of his own generation. An argument is presented below to show that MMBSC, MBZSC, and their reciprocals are structurally equated with BSC and FFSb and are therefore classified as <u>yenku</u>; but the <u>yenku</u> class is a special subclass of the PARALLEL-GRANDPARENT superclass, the designation for which is <u>kanini</u>, and therefore MMBSC, etc., may be designated also as _kanini_. Further still, the PARALLEL-GRANDPARENT class is a subclass of the SIBLING class, and so MMBSC, etc, may be designated also by sibling terms.

The FATHER and MOTHER classes of this system are similar to those of the Walbiri system. Here too a woman designates her BC by the same term as her brother does, _ngatamura_, so the FZ class is a subclass of the FATHER class. Apparently, a man does not designate his ZC by the same term as his sister does; he uses _tidnara_ instead of _ngatani_. This is not necessarily evidence that the MB class is not a subclass of the MOTHER class, for _tidnara_ may designate a subclass of the WOMAN'S CHILD class. We must assume that it does, because MMB is (or may be) identified terminologically with MM, and FMB with FM, and these terminological assignments must be governed by the parallel-cross neutralization rule.

The early sources are mostly silent on the possible classifications of parents' cross cousins and, reciprocally, cross cousins' children. Howitt (1904: 166) reports that a woman may designate her MBDS or FZDS as _paiara_ 'daughter's husband' (self-reciprocal) _if_ he is married to her daughter, but he says nothing about how kin of these types designate one another if they are not related also as in-laws. However, on one of the diagrams sent to Howitt by Siebert, it is shown that a person designates the daughter of his or her female cross cousin as 'father's sister'. According to Elkin

the designation for MMBC and MFZC is 'man's child', and
the reciprocals FZDC and MBDC are designated as 'father'
and 'father's sister'. Also, the designation for FFZC
and FMBC is 'sister's child' (or 'woman's child' for
female ego), and the reciprocals MBSC and FZSC are designated as 'mother' and 'mother's brother'. Elkin
(1931a: 493-6) indicates that the special in-law terms
paiara (WM, WMB), taru (WF, HF, WFZ, HFZ), and kalari
(HM, HMB), all self-reciprocal, are used only for actual
and "promised" (prospective) WF and WM and their siblings; potential in-laws who have not been singled out
as prospective in-laws are designated by the appropriate
consanguineal terms (see also Elkin 1938b: 57-8). Similarly, the special sibling-in-law terms are used only
for actual or prospective siblings-in-law, and not for
all potential siblings-in-law who are nadada or nadada
ngatata.

Equivalence rules

We may now consider the probable equivalence rules
of this system (see Table 9.3). Clearly, they are much
the same as the rules posited for the Walbiri system,
though the relations among them must be somewhat different. The designations assigned to parents' cross cousins and, reciprocally, cross cousins' children may be
accounted for only by positing the AGA and AGU rules
and their corollaries and by assuming that this system
does not feature the parallel-cross status-extension

rule. But it cannot be that the AGA rule is assigned
an auxiliary that permits it to determine the kin-class
statuses of cross cousins, for in this system cross
cousins are classified with FM and wSC rather than with
MF and mDC. That is, FZC are overtly identified with
their senior UTERINE SIBLINGS, ego's FM and FMB and
their MM and MMB; and, reciprocally, ego identifies
himself with his senior UTERINE SIBLING, his MM, and
classifies his MBC as equivalent to MMSC or ZSC. In
other words, in this system, the kin-class statuses of

Table 9.3. Dieri equivalence rules

I. 1. Half-sibling-merging rule

  2. Stepkin-merging rule

  3. Same-sex sibling-merging rule

  4. Parallel-cross neutralization rule

  5. Cross-stepkin rule

II. 6. AGU rule, subordinate to all the above,
    a. (MM → Z) = (wDC → wSb),
    b. corollary, (MMB → B) = (mZDC → mSb),
    c. auxiliary, has priority in relation to (d),
       (MF... → MMH...) = (...mDC → ...mWDC),
    d. auxiliary, (...mZ → ...mMD) =
            (wB... → wMS...).

III. 7. AGA rule, subordinate to all the above,
    a. (FF → B) = (mSC → mSb),
    b. corollary, (FFZ → Z) = (wBSC → wSb),
    c. auxiliary, (FM... → FFW...) =
           (...wSC → ...wHSC).

IV. 8. Spouse-equation rule, subordinate to all the
    above,
    a. (Sp → MO.FA..G=), self-reciprocal,
    b. (SpF → WOMAN'S SON) = (mCSp → MO. or MO.BRO.),
    c. (SpM → MAN'S DAU.) = (wCSp → FA. or FA.SIS.),
    d. etc.

---

cross cousins are determined by the AGU rule, and not
by the AGA rule. Therefore, we must suppose that the
AGU rule is assigned a special auxiliary that permits
it to determine the kin-class statuses of cross cousins,
and the AGA rule is not. The appropriate auxiliary is:

$$(\ldots mZ \rightarrow \ldots mMD) = (wB\ldots \rightarrow wMS\ldots).$$

Because this is an auxiliary of the AGU rule it applies
only when no other rule can be applied, and when it is
applied the AGU rule or its corollary is then applicable
and must be applied.

It appears that the AGU rule is assigned priority
over the AGA rule; it is possible to account for the
designation of MMBSC, FMBDC, etc., as yenku (FF), as
well as kanini (MM) or as 'sibling', only on this
assumption. By the AGU rule MMBSC is structurally
equivalent to BSC yenku; as noted above, the yenku
class is a special subclass of the KANINI class and the
KANINI class is a special subclass of the SIBLING class.
This means that, if the subclass distinctions are suspended, MMBSC, etc., are designatable as kanini or as
'sibling', as well as yenku. However, if the AGA rule
is applied to the kintype MMBSC, the result is MMSb.
The implication would be that MMBSC is designatable as
kanini or as 'sibling', but not as yenku - because yenku
in general are designatable as kanini but not vice
versa. Even so, MMBSS might be designatable as yenku
as structurally equivalent to MMB who may be so designated; but MMBSD would not be so designatable, because

she would be reckoned as structurally equivalent to MMZ who is not designatable as yenku.

Therefore, MMBSC, MFZSC, FMBDC, and FFZDC must be reckoned as structurally equivalent to FFSb and BSC, via the AGU rule (and its special stepkin-expansion auxiliary, rule 6c, Table 9.3), and the AGU rule must have priority over the AGA rule.

It follows that MMBDC and MFZDC are reckoned as structurally equivalent to BDC and MFSb, and FMBSC and FFZSC are reckoned as structurally equivalent to FMSb and ZSC, exactly as in the Walbiri system. Yet all these kintypes are designated by the same term as MF and mDC nadada, even though some of them are reckoned as structurally equivalent to FM and wSC kami. This is just the opposite of the problem we encountered in analysis of the Walbiri system, where all these kin-types are designated as jabala (FM, wSC), even though some of them are reckoned as structurally equivalent to the foci of the djamiri class (MF). Presumably, the solution to the problem in this case is similar to the solution offered in the Walbiri case. Here, of course, the special NADADA or MOTHER'S FATHER subclass of the KAMI or CROSS-GRANDPARENT class cannot include MF's AGNATIC SIBLINGS and, conversely, the DC of AGNATIC BROTHERS, because these are ego's cross cousins, who are his FM's UTERINE SIBLINGS or his UTERINE SISTER'S SC and who are equated with FM and wSC via the AGU rule. Instead, the special NADADA subclass consists of

all members of the KAMI class who are <u>not</u> FM or her
UTERINE SIBLINGS or, conversely, wSC or the SC of ego's
UTERINE SISTER.

Elkin's diagrams of the Dieri system show cross
cousins' spouses and spouses' cross cousins designated
as 'siblings', and in the text (1938b: 56) he says,
"cross cousins marry ego's tribal brothers and sisters."
The reference here is to ego's MMBSC, FMBDC, etc., who
are related as MMBDC, FFZSC, etc., to ego's cross cousins. That is, if ego's cross cousins marry their <u>nadada</u>
of their own generation, they marry persons whom ego may
designate as 'sibling' or as <u>yenku</u>, or even as <u>kanini</u>.
In fact, Elkin himself recorded all these designations
for cross cousin's spouse and spouse's cross cousin,
though not always for the same relative (Elkin's unpublished field notes and Elkin 1938a: 450). "Properly,"
then, ego's cross cousin's spouse or ego's spouse's
cross cousin is ego's kinsman of the YENKU class, whom
ego may designate also as <u>kanini</u> or as 'sibling'.

Kin classification and marriage

We saw that in Walbiri society classificatory
'cousins' are potential spouses - although in a tertiary
fashion. The Dieri arrangement appears to be the same.
Howitt says that the proper wife for a man is a woman
whose mother is his mother's 'cousin' <u>kami</u>; that marriage is not permitted between cross cousins; and that
classificatory 'cousins' "could not lawfully marry."

But he noted that, even so, such marriages do occur. He recorded one instance in which a marriage was arranged between the offspring of two opposite-sex parallel cousins (e.g., a man and his FFBDD or MFBSD). The mothers of the man and woman were mutually kamari (BW-HZ), rather than mutually kami 'cousin', but, Howitt says, "the respective kindreds got over the difficulty by altering the relationship of the two mothers from kamari to kami, by which change the two young people came into the noa [prospective spouse] relationship." Siebert (in Howitt 1904: 168) reported that the Dieri described this as an ancient practice. Of course, in an instance of this kind, a man's WM would be his 'father's sister' (FFBD) whose husband would normally be his 'mother's brother', or his WF would be his 'mother's brother' (MFBS) whose wife would normally be his 'father's sister'. If the kamari (BW-HZ) class is a subclass of the nadada class, and if the nadada class is a subclass of the KAMI class (as it is), then what is being done in this instance is to give recognition to the principal kin-class relationship of the two would-be mother's in-law. They are, in any event, kamari-nadada-KAMI, and if priority is assigned to their covert status as KAMI of one another, their offspring may be regarded as eligible to marry one another. Howitt (1904: 167-8) says this change "necessitated some consequential alterations of relationships" between the husband and some of his kin,

but he does not describe them in detail. Siebert wrote
to him that the change of the couple's relationship
from kami to nadada (and noa) did not affect the kin-
class statuses of their siblings in relation to one
another.

Elkin (1938b: 58) notes that Dieri men sometimes
"exchange" their ZDs in marriage. This also implies
the possibility of marriage between two persons who
classify one another as kami 'cousin' - because if one
man marries, in the orthodox manner, one of his own-
generation nadada and then gives his own ZD to his WMB,
the latter then marries his FZDDD or MBDDD, whom he
classifies as kami. Note that the mothers of such
kinswomen are ego's classificatory 'father's sisters'
(father's UTERINE SISTERS), and not his classificatory
'brother's daughters' (UTERINE BROTHER'S daughters),
who are his "proper" potential WMs; but their fathers
are still ego's classificatory 'sister's sons' (though
not his AGNATIC SISTER'S sons).

Some similar systems

Elkin (1931b, 1938b) provides data that indicate
that the systems of kin classification of the Piladapa,
Yantruwanta, Yaurawaka, Marula, northern Wongkonguru,
and Wilyakali, all located near the Dieri in the northwest of South Australia, are structurally similar to
the Dieri system. The main diacritical feature is that
cross cousins are classified with FM and wSC (not with

382

MF and mDC, as in Walbiri and Aranda). There are, of
course, many minor differences - mostly it seems in
subclasses - among these systems. So far as I have
been able to determine, systems of this kind are confined to northwest South Australia, but they are similar to the Murawari and Wongaibon systems of New South
Wales, in which it may be that the AGU rule also governs
the kin-class statuses of cross cousins (but not of
MMBC, MFZC, and their reciprocals, because the AGU rule
is subordinate to the parallel-cross status-extension
rule).

SUMMARY

This chapter has dealt with several systems of kin
classification whose principal structural difference
from the systems discussed in the preceding chapters is
that they do not feature the parallel-cross status-
extension rule. In the absence of this rule, the AGA
and AGU rules govern the kin-class statuses of all
second- and greater-degree collateral kinsmen related
as the descendants of opposite-sex siblings. In all
these systems mother's cross cousins are classified as
'man's child', father's cross cousins are classified as
'woman's child', and reciprocally, the offspring of
female cross cousins are classified as 'father' and
'father's sister', and the offspring of male cross
cousins are classified as 'mother' and 'mother's brother'. These classifications are sometimes obscured by

the presence of special subclass designations for
UTERINE BROTHER'S child (MMBC, MFZC) and, reciprocally,
father's UTERINE SIBLING (FZDC, MBDC), and for AGNATIC
SISTER'S CHILD (FFZC, FMBC) and, reciprocally, mother's
AGNATIC SIBLING (MBSC, FZSC). The classifications of
the offspring of parents' cross cousins depend on the
relationship between the AGA and AGU rules. In the
Walbiri and Aranda systems, the AGA rule is assigned
priority over the AGU rule and it is assigned an auxiliary that permits it to apply to cross cousins. Thus,
cross cousins are structurally identified with MF and
mDC; MMBDC, etc., are structurally identified with FM
and wSC; and MMBSC, etc., are structurally identified
with MM and wDC. But in the region of South Australia
north and west of Lake Eyre, the AGU rule is assigned
priority over the AGA rule and it is assigned an auxiliary that permits it to apply to cross cousins. Thus,
in the Dieri and several other closely related systems
cross cousins are identified with FM and wSC; MMBDC are
identified with MF and mDC; and MMBSC are identified
with FM and mSC.

There are some notable, if minor, differences in
subclass relations that correspond to this difference
in the relationship between the AGA and AGU rules. In
the Walbiri and Aranda systems the MF class is a special
subclass of the cross-grandparent class, and the
arrangement in the Dieri system is the same; but the
special subclass is defined differently in the Dieri

384

system. In the Walbiri and Aranda systems the MM class
is a special subclass of the parallel grandparent class
- which itself is a special subclass of the SIBLING
class; but in the Dieri system the FF class is the
special subclass of the parallel-grandparent class -
again, a special subclass of the SIBLING class.

Chapter 10

NGARINYIN

The focus of this chapter is the system of kin classification of the Ngarinyin language of the Kimberley
region of northwest Australia. Elkin (1932: 312-3)
describes this system as Aranda-like - because "descent
is reckoned through four lines, marriage with cross
cousins is prohibited, and the type-marriage is of the
second-cousin variety" - but he notes that it differs
from "other systems of the Aranda type" in at least
two ways. First, in Aranda-like systems ego's kin of
his MF's, FMB's, and MMB's descent lines are terminologically identified with one another in alternate
generations, but in the Ngarinyin system "the general
principle is that all the males of any one local horde
are classified together under one term, and so too are
their sisters. That is, a man and his brothers, his
and their children, and so on, are classified together
irrespective of generation levels." The second difference has to do with the "varieties of second-cousin
marriage which are possible." In "the ordinary Aranda

system" there are four types of second cousin whom a man may marry - MMBDD, MFZDD, FFZSD, FMBSD - and typically all such kinswomen are designated by the same term as FM and wSC or, alternatively, by special potential spouse and potential sibling-in-law terms. But in the Ngarinyin system only one of these four kintypes belongs to the class of marriageable kinswomen; this is FMBSD who, here too, is designated as 'father's mother'. None of the other three kintypes is so designated - MMBDD is 'mother', MFZDD is 'woman's child', and FFZSD is 'woman's son's child' (the reciprocal of 'father's mother' but not the same term) - and none is regarded as a potential wife.

Elkin (1932: 313; 1964: 79-80) attributed these differences to "the influence of local organization," and so did Radcliffe-Brown: "The principle that is obviously at work here is that of the solidarity of the local clan" (1930-31: 453). He argued that there is a "tendency" throughout Australia "for the individual to group together all the members of a clan other than his own and to regard his relationship to them as being determined by his genealogically close relationship to one member of the clan." Where this jural principle is expressed in kin classification, he said, it is most commonly through the terminological identification of agnatically related kin of alternate generations. But in the Ngarinyin case, as he interpreted it, "this tendency has . . . been given free play and has in a

certain sense over-come the division into generations
which elsewhere is so important." Elsewhere, Radcliffe-
Brown (1952: 83) described the "structural principle"
in question as: "For the outside related person the
clan constitutes a unity within which distinctions of
generation are obliterated." He noted also some of the
similarities between the Ngarinyin system and systems
of the so-called Omaha type found elsewhere in the
world, and he argued that they share this "structural
principle" with the Ngarinyin system.

It is shown below that the Ngarinyin system is,
indeed, structurally similar to systems of the so-called
Omaha type. Its most distinctive structural principle
is the covert structural equivalence of a woman's
brother with her father and, conversely, of a man's
sister with his daughter (cf. Lounsbury 1964b, 1965).
In this respect the Ngarinyin system is not wholly
unique in Australia, for the same principle occurs in
at least two neighboring cognate systems (Worora and
Wunambal) and in the wholly unrelated, or at best very
distantly related, system of the Yaralde of South Australia. It occurs also in a few other systems in Arnhem
Land, although in these systems its range is much more
limited than in the Ngarinyin system (see Hughes 1971;
Turner 1974). Terweil-Powell (1975) has recently shown
that the Koko Yimityirr system in southern Cape York
also features the "Omaha skewing rule." But the Ngarinyin system is similar also to a great many other Aus-

tralian systems of kin classification, for it too features a rule of structural equivalence of agnatically related kin of alternate generations. In the Ngarinyin system, however, this rule has a somewhat different form than it has in the other systems and this makes it possible for the two rules to complement one another in the same system.

The principal sources on the Ngarinyin system of kin classification are Elkin (1932, 1964), Jolly and Rose (1966), and Lucich (1968).[1] In addition Elkin permitted me to consult and to cite from an unpublished field report (1928d). Coate and Oates (1970) also provide some useful information. The terms and their assumed focal denotata are listed in Table 10.1. Additional data are introduced as they become relevant. The several sources are not entirely consistent with one another, but I believe it can be shown that most of the inconsistencies are only apparent.

Super- and subclasses

We may begin with the grandparent classes. Two of these, the MM class and the FM class, present no analytic difficulties. All sources agree that the designation for MM is kaiingi and that this term is extended to MMB and to all his agnatic descendants. The reciprocal junior category includes, first, wDC; it is expanded to include mZDC, FZDC, and the DC of ego's classificatory 'sisters' and 'father's sisters'. In Elkin's

published diagram <u>wolmingi</u> 'wife's mother' is given as
the junior reciprocal of <u>kaiingi</u> but in his field report
ZDC is shown as <u>kaiingi</u>. There is no indication why
<u>wolmingi</u> rather than <u>kaiingi</u> is shown on the published
diagram. Lucich (1968: 106) lists <u>mimingi</u> as the reciprocal of <u>kaiingi</u>. Presumably, the arrangement here is
similar to those we have encountered in other systems.
The MM term <u>kaiingi</u> is extended by the rule of self-
reciprocity to wDC, thus making wDC members of the
expanded 'mother's mother' class; but this junior subclass of the MOTHER'S MOTHER class is assigned the
special designation <u>mimingi</u>. Either term, <u>kaiingi</u> (MM)

Table 10.1 Ngarinyin reciprocal relations

1. kaiingi	MM / wDC	kaiingi or mimingi
2. maringi	FM / wSC	wuningi
waiingi	FMB / mZSC	
3. idje	F / mC	idje, amalngi
amalngi	FZ / wBC	
4. ngadji	M / wC	malengi
kandingi	MB / mZC	
mamingi	MBC / FZC	
5. ngolingi	Sb+ /	
	/ Sb-	margingi
lalingi	Z+ /	
6. wolmingi	WM / wDH	wolmingi
7. mariengi	HM / wSW	mariengi

or _mimingi_ (wDC), is then available to designate wDC
and structurally equivalent kintypes.

All sources agree that the designation for FM is
maringi and that the designation for FMB is _waiingi_.
The common reciprocal of both terms is _wuningi_ (wSC,
ZSC). Because both terms have the same reciprocal we
may assume that the _waiingi_ (FMB) class is a subclass
of the expanded _maringi_ (FM) class. The FM and FMB
terms are extended to all agnatic descendants of FMB;
reciprocally, the wSC term is extended to FZSC and to
the SC of ego's classificatory 'sisters' and 'father's
sisters'.

The data concerning FF and MF, their siblings, and
agnatic descendants are somewhat more difficult to
interpret. In all sources the designation for FF is
given as 'elder brother' and the designation for mSC
is given as 'younger sibling'. Most sources report FFZ
as 'elder sister' but Elkin gives _amalngi_ 'father's
sister' or 'man's daughter'. I am reluctant to dismiss
Elkin's report as an error, but because it is inconsistent with several other reports, and for other reasons
stated below, I assume that the designation for FFZ as
such is 'elder sister'. (The designation for wBSC is
not reported in any of the sources.) Apparently, in
this system, as in several of the systems analyzed in
preceding chapters, the potential FF subclass of the
parallel-grandparent class is evacuated and its members
are placed in the expanded 'elder-sibling' class;

conversely, the potential mSC subclass of the parallel-
grandchild class is evacuated and its members are placed
in the extended 'younger-sibling' class.

In Elkin's diagram the designation for MF is shown
as kandingi and the designation for MFZ is shown as
ngadji (M). The same terms are shown as the designations for all agnatic descendants of MF, including M
and MB. According to the other sources MF and MFZ are
designated as mamingi, and so are their agnatic descendants in alternate generations (MBC, MBSSC, etc.); M is
designated as ngadji and MB as kandingi, and these terms
are extended to MB's agnatic descendants in alternate
generations (MBSC, etc.). But Elkin himself describes
much the same arrangement in his field report; he notes
that MF, MB, MBS, MBSS, etc., may be designated as
kandingi, but the term mamingi "is sometimes applied"
to MF and MBS.

Apparently, there are two ways in which ego's
mother and her agnatic kin may be designated. (1) All
such kintypes may be designated as ngadji 'mother' or
kandingi 'mother's brother', depending on the sex of
alter; or (2) MF and MFZ and their agnatic descendants
in alternate generations may be designated as mamingi,
and M and MB and their agnatic descendants in alternate
generations may be designated as ngadji and kandingi.
To understand this arrangement we must consider how the
reciprocal relationships are designated. This is simple: All sources agree that the three terms ngadji (M),

kandingi (MB), and mamingi all have the same reciprocal
malengi 'woman's child'. Therefore, it must be that
the categories designated by the three senior terms are
related by class inclusion; they designate subclasses
of a higher-order class whose reciprocal class has no
specially designated subclasses. On the assumption
that the closest kintype within each superclass is the
focal member of that class, the focus of the senior
superclass is M and the focus of the junior superclass
is wC. Therefore, these are the familiar MOTHER and
WOMAN'S CHILD classes, with the difference that here
they include MF, MFZ, MBC, etc., and, reciprocally,
mDC, BDC, FZC, etc.

For reasons that will become apparent as the analysis proceeds and the equivalence rules of this system
are identified, the structure of these classes must be
as indicated in Figure 10.1. The hypothesis represented by this figure is that within the MOTHER class
two special subclasses are recognized. One is the
usual MB class, here designated as kandingi; as usual,
the reciprocal mZC class is not specially designated.
The other special subclass of the MOTHER class is a
'cousin' class with the focus MBC; the reciprocal FZC
class is not specially designated. The reasons for
choosing MBC, rather than MF, as the focus of the
mamingi class are presented below.

The FATHER class of this system is similar to the
FATHER classes of systems analyzed in preceding chap-

ters, except that here it is not expanded to include
mother's cross cousins (as in Kariera-like systems) or,
alternatively, the offspring of female cross cousins
(as in Aranda-like systems). F is designated as _ira_
or _idje_ and FZ as _amalngi_. There is a separate designation for mS and BS _ngongi_, but apparently there is no
separate designation for mD and BD, who are designated
as _amalngi_, the same as FZ. Similarly, however, mS and
BS may be designated as 'father'. As noted above,
similar arrangements are not uncommon in Australia,
though the absence of a separate designation for mD may
be unusual. At first glance, this case seems analyzable

Figure 10.1. Ngarinyin MOTHER-WOMAN'S CHILD class

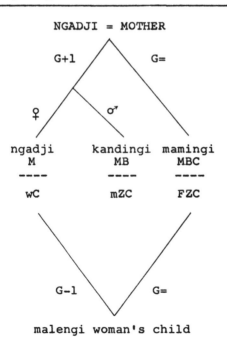

in the same way as the Karadjeri case (see Chapter 6),
with the difference that there is no special designation
for mC, but only for mS. But it is shown below that
this case is different from the Karadjeri case. Here,
the extensions of the F and FZ terms to mC are <u>not</u> based
on the rule of self-reciprocity applied to the F term
but on a rule of structural equivalence of agnatic kin
of alternate generations that is somewhat different from
the AGA rule posited for other systems in preceding
chapters.

The only remaining classes are those designated as
<u>wolmingi</u> and <u>mariengi</u>; these self-reciprocal terms
appear to designate in-law or potential in-law classes.
According to Elkin and others, <u>wolmingi</u> designates WM
and all members of her clan or agnatic line and, conversely, wDH and all members of his clan or agnatic
line. Similarly, <u>mariengi</u> designates HM and all members
of her clan or agnatic line and, conversely, wSW and all
members of her clan or agnatic line. We may postpone
further discussion of these classes until we have decided on the probable equivalence rules of this system.

Equivalence rules

Like most Australian systems of kin classification,
this system must also feature the same-sex sibling-,
half-sibling-, and stepkin-merging rules, and the
parallel-cross neutralization rule (see Table 4.4).
Clearly, however, it does not feature the parallel-

cross status-extension rule (whereby father's cross cousins are reckoned as members of the MOTHER class, and mother's cross cousins as members of the FATHER class), nor does it feature the cross-stepkin rule (whereby MBW and FZH are reckoned as members of the FZ and MB classes). And although this system must include some kind of rule of structural equivalence of agnatically related kin of alternate generations - for FF is designated as 'elder brother' and mSC is designated as 'younger sibling' - this rule does not operate, as it does in Aranda-like systems, to equate father's cross cousins with ZC and male cross cousins' children with M and MB. Also, as several observers have already noted, there are numerous similarities between the Ngarinyin system and systems of the so-called Omaha type elsewhere in the world. We may begin the process of specifying the equivalence rules of this system by considering these Omaha-like features.

The Omaha-like features of this system are especially clear in the cases of FM, FMB and MM, MMB, and their agnatic descendants. In all these cases, a female linking kinsman's brother's children are terminologically identified with that female linking kinsman's siblings and, conversely, FZ as a linking relative is treated as structurally equivalent to Z as a linking relative. That is, these terminological equivalences seem to be governed by the rule:

$$(FZ... \rightarrow Z...) = (...wBC \rightarrow ...wSb).$$

This is the rule identified by Lounsbury (1964b: 359) as the "Omaha Type I skewing rule." It is one of a series of rules found in different systems, all of which have the general property of equating a man's sister with his daughter and, conversely, a woman's brother with her father, in specified genealogical contexts.

This rule is adequate to account for the designations of the agnatic descendants of FMB and MMB, and it permits us to account for the designations of MBD and MBS as 'mother' and 'mother's brother'. By this rule MBD and MBS are reckoned as structurally equivalent to MZ and MB, and by the same-sex sibling-merging rule and the parallel-cross neutralization rule they are reckoned as structurally equivalent to M, the focus of the MOTHER class. Because of their structural equivalence to M and MB they may be designated as 'mother' and 'mother's brother', but as ego's kin of his or her own generation (rather than the first ascending generation) they are reckoned as members of the specially designated 'cousin' subclass of the MOTHER class. Conversely, by the Omaha skewing rule FZC is reckoned as structurally equivalent to ZC, and this relationship is reckoned as equivalent to wC (again via the same-sex sibling-merging rule and the parallel-cross neutralization rule). There is no special subclass designation for FZC; they are designated only as 'woman's child'.

So far we have considered only consanguineal relationships, but it is clear that the kin-class statuses of stepkin and in-laws also are affected by the Omaha skewing rule of this system. For example, FZH is designated as wuningi (ZH) and WBC are designated as maringi or waiingi (the same as WZ and WB). Although the rule posited above accounts for this, it does not account for some other terminological assignments, such as the designation of MBW as 'mother's mother', the designation of HZC as 'woman's daughter's child', and the designation of WBW and HZH as 'wife's mother' (or wDH). However, as Lounsbury (1964b: 378-9) notes, the four Omaha-type skewing rules specified in his general discussion of the subject were formulated to deal only with consanguineal relationships and it may be necessary to modify them appropriately when analyzing a system in which similar structural equivalences extend into the stepkin and in-law domains. The appropriate modifications may be made by altering the context restrictions on the rule most appropriate to the consanquineal domain. The Omaha type I rule specifies that FZ as a linking kinswoman is to be regarded as structurally equivalent to Z as a linking kinswoman; it is, in effect, restricted to the locus of a particular kind of linking kinsman, that is, any person's father. The rule may be made more generally applicable by substituting in the place of F a sign with the value "any male linking kinsman" (...m). When we do this, and adjust the other components of the rule accordingly, we may write the Omaha type I skewing rule, modified, as:

$$(...wB... \rightarrow ...wF...) = (...mZ... \rightarrow ...mD...),$$

that is, let a female linking kinsman's brother,
when considered as a linking kinsman, be regarded
as structurally equivalent to that linking kinswoman's father, when considered as a linking
kinsman; conversely, let a male linking kinsman's
sister, when considered as a linking kinswoman,
be regarded as structurally equivalent to that
male linking kinsman's daughter, when considered
as a linking kinswoman.

This rule accounts for the designation of MBW as
'mother's mother', because this is a case of a female
linking kinsman's (i.e., mother's) brother when considered a linking kinsman (to his wife). Thus, MBW → MFW
→ MM (by the stepkin-merging rule). Conversely, HZC →
HDC → DC. Also, WBW → WFW → WM and, conversely, HZH →
HDH → DH.

The probable form of the AGA rule of this system
is suggested by a comparatively odd feature of the system: It has no specially designated MF class. This is
shown by the designation of mDC as 'woman's child'.
Therefore, MF must be reckoned as a member of the MOTHER
class (that he is so reckoned is shown by his alternate
designation as 'mother's brother') and also as a member
of its special 'cousin' (MBC) subclass. More commonly,
in Australia, MBC is reckoned as a member of the MF
class, and sometimes as a member of a specially desig-

nated 'cousin' subclass of that class. But, of course, in these other cases neither the MF class nor its special 'cousin' class (self-reciprocal) is a subclass of the MOTHER class. Here we must suppose that MF, because he may be designated as 'cousin', is somehow reckoned as structurally equivalent to MBC (again just the opposite of the more common arrangement), and that mDC is somehow reckoned as structurally equivalent to FZC. The only way this can be done is via the rule:

(.mD... → .mFZ...) = (...wF. → ...wBS.),

that is, let male ego's daughter as a linking kinswoman be regarded as structurally equivalent to male ego's father's sister as a linking kins-woman; conversely, let a female linking kinswoman's father as a designated kinsman be regarded as structurally equivalent to that female linking kinswoman's brother's son as a designated kinsman. By this rule, MF → MBC, and, by the Omaha skewing rule, MBC → MSb. Reciprocally, mDC → mFZC → mZC. On the assumption that MBC is the focus of the _mamingi_ 'cousin' subclass, it follows that MF may be designated as 'cousin'; but the focus of this class, being a subfocus, is further reducible to MSb, so it follows also that MF may be designated as 'mother's brother' as a male member of the MOTHER class.

The corollary of this rule is, of course:

(.mS... → .mF...) = (...mF. → ...mS.),

and by positing this rule, as we must, we can account

for the designations of FF as 'elder brother' and mSC as 'younger sibling'. For by this rule FF → FS, and by the half-sibling-merging rule, FS → B; conversely, mSC → mFC → mSb. This places FF and mSC in the SIBLING class; FF is necessarily older than ego and is therefore designated as 'elder sibling'; mSC is necessarily younger than ego and is therefore designated as 'younger sibling'.

With regard to Elkin's report that FFZ is designated as 'father's sister' (and not as 'sister', as reported by Lucich and by Jolly and Rose), it may be noted that this kintype is not affected by the AGA rule formulated above or by the Omaha skewing rule (because Z in this context is not a linking kinswoman). Any variation on the Omaha skewing rule that would permit FFZ to be reckoned as equivalent to FZ would be too generally applicable and would specify terminological assignments not characteristic of this system. Therefore, it seems likely that this kintype reduces, via the parallel-cross neutralization rule, to FF; the AGA rule of this system equates FF with FS, equivalent to B by the half-sibling-merging rule. Of course, FFZ is a female, and not a male kintype, but even so this reduction is not inappropriate. Jolly and Rose (1966: 106) report ngolingi as the designation for an elder sibling of either sex and for FF and FFZ. Presumably, the designation for the elder-sibling class is ngolingi; the females of this class are singled out for special

designation as <u>lalingi</u>, thus leaving the nonfemales
(= males) of the class to be designated by the same
term as the class as a whole. That implies that FFZ
may be designated as <u>ngolingi</u> 'elder sibling' or as
<u>lalingi</u> 'elder sister' or 'female elder sibling'.

With only a small change in the context restrictions imposed on the AGA rule (as written above), it is
possible to use the same rule to account for the designation of mS as 'father' and mD as 'father's sister'.
This requires dropping the restriction "as a linking
kinsman," with the result that the rule may be expressed
in the formula:

$$(.mS \rightarrow .mF) = (mF. \rightarrow mS.)$$
$$(.mD \rightarrow .mFZ) = (wF. \rightarrow wBS.).$$

The rule then specifies that male ego's son is to be
regarded as structurally equivalent to male ego's
father, and male ego's daughter is to be regarded as
structurally equivalent to male ego's father's sister;
in addition, it implies the structural equivalences
accounted for by the weaker form of the rule as written
above. Positing this rule, rather than allowing the
extension of the father term to man's child to be
accounted for by the rule of self-reciprocity, has two
advantages. It eliminates one special rule (the rule
of self-reciprocity applied to 'father') and it permits
us to make the AGA rule more general in its applicability. Motivated by the rule of parsimony, we must posit
this form of the AGA rule.

There is one apparent (but only apparent) reason why we should not posit this form of the AGA rule. In the analyses presented in preceding chapters it was assumed that, in general, the foci of terminological classes are invariant to the equivalence rules of a system (unless they are foci of subclasses; see also Scheffler and Lounsbury 1971: 90-1). I have assumed that F is the focal denotatum of the _idje_ class, yet the rule posited above appears to imply that male ego's F as a designated kinsman is reckoned as structurally equivalent to male ego's son as a designated kinsman; that is, male ego's father as a designated kinsman appears not to be invariant to this rule. Moreover, the rule appears to imply a kind of circularity whereby male ego's father as a designated kinsman is reckoned as equivalent to male ego's son as a designated kinsman and vice versa. All of this would get us (and the Ngarinyin) precisely nowhere.

This situation may be clarified by noting that the assumption, that in general the foci of terminological classes are invariant to the equivalence rules of a system, is based on an empirical generalization. That is, this is the arrangement in the vast majority of cases and so it may be assumed - not dogmatically but only for heuristic purposes - to be the arrangement we are most likely to find when analyzing yet another system. It will come as something of a surprise to find an exceptional case, but the possibility of such exceptions must be recognized. Further, the invariant status of focal denotata of terminological classes may, in general, be attributed to the priority of definitional rules in relation to equivalence rules. This priority is a matter of logical necessity; classes must be defined before they can be expanded or widened. And it is for this reason that, in general, we do not expect to have to posit equivalence rules that apply to the foci of terminological classes (again, unless the classes in question are subclasses). Thus, in general, we do not have to make special note that definitional rules have priority in relation to equivalence rules; normally equivalence rules are not formulated that would apply to focal denotata. But in the odd or exceptional case it may be necessary to make this relationship explicit. This it seems is what we have to do in this case. The focal denotatum of the _idje_ class is F; the rule defining this class has priority over the equivalence rules of the system; the rule (.mS → .mF), etc., may be posited to account for the designation of mS as 'father', and it and its corollaries account for some other terminological assignments as well. One of the corollaries of this rule is (mF. → mS.) and it is applicable in some genealogical contexts, but not in the context of male ego's father as a designated kinsman, because F is the focus of the class _idje_ and is therefore invariant to this rule.

This completes our specification of the equivalence

rules of this system (see Table 10.2). These rules form an unordered set and, taken together with the specifications of class foci (Table 10.1), they account for the designations of all of any of ego's consanguineal relatives and for the designations of many stepkin and in-laws. We have yet to deal, however, with the in-law classes.

In-laws and marriage rules

In his published accounts Elkin gives the clear impression that in the "orthodox" case a man marries a kinswoman whom he designates as 'father's mother' maringi, whose father he designates as 'father's mother's brother' waiingi, and whose mother he designates as wolmingi. He says that a man designates all

Table 10.2. Ngarinyin equivalence rules

---

1. Half-sibling-merging rule

2. Stepkin-merging rule

3. Same-sex sibling-merging rule

4. Parallel-cross neutralization rule

5. Omaha skewing rule,
    $(...wB... \rightarrow ...wF...) = (...mZ... \rightarrow ...mD...)$

6. AGA rule,
    a. $(.mS \rightarrow .mF) = (mF. \rightarrow mS.)$,
    b. $(.mD \rightarrow .mFZ) = (wF. \rightarrow wBS.)$

7. Spouse-equation rule,
    a. $(.mW. \rightarrow FA.MO.) = (.wH. \rightarrow SON'S\ CH.)$,
    b. $(.mWM. \rightarrow FA.MO.MO.) = (.wDH. \rightarrow DAU.DAU.SON)$.

---

members of the clan or local horde of his WM as <u>wolmingi</u>
and all members of his WF's clan or local horde as
<u>maringi</u> (FM) or <u>waiingi</u> (FMB). Also, according to
Elkin, a woman designates her HM and all members of her
clan or local horde as <u>mariengi</u>. She designates her HF
as <u>malengi</u> 'woman's child'; she applies this term to
members of her HF's clan of his and alternate generations, and she designates other members of that clan as
<u>wuningi</u> 'woman's son's child'.

Most of these terminological assignments are
accounted for by the definitional and equivalence rules
posited above. We may begin with HF, designated as
'woman's child'. This follows from the AGA rule by
which HF → HS → S. Conversely, mSW → FW → M. HFF is
reckoned as equivalent to HFS and HB <u>wuningi</u>. The
relationships HM and wSW are invariant to the rules
posited above, but by those rules HMF is equivalent to
HMBS and HMB, and then HM via the parallel-cross
neutralization rule. Similarly, WF is reckoned as
equivalent to WBS and WB; WFF is reckoned as equivalent
to WFS and WB; WMF is equivalent to WMBS and WMB and
WM; and mDH is equivalent to FZH and ZH; and so on.

Several in-law relationships - WM, wDH, wSW, and
the sibling-in-law relationships - are not affected by
these rules. This suggests that they are the foci of
their respective classes. If so, these classes are not
subclasses of any consanguineal classes - although we
saw in preceding chapters that the usual arrangement in

Australian systems of kin classification is that in-law
or potential in-law classes are subclasses of certain
consanguineal classes. To appreciate this difference,
or to determine whether or not there really is a difference, we need to consider Ngarinyin marriage rules in
greater detail.

Elkin's two published accounts are brief and synop-
tic, and so it is perhaps not surprising that they do
not depict the complexity of the situation he observed
in the field. Briefly, the data presented in his field
report show that the most general rule is that a man
may not marry a kinswoman of his own patrimoiety; he
may marry any distant kinswoman of the opposite moiety
virtually without regard to her kin-class status,
provided that she is not also his WBW or ZHZ - thus two
men may not marry one another's sisters, but there is
no rule to prohibit a man from marrying his ZH's classificatory 'sister'. Elkin relates the rule against two
men being married to one another's sisters to the fact
that normally a man may marry his WBD whom he classifies as 'father's mother', but if his WBW were his own
sister he could not marry his WBD who would be also his
ZD. From Elkin's field report it appears that a man
has a rightful claim to his WBD as a second wife.

Elkin notes also that "wolmingi line and clans,
who provide WMF, WMB, and WM . . . may include persons
otherwise related as ngolingi [B], idje [F], lalingi [Z],
and amalngi [FZ], i.e., a distant F and FF line." Thus,

a man's WM wolmingi might be "otherwise" (that is, if she were not also his WM or regarded as a potential WM) designated as 'father's sister' or 'sister'. Or she might be his classificatory 'mother's mother' or 'sister's daughter's daughter'. Similarly, Elkin's genealogical data show that the kin-class status of WF is not invariably 'father's mother's brother' waiingi; some WFs are the husbands of classificatory 'sisters' or 'father's sisters', that is, wuningi (wSC); others are the husbands of classificatory 'mother's mothers', that is, classificatory 'mother's fathers' (or 'cousins') mamingi, or classificatory 'mother's brothers' kandingi. Even so, it seems that regardless of the premarital kin-class status of a man's wife, once she is betrothed or married to him he may designate her as maringi (FM), her father as waiingi (FMB), and her mother as wolmingi (WM).

As in his published accounts (see especially 1932: 313-4), Elkin says that here a man "looks to" his FMB, rather than to his MMB, to "give" him a wife. A man gives his classificatory 'daughter' or 'sister' to his ZSS as a wife. Elkin says, "in theory" a man may marry his FM, FMBD, FMBSD, but he recorded no instances of such marriages. Presumably, the 'daughter' or 'sister' whom a man's FMB gives to him is a woman of FMB's clan but not of his own line or lineage within that clan.

The information provided in Elkin's field report is confirmed by Lucich (1968: 33-4). Lucich says that,

in 1963, marriage choices "were restricted to persons
within specific categories," but he adds, "to express
this as a simple dichotomy between right and wrong marriages would be misleading." His informants agreed
that kin of the same patrimoiety should not marry and
neither should close kin of opposite moieties; a man
may not marry women of his mother's clan; nor should he
marry a woman he classifies as 'woman's son's child'
(e.g., ZHZ) or as 'woman's child' (e.g., ZD), although
"severe sanctions were not applied" against the partners
to such marriages. As for the other kinswomen of the
opposite moiety, Lucich's informants disagreed about
whether marriage was permitted with a classificatory
'mother' but they "classed as correct" marriages to
distant kinswomen of the 'father's mother' and 'cousin'
categories. Lucich says that, in 1963, 'father's
mother' was "the preferred category" and 'cousin' "was
regarded as a possible alternative if a 'father's
mother' was not available." In several places Lucich
implies, without quite asserting it, that whomever a
man marries he designates (or may designate) her as
'father's mother', her father as 'father's mother's
brother', and her mother as 'wife's mother' (1968: 44-
6, 149). Lucich says "terminological adjustments of
this type did not always extend completely to all clan
members, and a man might use one term for members of a
particular lineage, and another term, which reflected
previous usage, for the other members of that clan"

(1968: 46).

From these accounts it may be seen that the designation of FMB as <u>waiingi</u> - rather than as 'father's
mother' - is probably a special designation that implies
that he is a potential WB. Note that FMB rather than
WB is the focus of this class. The equivalence rules
of this system do not equate FMB with WB, nor do they
equate WB with FMB. The pragmatic marriage rules of
the society specify that FMB may be the classificatory
'brother' or the classificatory 'father' of male ego's
wife. As a potential WB he is designated as <u>waiingi</u>
rather than as 'father's mother'. This designation is
extended to ego's own WB in the event that ego happens
to marry a kinswoman of some class other than 'father's
mother'. The reciprocal potential ZH class is not distinguished from the wSC class <u>wuningi</u>, of which it is a
(covert) subclass.

We may now consider WM <u>wolmingi</u>. It was shown
above that the Omaha-like features of this system extend into the stepkin and in-law domains and, for example, that MBW is reckoned as equivalent to MFW and MM.
The rule that specifies these equivalences also specifies that FMBW is to be reckoned as equivalent to FMFW
or FMM; FMBW is designated as <u>wolmingi</u>, so it seems
highly probable that FMM also is designated as <u>wolmingi</u>
- though none of the sources provides information on
the designations of kin of the third ascending or
descending generations. Of course, <u>if</u> FMM is designated

as <u>wolmingi</u>, this is consistent with the fact that her daughter, ego's FM, is the focus of the kin class that includes ego's potential wives. And it is consistent with the probable designation of FMF as <u>waiingi</u>, the same as FMB. Thus, we may assume (until evidence to the contrary is produced) that FMM is the focus of the <u>wolmingi</u> class, the class of male ego's potential wives' mothers, and the class to which ego's actual WM is assigned in the event that she is not already a member of that class. Note, however, that because we have no information on how FMM may be designated other than as <u>wolmingi</u> (if, indeed, she may be so designated), we cannot say that the <u>wolmingi</u> class is a subclass of any other class. It appears to be a structurally independent class with the focus FMM, and its members are regarded as potential WMs.

Of course, not all relationship types reportedly designated as <u>wolmingi</u> can be equated with FMBW or FMM (or their reciprocals) using the rules posited above. For example, the kintypes representing the agnatic kin of WM are all equated with WM via these rules. To further equate them, and WM, with FMM, or rather to place them in the <u>wolmingi</u> class, we must posit a spouse-equation rule. We may postpone consideration of this rule until we have considered HM and wSW and the sibling-in-law relationships.

Although one of the relationships (FMBW) designated as <u>wolmingi</u> may be equated, under the rules that must be posited for this system, with a consanguineal kintype (FMM), this is not the case for any of the relationships designated as mariengi (HM, wSW). Therefore, the mariengi class also must be a structurally independent class (and not a subclass of any other class), but without a consanguineal focus. This is not inconsistent with the assumption that FMM is the focus of the wolmingi (potential WM) class; for it does not necessarily follow that if WM can be equated, in one way or another, with a consanguineal relationship, HM also can be equated with a consanguineal relationship. It does not ever happen in this society that a man's FMM is his WM; but, if she were, his wife would be his FM, and his wife's SW would be only her SW, not necessarily related to her in any other way. Conversely, for his wife's SW, his FM would be only her HM and not necessarily related to her in any other way.

All sources agree that WB, WZ, and BW are designated as 'father's mother's brother' and 'father's mother' and that HZ, HB, and ZH are designated as 'woman's son's child'. These relationships are not affected by the equivalence rules posited above, but their designations may be accounted for by positing a spouse-equation rule. Of course, if a man marries a woman whom he designates as 'father's mother', his WZ is already his 'father's mother's brother'. For his WZ and WB he (their ZH) is already their 'woman's son's child'. Should he marry a kinswoman of some other

category, it seems that he may designate his siblings-
in-law by these terms regardless of how he designated
them before the marriage. That is, he may treat them
terminologically as though he had made an orthodox
marriage. To account for this we may posit the spouse-
equation rule:

   (.mW. → FA.'S MO.) = (.wH. → SON'S CH.),
   that is, let male ego's wife as a designated relative be regarded as a member of the expanded
   father's mother class; conversely, let female
   ego's husband as a designated relative be regarded
   as a member of the expanded woman's son's child
   class.

Designation of WB as 'father's mother's brother' follows
as a corollary of this rule, but it would be redundant
to derive the designation of WF as 'father's mother's
brother' from a corollary of this rule because this
designation may be accounted for by the AGA rule that
equates WF with WB. Similarly the designation of HF as
'woman's child' need not be derived from a corollary of
the spouse-equation rule, because the AGA rule implies
HF is equivalent to wS. On the assumption that FMM is
the focus of the _wolmingi_ or potential WM class, designation of WM as _wolmingi_ may be derived from a corollary
of this rule:

   (.mWM. → FA.MO.MO.) = (.wDH. → DAU.DAU.SON).
But, again, it would be redundant to derive the designation of HM as _mariengi_ from a corollary of this rule,

because HM is _mariengi_ by definition.

There are several secondary in-law relationships
that are invariant to all these rules. These are WMM,
WFM, HMM, and HFM and their reciprocals. In his field
report (1928d) Elkin observes that, although he was
given _wolmingi_ (WM) consistently as the designation for
WMF (WMF → WMBS → WM), he was given three different
designations for different WMMs. These were 'father's
mother', 'sister's son's child', and 'mother', the
terms by which Elkin's informants designated their
respective WMMs before they married the DDs of these
women. Similarly he was given the same three designations, plus 'woman's daughter' for different WMBWs
(WMBW → WMFW → WMM). Similar data are not available
for WFM, HMM, and HFM, but we have no reasons to suppose
that the situation with regard to different relatives
of these kinds is not the same as that of different
WMMs. That is, it is probable that there are no standard designations for relatives of these kinds; these
relationships are invariant to the equivalence rules
of the system, and so the designations assigned to
relatives of these kinds depend entirely on their
previous consanguineal relationships to ego.

It seems worth noting, as Elkin does in his field
report (1928d), that these relationships between his
informants and their WMMs imply that many of his informants were married to kinswomen whom they did not designate as _maringi_ before marriage. Elkin notes that he

recorded eleven instances of marriage to classificatory 'woman's daughter' (which accounts for a number of cases in which WMM was designated as 'father's mother') and two instances of marriage to classificatory 'mother' (where WMF was 'mother's mother's brother', also 'wife's mother's brother', and the two WMMs 'father's mother' and 'mother'). Unfortunately, he does not report the number of instances recorded in which his informants designated their wives as 'father's mother' (or as 'cousin') before marriage. In any event, it appears that virtually all marriages recorded by Elkin were of men to women of the kin classes 'father's mother', 'woman's child', and 'mother' (in Elkin's report female 'cousins' are not distinguished from 'mothers'). It will be recalled that Lucich reports that his informants classified as "correct" marriages between men and women of the kin class 'father's mother' or 'cousin' (provided they were distant relatives); they disagreed about the propriety of marriages to women of the 'mother' class, and said that marriages to women of the 'woman's child' class were "incorrect" but not subject to strong negative sanctions. Even so, marriages of the latter type seem (from Elkin's data) to have been fairly common in the past. It is difficult to avoid the conclusion that it is not so much that such marriages are in any sense "incorrect" as it is that they are "extralegal"; they are instances of men marrying women to whom they have no rightful claim.

SOME SIMILAR SYSTEMS

The linguistically cognate systems of kin classification of the Worora and Wunambal (see Love 1941,
1950; Jolly and Rose 1966; Lucich 1968) appear to be
virtually identical to the Ngarinyin system and require
no additional comment. The linguistically unrelated
Yaralde system (Taplin 1879; Radcliffe-Brown 1918) is
quite similar to the Ngarinyin system but there are
some significant differences. There are a few apparently minor differences between Taplin's and Radcliffe-
Brown's data, but both sources agree on the following
features. There are four self-reciprocal grandparent
terms and the siblings of the four grandparents are
designated by the same terms as the linking grandparents. The FM (FMB) and MM (MMB) terms are extended to
FMB's and MMB's agnatic descendants in all generations,
but the MF (MFZ) term is not extended to MF's agnatic
descendants. In addition to the usual M and MB terms,
there is a distinct self-reciprocal 'cousin' term, but
MBSC are designated as 'mother' and 'mother's brother'
and, reciprocally, FFZC are designated as 'sister's
child' (by male ego; no data are given for female ego).
Another difference between the Yaralde and Ngarinyin
systems is that the Yaralde system does not distinguish
between man's child and woman's child. Also, the
Yaralde system features a full set of in-law terms; no
in-laws are terminologically identified with consan-
guines. The latter feature is consistent with Taplin's

(1879: 35) report that "near relatives" are not allowed
to marry and with Radcliffe-Brown's (1918: 238) report
that a man "may not marry a woman who is related to him
by any of the recognized blood relationships."

These data suggest that this system features the
Omaha-type rule (FZ... → Z...) = (...wBC → ...wSb).
This rule implies that MBC is reckoned as structurally
equivalent to M and MB and that FZC is reckoned as
structurally equivalent to ZC. That cross cousins are
not designated as 'mother' and 'mother's brother' (MBC)
and 'sister's child' (FZC) may be accounted for by
assuming that in this system, as in the Ngarinyin system, the 'cousin' class is a specially designated subclass of the MOTHER and WOMAN'S CHILD classes. The
difference is that in the Ngarinyin system only the MBC
subclass of the MOTHER class is specially designated.
On this hypothesis the kintypes MBSS and MBSD reduce
to MB and M, respectively, through the cross-cousin
types MBS and MBD, but they are not designated as
'cousin' because that term applies only to kintypes of
ego's own generation. That is, the 'cousin' class of
this system is like the 'cousin' classes of the Yir
Yoront and Murngin systems in that it does not focus
on MBC and FZC; it is a broadly defined class and consists of all kintypes reckoned as structurally equivalent to M, MB, and mZC but which are kin of ego's own
generation.

There is little or no evidence to suggest that this

system features an AGA rule or a spouse-equation rule.

SUMMARY

The Ngarinyin system of kin classification differs
significantly from the systems analyzed in previous
chapters, although there are some continuities. In
this system, as in most other Australian systems of kin
classification, the FZ and MB classes are subclasses of
the FATHER and MOTHER classes; that is, this system
also features the parallel-cross neutralization rule.
Like the Walbiri, Aranda, and Dieri systems it has no
parallel-cross extension rule, but it does feature an
AGA rule, although one that is different from the AGA
rules of the Walbiri, Aranda, and Dieri systems. The
most distinctive equivalence rule of the Ngarinyin
system is the Omaha-type skewing rule, and it shares
this rule with several neighboring systems and with the
Yaralde system of South Australia. It may be that the
Yaralde system does not feature an AGA rule; certainly
it does not feature a spouse-equation rule, although
the Ngarinyin system does.

Chapter 11

AN OVERVIEW

We have seen that much of the diversity among Australian
systems of kin classification may be accounted for as
the product of various combinations of a fairly small
stock of structural elements. These are, principally,
half-a-dozen or so dimensions of conceptual opposition
variously combined to yield a somewhat larger number of
principal kin classes and subclasses and a few rules of
genealogical structural equivalence. Because there is
relatively little variation at the level of principal
classes (especially if covert classes, indicated often
by reciprocal relations among terms, are taken into
account), the structurally most distinctive differences
among the systems are differences in their respective
sets of equivalence rules. These differences may be
summarized as follows.

The Pitjantjara system (Chapter 3) is structurally
quite similar to the so-called Hawaiian-type systems of
many Malayo-Polynesian languages. This similarity is
obscured by the presence, in the Pitjantjara system, of

a few specially designated subclasses. The principal
overt classes of this system are sexually differentiated
subclasses of covert grandkin, parent, child, and sibling superclasses. These classes are expanded by the
"sibling-merging rule," (PSb → P) and (SbC → C), that
is, "let one's parent's sibling be regarded as structurally equivalent to one's own parent and, conversely,
let one's sibling's child be regarded as structurally
equivalent to one's own child." In other words, the
opposition between lineal (and colineal) and collateral
kin relevant to the definitions of the primary senses
of the kin terms are neutralized, and the classes they
designate are expanded to include all collateral kin of
the same generations as the class foci. In addition,
this system features the half-sibling and stepkin-
merging rules (nos. 1 and 2, Table 4.4), but it shares
these with other Australian systems of kin classification; therefore, the sibling-merging rule is the most
distinctive equivalence rule of this system. The
PARENT, CHILD, and SIBLING (expanded parent, child, and
sibling) classes generated by these rules contain a few
additional specially designated subclasses. Within the
FATHER class there is a specially designated MB subclass; within the MOTHER class a FZ subclass; and
within the CHILD class a mZC, wBC subclass. Within the
SIBLING class there is a specially designated 'distant-
sibling' class. There are a few distinct in-law classes
but these, it seems, are not subclasses of any of the

kin classes. It may be that this system, unlike many
other Australian systems, does not feature a spouse-
equation rule, that is, a rule that assimilates specified in-law relations to specified kin classes.

The systems discussed in Chapter 4 - Kariera-like
systems - differ from the Pitjantjara system (and the
many others that share its main features) in a number
of ways. At the level of principal classes there are
both similarities and differences. Here, too, there
are covert grandparent, parent, child, and sibling
classes; but whereas the parent class is, again, divided
into a father and a mother class, the reciprocal child
class is divided into a man's child and a woman's child
class. Further, the grandparent class is divided into
a parallel-grandparent (FF, MM) and a cross-grandparent
(MF, FM) class; in some of these systems one or the
other (or both) of these two classes is divided into
two sex-specific classes. So the number of grandparent
classes varies from two to four. There are no distinct
grandchild terms, for the grandparent terms are extended
by the rule of self-reciprocity. A further notable
difference is that these systems contain distinct cousin
(cross-cousin) classes and terms. This class is not a
subclass of the SIBLING class. Again, there are distinct MB and FZ terms, but the MB class is not a subclass of the FATHER class, and the FZ class is not a
subclass of the MOTHER class. Here the MB and FZ terms
do not share a common reciprocal term (or terms). The

reciprocal of the MB term is 'woman's child' and the reciprocal of the FZ term is 'man's child'. This indicates that the MB class is a subclass of the MOTHER class and the FZ class is a subclass of the FATHER class.

Typically, in these Kariera-like systems, the FATHER'S SISTER class includes a specially designated potential or prospective WM subclass. Correspondingly, the FATHER class may include a specially designated potential or prospective WMB subclass, and the reciprocal MAN'S CHILD class may include a specially designated potential or prospective wDH, ZDH subclass. Less often, the MOTHER'S BROTHER and MOTHER classes include specially designated potential or prospective WF and WFZ subclasses.

Like the Pitjantjara system, these systems feature the half-sibling- and stepkin-merging rules, but in place of the general sibling-merging rule they substitute the same-sex sibling-merging rule, and they add two others for reckoning the kin-class statuses of collateral kin. These two others - the most distinctive equivalence rules of these systems - are the parallel-cross neutralization rule and the parallel-cross status-extension rule. The first equates FZ with FB (equivalent to F under the same-sex sibling-merging rule), MB with MZ and, conversely, wBC with mBC, and mZC with wZC. But it is weaker than the same-sex sibling-merging rule and does not apply when FZ and MB are considered

as linking kin. Therefore, MBC and FZC are not reckoned
as structurally equivalent to MZC and FBC and are not
designated as 'brother' or 'sister'; MBC, FZC are invariant to the equivalence rules of these systems and are
left to be designated by separate terms. The parallel-
cross neutralization rule governs also the collateral
extensions of the grandparent terms. Where there are
more than two such terms, this results in the possibility of alternative designations for kintypes of the
second ascending and descending generations. In some
societies the alternatives are governed by considerations of 'tabu' (especially potential or prospective
marital) social relations, which require designation at
the appropriate superclass level, rather than at the
subclass level appropriate to the kintype being denoted
(see Chapter 4, Ompela and Mongkan). Because of all
this, patterns of classification of grandparents and
their siblings which Radcliffe-Brown and others regarded
as virtually definitive of Aranda-type systems may occur
in association with systems that are otherwise entirely
Kariera-like.

The parallel-cross status-extension rule is so
called because it permits the kin-class statuses of
relatively distant collateral kintypes to be governed
by the same-sex sibling-merging rule and the parallel-
cross neutralization rule. It does this by equating
one's father's cross cousins with one's mother's siblings and one's mother's cross cousins with one's

father's siblings; and, conversely, it equates the children of one's male cross cousins with one's sister's children, and the children of one's female cross cousins with one's brother's children. The same-sex sibling-merging and parallel-cross neutralization rules then become applicable.

In addition to these five equivalence rules, Kariera-like systems feature a cross-stepkin-merging rule (which governs the kin-class statuses of relationships such as FZH, MBW, and their reciprocals), and a spouse-equation rule, which specifies that a person's spouse is to be regarded as a member of the COUSIN (expanded cross-cousin) class in the event that he or she is not already a member of that class. The various corollaries of this latter rule govern also the kin-class statuses of ego's in-laws (again, if they are not already members of the culturally appropriate potential or prospective in-law subclasses). For example, one such corollary specifies that WM is to be regarded as a member of the FATHER'S SISTER (expanded FZ) class. This rule does not imply, however, that WM is to be designated as 'father's sister'. Again, in most of these systems the FATHER'S SISTER class includes a potential or prospective WM subclass, and typically the designation for this subclass is the required designation for WM. The kintypes included in this subclass vary with the local rules of interkin marriage. Where marriage between cross cousins is permitted, a man **may**

designate his own FZ or MBW, as well as more distant
kinswomen of the FATHER'S SISTER superclass, as 'wife's
mother'; but where marriage between cross cousins and
close classificatory 'cousins' is prohibited, only relatively distant kinswomen of the FATHER'S SISTER class
may be designated as 'wife's mother'.

The Nyulnyul and Mardudhunera systems discussed in
Chapter 5 have the same principal classes and equivalence rules as the Kariera-like systems discussed in
Chapter 4. They differ only in having more specially
designated potential or prospective in-law subclasses.
They required special notice, however, because Elkin and
Radcliffe-Brown misinterpreted certain practical consequences of a form of intergenerational marriage as
though they were "basic" structural features of these
systems and therefore described them (erroneously) as
Aranda-type systems.

The Karadjeri system (Chapter 6) shares most of the
basic categories and all of the equivalence rules of the
Kariera-like systems discussed in Chapter 4. But it
adds to those rules a rule of structural equivalence of
agnatic kin of alternate generations, that is, the AGA
rule (FF → B) and (mSC → mSb). Because all other equivalence rules have priority in relation to the AGA rule
(and its corollary and auxiliary), the AGA rule affects
the classification of kin in only two notable ways.
First, it determines that the FF class has a dual
status; it is both a subclass of the parallel-grandpar-

ent class and a subclass of the SIBLING class. Second, coupled with a special auxiliary, it determines that cross cousins are classified as 'cross grandkin', and this leaves the system without a terminologically distinct 'cousin' class; the COUSIN class becomes a covert subclass of the CROSS-GRANDKIN class. Despite ethnographic reports to the contrary, MBC and FZC as such are not terminologically distinguished; there is no basic asymmetry in the classification of cross cousins, and there is no "rule of matrilateral cross-cousin marriage." A man's FZ may not be his prospective WM, but his MBW may be if she is not also a close 'father's sister'. Therefore, a man <u>may</u> regard a MBD as a prospective wife and designate her as such, but he may not so regard or so designate any of his FZDs. Thus men do distinguish terminologically among their female COUSINS, but the distinction is between those who are prospective wives and those who are not; it is not between MBD and FZD as such.

The Arabana system (Chapter 7) adds yet another equivalence rule to the Kariera-like base. In addition to the AGA rule, it features a rule of structural equivalence of uterine kin of alternate generations, the AGU rule (MM → Z) and (wDC → wSb). Again, the other equivalence rules have priority in relation to the AGA and AGU rules (and their corollaries and auxiliary). The AGA and AGU rules determine that the FF and MM classes are subclasses of the SIBLING class. In this system

the AGA rule is highly restricted and has no special
auxiliary but the AGU rule does. The kin-class statuses
of cross cousins are determined by the AGU rule (rather
than by the AGA rule, as in the Karadjeri system) and
cross cousins are equated both structurally and terminologically with FM and wSC (rather than with MF and mDC,
as in the Karadjeri system).

In the Karadjeri and Arabana systems the AGA and
AGU rules are subordinate to the parallel-cross status-
extension rule (as well as to the other equivalence
rules). Therefore, although they affect the kin-class
statuses of cross cousins and of kintypes structurally
equivalent to them, they do not affect the kin-class
statuses of parents' cross cousins, cross cousins'
children, or parents' cross cousins' children. In the
Yir Yoront system (Chapter 8), however, the AGA rule is
assigned priority over the parallel-cross status-
extension rule (the AGU rule is absent). The result is
that, in addition to cross cousins being reckoned as
structurally equivalent to MF and mDC, FFZC is reckoned
as structurally equivalent to ZC and, reciprocally, MBSC
is reckoned as equivalent to M or MB. Further, FFZSC →
ZSC, FFZDC → ZDC, FMBSC → FMSb, and MMBSC → MMSb. No
other kintypes in G+1, G-1, or G= are affected; they
are classified in the same ways as in simple Kariera-
like systems, except, of course, for a few structurally
minor differences in modes of subclassification.

The Murngin system of kin classification (also

Chapter 8) is similar to the Yir Yoront system except
for the presence of the AGU rule. Here, however, the
AGU rule is highly restricted and determines only that
the MM class, like the FF class, is a subclass of the
SIBLING class. It has no effects on the kin-class
statuses of collateral kin. The Murngin system differs
from the Yir Yoront system also in that it has a few
more specially designated potential in-law classes.

The Yir Yoront and Murngin, like the Karadjeri,
prohibit a man's FZ from being his prospective WM, but
a man's MBW may be his prospective WM if she is not also
a close 'father's sister'. Thus a man may marry a MBD
but not a FZD. However, in all three societies a man's
potential WMs include his MMBD, who is a member of the
potential or prospective WM subclass of the FATHER'S
SISTER class. Whomever a man marries, he classifies
his WM as a special kind of 'father's sister'; so,
although he also classifies his WF as 'mother's brother', there is no "rule of matrilateral cross-cousin
marriage!" Because the structures of the COUSIN classes
of the Yir Yoront and Murngin systems are somewhat
different from those of the other systems noted so far,
the spouse-equation rule of the Yir Yoront and Murngin
systems is correspondingly slightly different.

All the systems so far noted (except Pitjantjara)
share most of their basic categories and many of their
equivalence rules, perhaps most notably the parallel-
cross status-extension rule. Some add the AGA or the

AGU rule. or both, usually subordinating them to the other equivalence rules; but in some systems the AGA rule is assigned priority over the parallel-cross status-extension rule. There are, however, a number of Australian systems of kin classification from which the parallel-cross status-extension rule is absent.

The Walbiri and Aranda systems (Chapter 9) feature the AGA and AGU rules, but not the parallel-cross exten- sion rule. Both assign priority to the AGA rule in relation to the AGU rule. In both systems the FF and MM classes are subclasses of the SIBLING class (AGNATIC SIBLINGS and UTERINE SIBLINGS), and the COUSIN class is a subclass of the MOTHER'S FATHER class (MF's AGNATIC SIBLING, under the AGA rule). Coupled with a special auxiliary, the AGA rule affects the kin-class statuses of father's cross cousins, who are classified as a special kind of 'woman's child' (AGNATIC SISTER'S child). Conversely, the children of one's male cross cousins are classified as a special kind of 'mother' and 'mother's brother' (mother's AGNATIC SIBLING). The AGU rule, also coupled with a special auxiliary, deter- mines the kin-class statuses of mother's cross cousins, who are classified as a special kind of 'man's child' (UTERINE BROTHER'S child). Conversely, the children of one's female cross cousins are classified as a special kind of 'father' and 'father's sister' (father's UTERINE SIBLING). All these classifications are just the oppo- site, at the superclass level, of those that characterize Kariera-like systems. These rules further affect the kin-class statuses of many G= kintypes.

The Dieri system (Chapter 9) is similar to the Walbiri and Aranda systems, but assigns priority to the AGU rule in relation to the AGA rule (rather than vice versa). Because of this difference, the Dieri COUSIN class is a (covert) subclass of the FM class (rather than of the MF class). Other differences in the classifications of G= kin also are attributable to this simple difference in the relations between the AGA and AGU rules, as well as to slight differences in the structures of the CROSS-GRANDPARENT classes of these systems. One interesting difference is that in the Walbiri and Aranda systems a man's potential wives are his classificatory 'father's mothers' and 'sister's sons' daughters' of his own generation, whereas in the Dieri system they are his classificatory 'daughter's daughters' and 'mother's father's sisters' of his own generation. However, the kintypes included in these potential wife classes are the same in all three systems.

Finally, several Australian systems of kin classification feature one or another form of the Omaha-type skewing rule (Chapter 10). Characteristically, such rules specify that a man's sister is to be regarded as structurally equivalent to his daughter and, conversely, a woman's brother is to be regarded as structurally equivalent to her father, in certain genealogical contexts. In some systems, it seems, this is a relatively low-order rule, normally subordinate to the parallel-cross status-extension rule but available for reckoning alternative kin-class statuses in certain social contexts. However, the parallel-cross status-extension rule is absent from the Ngarinyin and several other systems, and the Omaha skewing rule is complemented by an AGA rule that is somewhat different from the AGA rule found in the systems noted above. This has the effect of extending the FM, FMB and MM, MMB terms to all agnatic descendants of FMB and MMB. Also, these rules evacuate the potential MF class, making MF, MFZ structurally equivalent to MBC, the focus of a specially designated subclass of the MOTHER class. Here, MF, MFZ are classificatory 'cousins' (rather than MBC being a classificatory 'mother's father', as in many of the systems noted above). In the Ngarinyin system a man's potential wives are his classificatory 'father's mothers' but not his classificatory 'sister's son's daughters'. Correspondingly, the spouse-equation rule of this system is somewhat different from the spouse-equation rule of the Walbiri and Aranda systems. Apparently, the Yaralde system is similar to the Ngarinyin system but lacks a spouse-equation rule.

I have argued throughout this study, contrary to Radcliffe-Brown and others, that rules of kin-class expansion that express rules of interkin marriage are not among the basic structural features of Australian

systems of kin classification. The spouse-equation
rules posited in this study are not structurally basic.
They do not govern the kin-class statuses of collateral
kintypes, for they are low-order rules that take the
expanded kin classes as given. Their sole function is
to assign in-law relationships to these expanded classes
only when those in-laws are not already kin of the culturally appropriate classes. Rules of interkin marriage
do govern the kin-class statuses of some collateral kintypes, but only insofar as they govern inclusion in
specially designated potential or prospective in-law
subclasses. Those patterns of collateral kin classification that appear to be determined by rules of interkin
marriage are instead determined, in some cases, by the
parallel-cross status-extension rule, in other cases by
this rule in conjunction with the AGA rule or the AGU
rule, and in yet other cases by AGA and AGU rules alone.

This understanding of the structures and relations
among the structures of Australian systems of kin classification makes it possible to demonstrate (in the next
chapter) that Radcliffe-Brown was correct when he argued, contrary to prevailing scholarly opinion at the
time, that section and subsection systems are derived
from systems of kin classification, rather than vice
versa.

Chapter 12

KIN CLASSIFICATION AND SECTION SYSTEMS

The thesis of this chapter is that the categories and
intercategory relations of section and subsection systems are derived from the highest-order superclasses of
systems of kin classification. As Elkin (1933b: 90) put
it, section and subsection systems "systematize" and
"summarize" the more specific kin categories. Or as
Radcliffe-Brown said (1951: 39): "The 'classes' [4 or
8] result from the giving of names to kinship divisions
which can quite well exist, and do exist, without names
as part of the kinship system organized by means of the
kinship terminology." This interpretation has not been
generally understood, much less generally accepted, but
it is substantially correct. Moreover, all other theories of the structures of section and subsection systems
are deficient and defective in a number of ways.

THE STRUCTURE OF SECTION AND SUBSECTION SYSTEMS
The predominant opinion among anthropologists is
that section (four-class) systems result from division

of exogamous moieties (patrilineal or matrilineal) into
two subclasses each, and that subsection (eight-class)
systems result from division of sections into two subclasses each (see Thomas 1966 [1906]; Lawrence 1937;
Berndt and Berndt 1964; Dumont 1966; Maddock 1973).
There is disagreement, however, about the principles of
subdivision and therefore about the structures of section and subsection systems.

According to one theory, sections arise from and
are based on division of moieties (patrilineal or matrilineal) into alternate-generation subclasses. According
to another (the so-called double-descent) theory, sections are the logical product of the intersection of
patrilineal _and_ matrilineal moieties. Both theories
founder on the ethnographic facts and on faulty logic.
Not all societies that have section systems also have
both patrilineal and matrilineal moieties - indeed, few
if any Australian societies do. Therefore, although it
is conceivable that the section system originated in a
society that did have both kinds of moieties, it is not
now structurally dependent on the presence of both kinds
of moieties. In the absence of one or the other of the
required sets of moieties, some proponents of the
double-descent theory posit a set of "implicit" or
"unnamed" moieties, in addition to the set of overt or
named moieties. In response to this it has been pointed
out that these hypothetical constructs are quite unnecessary, because the hypothetical implicit or unnamed

moieties may be accounted for as logical products of
division of the known moieties into alternate-generation
subclasses. That is, given a set of (for example) exogamous patrimoieties, A and B, if A is divided into two
alternate-generation levels or sections A1 and A2, and
B is divided into B1 and B2, then A1 and B2 are in
effect "matrilineally" related and constitute a de facto
matrimoiety, and similarly for A2 and B1.

Because overt terminological recognition of alternate-generation levels is very widespread, the "moiety
plus alternate-generation levels" hypothesis is more
acceptable than the double-descent hypothesis, but it
too is deficient and defective. Section systems are
not invariably associated with moieties of either kind,
so it is not true that sections everywhere are subdivisions of moieties. Again, it is conceivable that the
system originated in a society with moieties, but it is
not now structurally dependent on the presence of moieties of any kind. This conclusion cannot be avoided by
positing a set of "unnamed" or "implicit" moieties.
The only reason to posit such entities is that any section system contains two "patricouples" (two sections
related by patrifiliation, A-D and B-C in Figure 12.1)
and two "matricouples" (two sections related by matrifiliation, A-C and B-D in Figure 12.1). If we say that
the two patricouples (or matricouples) constitute a
pair of moieties, we cannot then account for the sections by deriving them from the moieties (so-called),

because we have in effect already derived the moieties
from the relations among the sections.

It is sometimes argued that the absence of proper
names for moieties is not decisive evidence against the
presence of moieties or against the claim that section
systems are based on moieties. According to Radcliffe-
Brown (1913), the Kariera have a section system but no
moieties. However, the Kariera language has two expressions, ngaju maru (or maman 'father' maru) and balu
maru, the first used by a person to designate members
of his or her own patricouple of sections, and the second to designate members of the other patricouple. It
might be argued that although these expressions are not
proper names the categories they designate are nonetheless moieties. But to so argue we would have to over-
look a significant structural difference between the

Figure 12.1. Conventional representation of a
section system

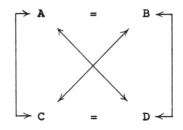

	mother-child relations
X	father-child relations
=	marital relations

categories designated by these Kariera expressions
and genuine moieties designated by proper names.
Unlike genuine moiety names, these Kariera expressions
signify relationships; the categories they designate
are relationally defined. It is sufficient to say
that, from the perspective of some specified or
specifiable person, members of two sections related
by patrifiliation (A and D, or B and C) are classified
together in opposition to members of the other two
sections, to which members of the first two sections
are related by matrifiliation (see also Epling 1961).
The existence of "moieties" that are anything other
than the sums of A+D and B+C cannot be deduced from
these two expressions and their meanings. It is
circular and explains nothing to argue, as Dumont
(1966: 236) does, that the sums of A+D and B+C,
when divided in two, result in the classes A, B, C,
and D.

Of course, these observations do not disprove the
hypothesis that the section system originated, sometime
in the distant past, from division of moieties into
alternate-generation levels. However, they do demonstrate the falsity of the thesis that moieties of one
kind or another are a basic underlying structural feature of the system wherever it now occurs. This suggests the possibility that the section system is not now
and never was structurally dependent on moieties, and
that where the system occurs in association with moie-

ties that association is only accidental. It is fair
to say that, among the many anthropologists who have
puzzled over this problem, Radcliffe-Brown is the only
one who has given the possibility the consideration it
deserves.

Radcliffe-Brown recognized that section and subsection systems are not everywhere (if anywhere) based
on moieties, much less on a combination of two kinds
of moieties. He recognized also that Australian systems of kin classification are not based on moiety,
section, or subsection systems. Further, he recognized
that the modes of kin classification allegedly based on
section and subsection systems are more widespread than
are section and subsection systems. Therefore, he proposed to reverse the conventional explanation and to
account for section and subsection systems by deriving
them from systems of kin classification. He argued
that sections and the relations (of "descent" and "marriage") among them are nothing more than reifications
of aggregates of kin classes and the relations (again,
of "descent" and "marriage") among them. Unfortunately,
he obscured his argument and its radical departure from
the then (and still) generally accepted theory by con-
tinuing to describe these aggregates of kin classes as
"moieties." As a consequence, he sometimes appeared to
contradict himself and to support the "moieties plus
alternate-generation levels" hypothesis (cf. Dumont
1966).

In 1930 Radcliffe-Brown wrote: "An examination of
the simplest of the normal Australian kinship systems,
that of the Kariera, based on first cross-cousin marriage, shows that such a system if logically and consistently carried out must inevitably result in the
formation of moieties" (1930-31: 440). That is, if all
marriages are terminologically regular (or are regular-
ized in some way), the practical consequence is that
there are two opposed sets of individuals. The members
of one set all designate one another as 'father's
father', 'father', 'brother', 'man's child', or 'man's
son's child' (to mention only the terms for male relatives), and they designate all members of the other set
as 'mother's father', 'mother's brother', 'cousin',
'woman's child', or 'man's daughter's child'. These two
sets, Radcliffe-Brown argued, could exist and do exist
in many Australian societies quite independently of the
existence of patrimoieties. Similarly, there are two
other sets, the so-called matrimoieties. One consists
of individuals who designate one another as 'father's
father', 'mother's brother', 'brother', 'woman's child',
or 'man's son's child'; the other set consists of individuals whom members of the first set designate as
'mother's father', 'father', 'cousin', 'man's child',
or 'man's daughter's child'. Again, these two sets
could exist and do exist in many Australian societies
quite independently of the existence of matrimoieties -
or so Radcliffe-Brown argued. Now the existence of such

sets, if recognized by the people who constitute them, could be conceptualized, and the result would be a set of categories like those designated by the reciprocal Kariera expressions ngaju maru (or maman maru) and balu maru. As Spencer and Gillen (1927: 41) observed about the semantically equivalent Aranda expressions nakrakia and mulyanuka, "at first sight" such expressions "might be mistaken for moiety names."

As we have already seen, they are not moiety names. Like the designations of kin classes, these expressions are reciprocals of one another and, surely, it is not accidental that one of them is based on a kinship term, maman 'father'. As Radcliffe-Brown interpreted these expressions, they designate conceptual aggregates of persons who stand uniformly in certain kin-class relationship vis-à-vis one another. Because these aggregates are "patrilineally" constituted, each is composed in effect of the members of one patricouple of sections. Thus, these expressions may be used not only to describe certain aggregates of kin (related "patrilineally" among themselves), but also to describe relationships within the section system. It is somewhat misleading, however, to describe these aggregates as "patrilineal." Each is composed of kin of several classes, and the members of the classes of one set are related to one another as 'father' and 'man's child', whereas the members of classes of opposite sets are related to one another as 'mother's brother' and 'woman's child'. It is sufficient to say that the several classes of one set are related to one another through <u>patrifiliation</u> and to the classes of the opposite set through <u>matrifiliation</u>. We could, and many anthropologists do, describe such relationships between kin classes in terms of patrilineal descent, but this may and often does result in the confusion we are trying to avoid - confusion between moieties and categories that are only aggregates of kin classes - and in the totally unwarranted analytic reduction of kin classes (and relations of kinship in general) to subclasses of descent systems (cf. Scheffler 1973).

Radcliffe-Brown, however, chose to describe these aggregates of kin classes as moieties, and to describe the relations among the several kin classes of which they are composed in terms of patrilineal descent. He noted also that there is another such kind of "kinship divisions," the division of kintypes and kin classes "into two alternating generation divisions," and he described these, too, as "endogamous moieties." They may be so called, he said, because "marriage systems are as a rule such as only to permit marriage between persons of the same generation division" (1951: 39). These divisions may or may not be "named," but where they are, the "names" (so-called) are relative appellations, as in the Kukata language where <u>nganandaga</u> designates the category of a person's relatives of his or her own generation and generations alternate to it (G=,

G+2), and <u>tanamildjan</u> designates the category of a person's kin of adjacent generations and generations alternate to them (G+1, G+3). The Kukata do not have or use a section or subsection system, so this conceptual opposition is structurally independent of such systems. However, if this opposition were combined with an opposition like Kariera <u>ngaju</u> <u>maru</u> versus <u>balu</u> <u>maru</u>, the logical product would be a system of four interrelated classes. For any speaker, these four classes would consist of the following classes of kin:

1. his 'brothers' and 'father's fathers',
2. his 'fathers' and 'man's children',
3. his 'mother's fathers' and 'cross cousins',
4. his 'mother's brothers' and 'woman's children'.

Of course, these four classes are composed of exactly the same kin classes as are the four categories of a section system, and, further, the patrifilial and matrifilial relations among them are the same. These four higher-order or superclasses could easily be transformed into a section system. This could be done by taking some person arbitrarily as the starting point and by giving each of the four superclasses of his relatives a proper name; it would then be possible to assign one of the four names to any member of the society, past, present, or future, given nothing more than a knowledge of how the party to be designated is related (by birth or by marriage) to the arbitrary starting point, or to yet some other person whose designation has already been determined.

The result would be a section system, and the
patrifilial, matrifilial, and marital relations among
the four sections could be specified in kinship terms.
Sections A and C (see Figure 12.1) are related self-
reciprocally as 'father' and 'man's child', and so are
sections B and D. Sections A and B are related self-
reciprocally as 'cross grandparent' and 'cross grandchild', and so are sections C and D. Of course, it may
be said also that A-B and C-D are related "by marriage,"
that is, self-reciprocally as 'husband' and 'wife' or
as 'siblings-in-law', because many (but not all, cf.
Radcliffe-Brown 1930-31: 55) of their respective members
are potential spouses of one another. However, for
reasons stated below, these "by marriage" relationships
among the sections must be regarded as structurally
derivative.

Subsections, according to Radcliffe-Brown, are
derived from Aranda-type systems of kin classification.
Here the rule is that one marries a "mother's mother's
brother's daughter's daughter," rather than a "mother's
brother's daughter," and as a consequence four, rather
than only two, descent lines are recognized in ego's
genealogy. Again, if all marriages were terminologically regular (or made regular, as it were, after the
fact), the practical result would be that the society
as a whole would be divided into four sets of persons.
Each set would be composed of persons who designate one

another as 'father's father' and so on, and the members
of any one set would designate the members of the other
three sets as 'mother's father', etc., or as 'father's
mother', or as 'mother's mother', etc. If these four
sets of classes of kin were each divided into two
alternate-generation classes, the result would be a set
of eight superclasses of kin classes. These eight
superclasses could then be reified as objective, nonego-
centric categories and assigned proper names, and the
result would be a subsection system (Radcliffe-Brown
1930-31: 56-8).

Radcliffe-Brown's interpretation of the structural
relations between systems of kin classification and
section and subsection systems is fundamentally sound,
despite the limitations of his interpretation of the
structures of Australian systems of kin classification.
There are, however, several difficulties that should be
discussed. First, his account of the derivation of the
necessary "patrilineal moieties" and "endogamous alternate-generation moieties" requires an unrealistic degree
of interindividual "consistency" in the designations of
other individuals. It requires not only that, in
general, if individuals X and Y designate one another
as, say, 'brother' and X designates Z as 'cousin' then
Y does the same, but also that there are few if any
exceptions to such arrangements. But Rose (1960) and
Hamilton (1971) have established that this is not the
modern-day arrangement in some Australian communities,

and there are no good reasons to suppose that it ever
was the typical arrangement in any Australian community.[1] This, however, is not a serious objection to
Radcliffe-Brown's analysis, because the necessary aggregates of kin classes already exist, as superclasses,
within many Australian systems of kin classification,
and their existence is not in the least dependent on
this kind of interindividual "consistency" in the designation of kin.

The second difficulty is more serious. Although
Radcliffe-Brown derived subsection systems from Aranda-
type systems of kin classification, he stated also that
subsections are "subdivisions" of sections (1930-31:
39). Given his analysis of Aranda-type systems of kin
classification, the two propositions are mutually contradictory. They are contradictory because in his model
of Aranda-type systems, the four descent lines are not
related to one another by class inclusion; the MMB line
is not so related to the FF line, and the FMB line is
not so related to the MF line. Instead, in the model,
these lines are radically distinguished as an implication of the rule that "a man may not marry his 'mother's
brother's daughter' but must marry a 'mother's mother's
brother's daughter's daughter'." But in a subsection
system, the two subsections that contain a person's FF
and MMB correspond to a section; these two subsections
are related by class inclusion, and if they were not it
would make no sense to describe them as "subdivisions

of a section." There is abundant evidence to show that subsections are generally regarded by Australians as subdividions of sections (see below, Subsection systems). Therefore, the systems of kin classification from which subsection systems are derived must contain, at the highest level of class inclusion, four superclasses that may be objectified as sections and subdivided to form subsections. Radcliffe-Brown's model of Aranda-type systems does not allow for this possibility.

We have seen, however, that his model of Aranda-like systems is inadequate and that such systems do contain four superclasses; he was prevented from recognizing this by his overdependence on the concept of descent lines and on differences in rules of interkin marriage as determinants of the number of descent lines recognized in a system. It remains to be shown how these four superclasses correspond to the four sections of a subsection system. We should begin, however, with the less complex problem of the derivation of section systems. These are derived not from simple Kariera-like systems (Chapter 4) but from Kariera-like systems on which the AGA and AGU rules are superimposed (Chapter 7). This is not inconsistent with the frequent occurrence of section systems in association with simple Kariera-like systems, because it is fairly clear (Elkin 1970) that the section system originated in, at most, one or two places, and diffused widely across the continent, being accepted and used (even if sometimes with difficulty) by peoples whose systems of kin classification do not provide the most probable basis for the development of such a system.

Section systems

A section system consists of four classes whose interrelationships are conventionally represented as in Figure 12.1. For any member of section **A**, his or her kin may be distributed among the sections as indicated in Figure 12.2. Given a simple Kariera-like system of kin classification, a person does not designate all of

---

Figure 12.2. Normal distribution of kintypes by sections

A	=	B
FF, MM		FM, MF
Ego, Sb, FBC, MZC		MBC, FZC, SpSb
MFZSC, MMBSC,		MMBDC, MFZDC,
FMBDC, FFZDC		FMBSC, FFZSC
mSC, wDC		mDC, wSC
M, MB, SpF		F, FZ, SpM
FFZC, FMBC		MMBC, MFZC
wC, mZC, mCSp		mC, wBC, wCSp
MBSC, FZSC		MBDC, FZDC
C	=	D

his kin of any one section by the same kin term, even
allowing for differences of sex of ego and sex of alter.
Instead he has several terminological classes of kin in
each section. We have seen, however, that many terminologically distinct kin classes are related by class
inclusion. It is a plausible hypothesis that, at a
higher or superclass level, each section contains ego's
kin of only one class. Indeed, it is clear that some
of the required superclasses do exist even in simple
Kariera-like systems. We saw in Chapter 4 that the
FATHER'S SISTER class is a subclass of the FATHER class
and that there is a single reciprocal class, the MAN'S
CHILD class. These two reciprocal classes, taken
together, may be said to constitute a still-higher-order
superclass, the FATHER-MAN'S CHILD class. We saw also
that in the Karadjeri language it is permissible to
designate any member of this class as 'father' (Chapter
6). Similarly, the MOTHER'S BROTHER class is a subclass
of the MOTHER class and there is a single reciprocal
class, the WOMAN'S CHILD class. Again, these two reciprocal classes, taken together, may be said to constitute
a superclass, the MOTHER-WOMAN'S CHILD class. According
to Elkin (1964: 66-7), there are Australian languages
in which this opposition also may be neutralized (although perhaps not to the same extent as the opposition
between the FATHER and MAN'S CHILD classes). Of course,
these superclasses could be objectified to form categories C and D in a section system. But for a section

system to be derived in this way there would have to be four, rather than just two, such superclasses. The other two superclasses would have to consist of (1) the SIBLING class plus the PARALLEL-GRANDKIN class, which could form the basis of section A; and (2) the COUSIN class plus the CROSS-GRANDKIN class, which could form the basis of section B. In a simple Kariera-like system there are no such superclasses, because the PARALLEL-GRANDKIN class is not related to the SIBLING class by class inclusion, nor is the COUSIN class so related to the CROSS-GRANDKIN class. Because a simple Kariera-like system lacks these two superclasses, its structure provides an inadequate basis for the derivation of a section system.

It is, however, a relatively simple matter to generate the required superclasses. This may be done by superimposing the AGA and AGU rules on a simple Kariera-like system. If neither of these rules is assigned priority over the parallel-cross status-extension rule, it makes no difference how they are ordered in relation to one another. The consequence, in any event, would be to make the PARALLEL-GRANDKIN class a subclass of an even wider SIBLING class, and to make the COUSIN class a subclass of an even wider CROSS-GRANDKIN class - and thereby to establish the other two superclasses essential to the derivation of a section system.

The four superclasses of such a system of kin

449

classification would then be the FATHER, MOTHER,
SIBLING, and CROSS-GRANDKIN classes, and these would
contain more or less the following terminological
subclasses:

FATHER   : 'father', 'father's sister', 'man's child';
MOTHER   : 'mother', 'mother's brother', 'woman's child';
SIBLING : 'sibling', 'parallel grandparent';
CROSS
GRANDKIN: 'cross grandparent', 'cousin'.

Of course, further subclasses might be distinguished
within each of these subclasses.

We may now note how these four superclasses are
related to one another. These relationships may be
specified in terms of the relations among the respective
class foci (F, M, Sb, and MF, FM), and all the interclass relationships are specifiable solely in terms of
father-child (F-C) and mother-child (M-C) relationships
- the two kinds of parent-child (P-C) relationship.
We may specify the pair relations among the several
classes as follows, where (m) stands for M-C or
matrifilial relationship, and (p) stands for a F-C
or patrifilial relationship.

SIBLING-MOTHER	Ego's Sb is ego's M's C	=(m)
	Ego's M is ego's Sb's M	=(m)
SIBLING-FATHER	Ego's Sb is ego's F's C	=(p)
	Ego's F is ego's Sb's F	=(p)
MOTHER-CROSS GR.KIN	Ego's M is ego's MF's C	=(p)
	Ego's MF is ego's M's F	=(p)

```
FATHER-CROSS GR.KIN  Ego's F is ego's FM's C      =(m)
                     Ego's FM is ego's F's M      =(m)
```

These interclass relations may be described as elementary because they are simple matrifilial or patrifilial
relationships. The remaining interclass relationships
are:

```
SIBLING-CROSS GR.KIN Ego's Sb is ego's MF's DC    (p + m)
                     Ego's MF is ego's Sb's MF    (m + p)
                     Ego's Sb is ego's FM's SC    (m + p)
                     Ego's FM is ego's Sb's FM    (p + m)
MOTHER-FATHER        Ego's M is ego's F's CM      (p + m)
                     Ego's F is ego's F's CM      (m + p)
```

These interclass relationships may be described as complex and derivative because they are relative products
of simple matrifilial and patrifilial relationships.

It is not stretching a point to describe the relationship between the FATHER and MOTHER classes in this
way. This is just another way of saying that F and M
are the two kinds of parent, the two subclasses of the
parent class, so their relationship to one another
vis-à-vis ego is as coparents (CP). This relationship,
like the cross-grandparent and cross-grandchild relationship, is a relative product of P-C relationships.

Now if any system features these four superclasses
- and many Australian systems do - the superclasses
may be objectified to form a section system. As noted
above, this may be accomplished by taking some person
arbitrarily as the starting point and by giving each of

the superclasses of his relatives a proper name (in place of the kin category designations SIBLING, FATHER, MOTHER, CROSS GRANDPARENT). These names would then bear the same relationships to one another as the several kinship superclasses do. Therefore, the name to be assigned to any member of the society, past, present, or future, could be deduced from the name assigned to any one of his relatives, given a knowledge of the genealogical links between them.

Of course, the relationships between certain sections may be described not only in terms of relative products of patrifilial and matrifilial relation but also in terms of intermarriage. That is, "A marries B, and C marries D." The possibility of describing sections A-B, and C-D, as related in this way depends on certain general structural features of Australian systems of kin classification (which they share with the systems of many other peoples.) These include division of the parent class into two subclasses on the basis of sex of alter, and use of the same-sex sibling- and stepkin-merging rules. The use of these rules establishes a further relationship between the 'father' and 'mother' terms, in addition to their relationship as the designations of two subclasses of the parent class. That further relationship is that the spouse of a kinsman designated as 'father' is normally a woman designated as 'mother' and vice versa. A similar relationship exists between the 'man's child' and

'woman's child' terms of a Kariera-like system, for the
extension rules of the system establish these two terms
as a normal marital pair. Therefore, it may be said
that there is an additional derivative relationship
between the FATHER-MAN'S CHILD and MOTHER-WOMAN'S CHILD
superclasses: They are related also as a spouse pair.
Normally, a member of one kin class would have as his
or her spouse a member of the other kin class. It
follows logically from this that the same relationship
exists between the two sections based on these two
superclasses: A member of C is normally the spouse of
a member of D, and vice versa.

The same relationship exists between the SIBLING
and CROSS-GRANDKIN classes, and, therefore, between
sections A and B. The rules of kin-class definition
and expansion establish the cross-grandkin and parallel-
grandkin terms as a normal marital pair; normally a
member of one of these categories is married to a member
of the other. These same rules also establish the
sibling and cross-cousin terms as a normal marital pair.
The PARALLEL-GRANDKIN class is a subclass of the SIBLING
class, and the COUSIN class is a subclass of the CROSS-
GRANDKIN class; it follows that the two superclasses
SIBLING and CROSS GRANDKIN also form a normal marital
pair. The same relationship must exist between the two
sections based on these two superclasses: A member of
A is normally a spouse of a member of B, and vice versa.

Subsection systems

We have seen that the section systems are derived
from the superclasses of systems of kin classification
in which the AGA and AGU rules are subordinate to the
parallel-cross status-extension rule. Virtually all
anthropological observers agree that subsection systems
are derived from section systems by dividing each of
the sections in two. There is strong comparative and
historical evidence to support this analysis.

As noted in Chapter 1, the Northern Aranda have a
subsection system but the Southern Aranda have a section
system, although their systems of kin classification are
virtually identical. The Southern Aranda section system
is illustrated (following Spencer and Gillen 1927: 43)
in Figure 12.3. In this figure the patrifilial relations are represented vertically and the matrifilial relations are represented diagonally. Apparently Spencer

---

Figure 12.3. Southern Aranda section system

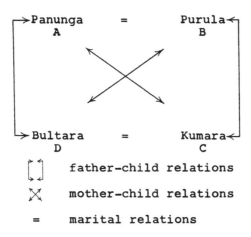

```
      ┌─→Panunga      =      Purula←┐
      │    A                    B   │
      │                             │
      │          ↖      ↗           │
      │            ╲  ╱             │
      │             ╳              │
      │            ╱  ╲             │
      │          ↙      ↘           │
      │                             │
      └─→Bultara      =      Kumara←┘
           D                    C
```

[]   father-child relations

✕   mother-child relations

=   marital relations

and Gillen wanted to stress that in the Aranda language
the set A+D is conceptually opposed to the set B+C.
They described the two sets as "moieties," although
"there are no distinctive names in the Aranda for the
moieties" (1927: 41). They added, "there are, however,
two terms employed that, at first sight, might be mistaken for moiety names." The two terms are nakrakia
and mulyanuka. An individual who belongs to one set
uses the expression nakrakia to designate his own set
and the expression mulyanuka to designate the other set.
So these are not proper names; they signify relations
among kin classes and then sections; sections related
by patrifiliation are classified together in relation
to sections to which they are related by matrifiliation.
In the Northern Aranda subsection system these designations are retained, but they signify the relationship
between two sets, each of which consists of four subsections arranged as indicated in Figure 12.4. Again,
Spencer and Gillen described the two sets as "moieties"
(patrilineal), but they did not specify how the two
subsections of any section may be regarded as "patrilineally" related to one another. However, they did
point out that in the North, "the old names" for the
sections are retained as names for subsections, and
each name remains associated with one of the subsections of the section originally so named. A new name
is added, as it were, for the new subsection. For
example, the name for section A in the South is Panunga;

in the North Panunga is retained as the name for subsection A1 and Knuraia is added as the name for subsection A2. According to Spencer and Gillen's informants,
the new names for the subsections were taken over from
the Walbiri to the north (1927: 42). (On the Walbiri
subsection system see Meggitt 1962: 165 ff.) Indeed,
the process of the introduction of the subsection system among the Northern Aranda was observed by Gillen
himself. In his unpublished notes for the period
1894-96, he wrote, "all tribes I know anything of are
divided into four castes [sections]," but he added that
the Ilipirra (Walbiri) use a system with eight classes.
This system, he said, was rapidly being adopted by the
Aranda and he knew families who had only recently
adopted it (Gillen Ms.). Many Aranda statements recorded by Gillen provide valuable information on how

---

Figure 12.4. Northern Aranda subsection system

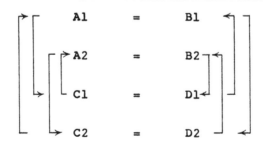

Patricouples: A1-D2, A2-D1, C1-B1, C2-B2

Matrifilial cycles: A1-C1-A2-C2-A1

B1-D2-B2-D1-B1

the two subsections of a section were distinguished from one another, that is, on how the sections were divided into subsections.

Gillen's informants explained this by referring to kin classes. They pointed out that in the section system any person's 'siblings', 'father's fathers', and 'mother's mothers' all belong to the same section as he does; but in the subsection system his kin of the latter category belong to a different subsection. The members of these two subsections "stand respectively in the relationship of ipmunna ['mother's mother'-'woman's daughter's child'] to each other" (Spencer and Gillen 1927: 44). That is, each section is divided into two subsections by systematically placing in different subsections those persons who are related as 'mother's mother'-'woman's daughter's child' to one another. The name for the original section is retained as the name for one of the subdivisions and a new name is assigned to the other. One of the consequences of subdividing the sections in this way is that no two subsections can be related self-reciprocally as 'mother'-'woman's child'. This does not do away with the apparent matrimoieties, for another consequence is that two four-generation "matrilineal" cycles are thereby established - a woman's DDDC belongs to the same subsection as she does. These two sets of "matrilineally" related subsections are often described (by anthropologists) as matrimoieties or matrilines, but they are only by-

products of the subsection system, and not at all basic
to its structure. The division of sections into subsections is not the product of the recognition of
matrimoieties; it is instead the product of assigning
priority to the AGA and AGU rules in relation to the
parallel-cross status-extension rule, and of providing
these rules with certain auxiliary rules, as in the
Walbiri and Aranda systems of kin classification
(Chapter 9). Apparently, the Aranda system of kin
classification already had this structure when the
Northern Aranda took over the subsection system from
the Walbiri; they were preadapted to understand and
make use of the subsection system.

We saw in analyzing the Walbiri and Aranda systems
that they too contain the four superclasses SIBLING,
FATHER, MOTHER, and CROSS GRANDKIN. Their respective
superclasses, however, are differently constituted than
are the subclasses of systems in which the AGA rule has
less scope. To describe this arrangement it was necessary to distinguish between two kinds of SIBLINGS in
$G\pm2$, namely AGNATIC SIBLINGS (focus = FF) and UTERINE
SIBLINGS (focus = MM). The opposition between AGNATIC
SIBLINGS and UTERINE SIBLINGS is a recurrent feature
of the structure of all four superclasses.

In the case of the SIBLING class itself, the
AGNATIC SIBLING subclass is the PARALLEL-GRANDFATHER
class, and the UTERINE SIBLING subclass is the PARALLEL-
GRANDMOTHER class. These two subclasses constitute the

458

PARALLEL-GRANDPARENT subclass. Within the widest
SIBLING class, the PARALLEL-GRANDPARENT class is a
special subclass, and within the PARALLEL-GRANDPARENT
class the PARALLEL-GRANDMOTHER class is a special
subclass. That is, the order of neutralization is
'parallel grandmother' → 'parallel grandfather' →
'sibling' (or UTERINE SIBLING → AGNATIC SIBLING →
'sibling').

In the case of the MOTHER class, mother's AGNATIC
SIBLINGS are MBSC, and mother's UTERINE SIBLINGS are
wC and ZC. The mother and woman's child classes are
distinguished to begin with, as reciprocals. Mother's
AGNATIC SIBLINGS constitute a specially designated
potential in-law subclass of the MOTHER class, but
otherwise members of this subclass are designatable as
'mother' and 'mother's brother'.

In the case of the FATHER class, father's AGNATIC
SIBLINGS are mC and BC, and father's UTERINE SIBLINGS
are FZDC. The father and man's child classes are
opposed and distinguished as reciprocals of one another.
Father's UTERINE SIBLINGS constitute a specially designated potential in-law class, but otherwise members of
this subclass of the FATHER class are designatable as
'father' and 'father's sister'.

The arrangement within the CROSS-GRANDKIN class is
slightly different. This class is divided initially
into two subclasses that focus on FM and MF, but again
the UTERINE SIBLINGS and the AGNATIC SIBLINGS of the

foci are in different subclasses. FM's AGNATIC SIBLINGS
are FMBSC, and FM's UTERINE SIBLINGS are FZC; FMBSC is
in the same subclass as FM, and FZC is in a different
subclass. MF's AGNATIC SIBLINGS are MBC, and MF's
UTERINE SIBLINGS are MFZDC; MBC are in the same subclass
as MF, and MFZDC are in a different subclass. Because
MBC and FZC are reciprocals they are in the same subclass, which is the AGNATIC SIBLING subclass of the MF
class. All kintypes that fall into the CROSS-GRANDPAR-
ENT class but which are not MF or his AGNATIC SIBLINGS
fall into the CROSS-GRANDMOTHER class (= FM, etc.).

If we divide each of these four superclasses in two
by merging the AGNATIC SIBLINGS of the focus of the
class with the focus of the class, thus retaining the
UTERINE SIBLINGS of the class focus as a special subclass, the result is:

SIBLING   → Sb + FF   vs. MM
CROSS GRANDKIN   → FM + FMBSC   vs. MF + MBC
MOTHER   → M + MBSC   vs. wC + FFZC
FATHER   → F + mC   vs. FZDC + MMBC

These eight categories and the matrifilial, patrifilial,
and marital relations among them correspond exactly to
the categories and intercategory relations of a subsection system (see Figure 12.5). They may be converted
into a subsection system in the same way as the four
superclasses may be converted into a section system
(see above, Section systems). It seems probable, however, that the subsection system developed out of the

section system, so that historically the four sections did indeed subdivide into two subsections each, four new names being added to an already existing stock of four names - the same as when the Aranda, who already had the section system, adopted the subsection system from the Walbiri.

ALTERNATE REALIZATIONS

The most direct way in which kinship superclasses are verbally realized in Australian languages is by suspending the terminological distinctions (if any are present) among the various subclasses of a superclass and designating the superclass as a whole by the designation of its focal member - as where any member of the FATHER class, including its specially designated subclasses, may be designated as 'father', and any member

Figure 12.5. Distribution of kin by subsections

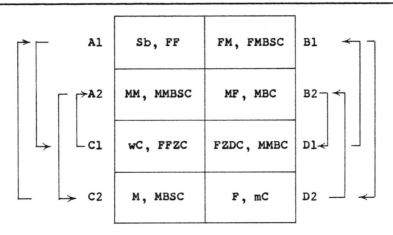

of the reciprocal MAN'S CHILD class, including its specially designated subclasses, may be designated as 'man's child'. The still higher-order class that includes these reciprocal classes may be directly realized by suspending the opposition between them and designating the class as a whole by the designation of the focal member of its senior subclass, again 'father'. Although this form of direct expression of superclasses is not always available - because of cultural rules that may require the use of subclass designations (especially those for potential or prospective in-law classes) - there are many other ways in which kinship superclasses may be and are expressed in Australian languages.

Another relatively direct way to express a superclass is through the union of the expressions for its subclasses, as when Beckett's Wongaibon informant described the SIBLING class (in part anyway) by the expression 'grannies-sisters' (see Chapter 7, Murawari and Wongaibon). Kinship superclasses are often realized also in specialized indexical vocabularies such as the Djiribal "mother-in-law" language (Dixon 1971a, 1971b) or in the "mourning terminologies" of many Australian languages (Thomson 1972, and Chapter 4, Ompela and Mongkan), wherein the number of expressions designating kin classes is greatly reduced in comparison with the number available in the "ordinary" or "everyday" vocabulary. Further, kinship superclasses are frequently

realized in sign language or gesture systems, wherein
all kintypes of a pair of reciprocal superclasses are
denoted by one sign or gesture (see e.g., Meggitt
(1954). Finally, although probably not to exhaust the
list of possibilities, kinship superclasses are sometimes realized in pronominal systems.

Indeed, those relational designations frequently
misrepresented in the ethnographic literature as "names"
of patrimoieties, matrimoieties, or alternate-generation
moieties are perhaps best understood as pronominal forms
(see also Hale 1966, Ms. 1, Ms. 2). The Kariera expressions ngaju maru and balu maru probably are third-
person-plural pronouns, for which the gloss 'they' is
inadequate only because it is not sufficiently specific.
Both forms signify that the parties referred to are
related among themselves as 'father' and 'man's child'
(or through kin-class relations that may be described
as relative products of the 'father'-'man's child'
relationship). They differ in meaning in that one
(ngaju maru) signifies that the speaker or propositus
is related in the same way to the designated parties,
whereas the other (balu maru) signifies that he or she
is not so related to the designated parties. It is not
the least puzzling that maman 'father' may be substituted for ngaju. A reasonable gloss for maman maru may
be 'they, related among themselves as "father"-"man's
child" (etc.) and including the father of the speaker
or propositus'.

In some Australian languages pronominal systems based on kin classes have much the same structure as a section system, although the section system is either unknown to or not used by the speakers of these languages. It can only be concluded from this that these pronominal systems are alternate realizations of the relations among the superclasses of systems of kin classification.

Consider the case of the Adnjamathanha (or Wailpi) of South Australia, whose personal pronouns and their relations to kin classification are the subject of brilliant studies by two linguists and an anthropologist (Schebeck 1973, 1974; Hercus and White 1973). The Adnajamathanha system of kin classification is described by Schebeck (1973: 5) as "a slightly modified Kariera type" system. One of the "modifications" is that cross cousins are designated by the cross-grandparent terms 'mother's father' and 'father's mother'. This indicates that this system features the AGA or the AGU rule, or both. Although this is not apparent from the information given by Schebeck, and Hercus and White (or by Elkin 1938a), it features both. My informants (who were some of Hercus and White's informants too) indicated that FF is "like a brother" and MM is "like a sister." Kin of these types are not addressed or referred to by sibling terms, but their siblings and classificatory 'siblings' may be. This is another case in which it is difficult to know which rule, the AGA or the AGU rule, governs the classification of cross cousins, but it is clear that here the parallel-cross status-extension rule has priority. This system, then, is structurally similar to the Murawari and Wongaibon systems (see Chapter 6), and its superclasses are the same as those on which section systems are based. A few additional complications are noted as they become relevant to the discussion of the pronominal system.

In addition to a set of eight "basic" pronouns, there are ten sets of pronouns, the use of which "depends both on the kin relationship of the speaker to the referent, and the relation of the referents to each other." One way to describe these ten sets is *as if* they were based on a section system. Assuming that the speaker is a member of section A, and considering only the third-person-plural forms, the first four pronouns are used as follows.

1. A refers to other members of A.

2. A refers to members of B.

3. A refers to members of C.

4. A refers to members of D.

If the referents belong to two sections, the possibilities are:

5. A refers to members of A and C.

6. A refers to members of B and D.

7. A refers to members of A and B.

8. A refers to members of C and D.

9. A refers to members of B and C.

10. A refers to members of A and D.

With a few exceptions (noted below) the pronouns are used in this way, or, more accurately, they could be said to be so used were it not that the Adnjamathanha do not have a section system. Therefore, the relationships signified by the pronouns are not relations between persons as members of sections; they can only be relations between persons as members of kin classes. The pronominal system - this much of it, anyway - can be understood only by relating it to the system of kin classification. For example, the speaker's 'father's fathers' and 'brothers' are designated by the same pronoun not because they belong to the same section as the speaker (they belong to no section at all!), but because they stand in the same kin-class (superclass) relationship to one another as either does to the speaker. The terminological opposition between 'brothers' and 'father's fathers' is neutralizable only under certain conditions in ordinary use of the kinship terms themselves, but in the pronominal system this opposition (within the broadest SIBLING class) is neutralized and all the speaker's SIBLINGS (also SIBLINGS of one another) are denoted by the same form.

The first four pronouns denote sets of persons who are SIBLINGS of one another. That is, the designated parties designate one another as 'father's father', 'mother's mother', 'brother', 'sister', etc. The differences among the four sets are in how the denotata are related to the speaker.

The first four third-person-plural pronouns may therefore be defined in this way.

1. They, who are related as SIBLINGS to one another and to the speaker.

2. They, who are related as SIBLINGS to one another, but as CROSS GRANDKIN to the speaker.

3. They, who are related as SIBLINGS to one another, but as MOTHER or WOMAN'S CHILD to the speaker.

4. They, who are related as SIBLINGS to one another, but as FATHER or MAN'S CHILD to the speaker.

The remaining forms refer to sets of persons not related to one another as SIBLINGS.

5. They, who are related to one another and to the speaker as MOTHER-WOMAN'S CHILD.

6. They, who are related to one another, but not to the speaker, as MOTHER-WOMAN'S CHILD.

7. They, who are related to one another and to the speaker as CROSS GRANDKIN.

8. They, who are related to one another, but not to the speaker, as CROSS GRANDKIN.

9. They, who are related to one another and to the speaker as FATHER and MAN'S CHILD.

10. They, who are related to one another, but not to the speaker, as FATHER and MAN'S CHILD.

The definitions given above for categories 7, 8, 9, and 10 are not quite correct. To understand these departures from the pattern we must, as Schebeck and Hercus and White point out, take into account certain marriage practices and their expression in kin classification.

The orthodox marriage is between COUSINS; a man may marry his MBD or FZD or a classificatory 'cousin', provided she is younger than he. A man designates a junior female COUSIN whom he regards as a prospective wife as 'wife' and she designates him as 'husband'. If a man does not designate a particular junior female COUSIN as 'wife', he and others do not regard her as his prospective wife and he may not marry her. This distinction is relevant to the classification of father's 'brothers' and mother's 'cousins'. A person designates his or her father's 'brothers' and mother's male 'cousins' as <u>vapi</u> 'father' or as <u>ubmali</u> <u>vapi</u>, but mother's COUSINS who do not designate her as 'wife' may be designated also by the special expression <u>ubmali</u> <u>ngamana</u> (<u>ngamana</u> = 'mother's brother'). Kinsmen designated as <u>ubmali</u> <u>ngamana</u> may, reciprocally, designate ego as <u>ubmali</u>. In addition to those COUSINS whom he designates as 'wife', a man may marry the daughter of a COUSIN whom he does not designate as 'wife', she would be a kinswoman he designates as <u>ubmali</u>, rather than as 'man's daughter'. Also, he may marry a kinswoman whom his father designates as 'younger sister' (though not of course his FZ-).

Usage of the pronominal forms 7, 8, 9, and 10 may now be explained in this way. There are two forms in each of the categories 7 and 8.

7a. The speaker refers to several of his SIBLINGS and
CROSS GRANDKIN (who are related as CROSS GRANDKIN
to one another) who are actual or prospective
spouses and siblings-in-law of one another. In
the event that one of the speaker's SIBLINGS or
CROSS GRANDKIN marries a classificatory 'man's
daughter' of the ubmali subclass, this couple and
their siblings are denoted by this form.

7b. Like 7a this form refers to actual or prospective
spouse and sibling-in-law sets, but the referents
are related to ego as MOTHER and FATHER (including
'mother's brother' and 'father's sister') or as
MAN'S CHILD and WOMAN'S CHILD.

8a. The speaker refers to several of his SIBLINGS and
CROSS GRANDKIN (again, they are CROSS GRANDKIN of
one another), but none is married or 'promised'
to one of the others.

8b. The speaker refers to several of his kin who are
related as CROSS GRANDKIN to one another but as
MOTHER or WOMAN'S CHILD or as FATHER or MAN'S
CHILD to him, none of whom is married or 'promised'
to one of the others.

9. The speaker refers to several of his kin who are
related to one another as FATHER and MAN'S CHILD,
except for instances in which form 10 is appropriate. It does not matter how the speaker is related
to the referents.

10. The speaker refers to a man and his classificatory

'father's sister' or, conversely, to a woman and her classificatory 'brother's child' who are married to one another.

It may aptly be said of this pronominal system, as Elkin has said of section and subsection systems, that it "systematizes" or "summarizes" the categories of the system of kin classification. It does so by predicating pronominal categories on kinship superclasses, and this is why its structure is similar (though by no means identical) to the structure of a section system.

REGULATION OF MARRIAGE

So far we have focused on the formal properties of section and subsection systems. We have seen that they correspond to and are structurally derived from the formal properties of systems of kin classification. One of these formal properties is that certain categories of kin, and certain sections or subsections, are, as it were, "related by marriage." These relationships be-tween sections or subsections are structurally deriva-tive; they are consequences of the rules of kin-class definition and expansion whereby certain kinship terms become normal marital pairs. In reciting how sections or subsections are related formally to one another, Australians often refer to these structurally derivative relationships and say, for example, "A (male) marries B (female) and her children are D; D (male) marries C (female) and her children are A," or (if certain ethno-

graphic accounts are reliable) even "A should marry B,"
and so on. Not surprisingly, such statements have been
misunderstood by many observers not only as formal
descriptions of these systems of classification but
also as assertions of pragmatic rules that govern
individual marriages (see also Meggitt 1962: 185 on
this point). Thus section and subsection systems are
often described functionally as "marriage class" systems, although this practice is no longer as common as
it used to be. It has become relatively uncommon for
the good reason that innumerable studies have shown
that section or subsection affiliations are not in and
of themselves determinants of eligibility or ineligi-
bility to marry. They may appear to be, however,
because in general the women whom a man may claim as
wives by right of kinship belong to the section or subsection formally paired "by marriage" with his own.
Conversely, the types and classes of kinswomen whom a
man may not claim by right of kinship belong to other
sections or subsections, and he is expressly prohibited
from marrying many (though not all) of these women by
reason of kin-class status, genealogical closeness, and
so on. Therefore, it is approximately correct to say
that a man of section A, for example, may not marry
women of sections C and D, but only women of section B.
But this is only a summary and somewhat inexact way of
stating kinship-based prohibitions and rights (again,
see also Meggitt 1962: 185).

It is somewhat inexact because sections and subsections are not unitary wholes with regard to the
marriage possibilities of their members; not all members
of a specified section or subsection have the same
potential spouses. Formally, "A marries B," or "Al
marries Bl," but for an A or Al man B or Bl includes
kinswomen who are two generations removed from him as
well as kinswomen of his own generation. Most (if not
all) of the former are unmarriageable or marriageable
only if distantly related and only if there are no other
men who may and do claim them by right of a particular
genealogical or kin-class relationship; and many of the
women of a man's own generation may be unmarriageable
because regarded as too closely related. At the very
least, it has to be acknowledged that kin class, relative generational status, and genealogical and social
distance are additional determinants of eligibility to
marry. This, however, would be to concede entirely too
much to sections and subsections as "marriage classes."

In a series of papers published in the 1890s and
1900s R. H. Mathews demonstrated that, in a great many
societies scattered all across the continent, it was
entirely permissible for a man to marry women of one or
more sections or subsections other than the one formally
paired "by marriage" with his own. In some cases, he
found, a man might without impugnity marry a woman of
any section, even his own; and in others he might marry
women of as many as four subsections, again including

his own. In all cases he knew of, there was no rule
that a man was obliged to take his wife from among the
women of one and only one other specified section or
subsection. Mathews concluded that sections and subsections are not properly described as "marriage
classes," and they "have not been deliberately formulated with intent to prevent consanguineous marriages
and incest, but have been developed in accordance with
surrounding circumstances and conditions of life" (1905:
123). Mathews's reports have since been confirmed by
many other anthropologists, some of whom have gone on
to give substance to Mathews's "surrounding circumstances and conditions of life" by showing that, functionally, section and subsection systems are far more
important in the organization of ritual than they are
(and probably ever were) in the regulation of marriage
(see especially Elkin 1933b).

Mathews observed that in practice marriages between
men and women of sections or subsections formally paired
"by marriage" are much more common than the permitted
alternatives; and among the "alternatives" some are much
more common than others. For example, A1 is formally
paired with B1, but an A1 man might also marry a B2,
A2, or A1 woman; in practice A1-B1 unions are most
common, then A1-B2 unions, with the others being quite
rare.[2] This does not demonstrate that in practice
subsection affiliations are indeed critical and that,
by and large, people tend to follow the rules of the

subsection system - for there are no such practical rules for them to follow or not to follow. What it does show is that, by and large, women are claimed and taken as wives (at least in their first marriages or while they are still young) by kinsmen who are rightfully entitled to them (Goodale 1962, 1971; Falkenberg 1962). By and large, but not invariably, these men belong to the subsections formally paired "by marriage" with the subsections of these women. In other words, the structures of section and subsection systems are consistent with the ways in which marriage is regulated (by kinship), and so marriage may appear to be regulated by such systems even though it is not.

Perhaps the most conclusive evidence that genealogy, rather than "category" (in this instance section or subsection affiliation), is the critical determinant of marital eligibility is this. Because of previous irregular or even wrong marriages, it sometimes happens that a woman X to whom a man Y would normally have a marital right (such as his MMBDD) is not in the section or subsection formally paired with his own "by marriage," sometimes not even in one of the customary recognized "alternate" sections or subsections. If eligibility to marry were contingent, first, on section or subsection affiliations and only secondarily on genealogical or kin-class relations, it should follow that some other man Z, who belongs to the section or subsection formally paired "by marriage" with that of

woman X, would have the stronger right to her. If Y
were to press a claim to the woman X on genealogical
grounds and Z were to press a claim to her on section
or subsection grounds, Z would be given the woman or
would have a legitimate complaint and could take action
(through self-help) if she were given instead to Y.
The evidence, however, is all to the contrary. There
is little or no evidence in the ethnographic literature
that section or subsection affiliation entails a preferential right to the women of another specified section
or subsection. Therefore, men cannot rightfully claim
women on such grounds, although of course this does not
prevent men from occasionally justifying a bestowal or
acquisition at least partly in terms of the formal
properties of the section or subsection system itself.
Typically, then, if a man has a right (by kinship) to
claim a woman who does not happen to belong to the
section or subsection formally paired "by marriage"
with his own, her section or subsection affiliation is
simply ignored and she is given to him anyway.[3]

In short, section and subsection systems are not
formally or functionally "marriage class" or "prescriptive marriage systems." They are, as Elkin has said,
not only "mechanisms for summarizing kinship and the
social behavior of which it is the determining factor"
but "also, and perhaps primarily, forms of totemism"
(1933b: 71, 90). But more of this below (Chapter 14).

IRREGULAR MARRIAGES AND THE "DESCENT" OF SECTIONS

Section and subsection systems are formally closed.
Therefore in principle a person's affiliation is the
same whether reckoned through his father or through his
mother. As we have seen, many anthropologists have
sought to account for these features of section and
subsection systems by deriving them structurally from
moieties. From this perspective, either the patrifilial
or the matrifilial relations among the sections or subsections are taken to be structurally most basic,
depending on whether the analyst assumes the underlying
moieties to be patrilineal or matrilineal, and the other
set of filial relationships is regarded as structurally
derivative. In support of such analyses anthropologists
have sometimes cited the rules for section or subsection
assignment in cases of structurally irregular marriage.

Maddock (1973: 74), for example, says that in cases
of irregular marriage "the usual practice is to 'throw
away the father' and decide the child's class by its
mother's"; that is, the section or subsection affiliation of the father is ignored and the child is assigned
to the category to which its mother's category stands
in the 'mother' relationship. According to Maddock,
this shows "that Aborigines prefer to take the mother
as a point of reference." Therefore, he argues, we may
"define the rule of descent by which classes of any
kind [moieties, sections, subsections, semimoieties]
are governed as a version of the principle of matri-

liny." This requires specifying "the rule for patrimoieties" as "indirect matrilineal descent: a child
belongs to the class to which its mother does not
belong." According to Maddock, this "formulation . . .
has the advantages of economy and consistency." Aside
from being a shade devious and resting on the false
assumption that section and subsection systems are
based on "a rule of descent," this interpretation
requires the postulation, in many instances, of implicit
or unnamed moieties that we would not otherwise be
forced to posit. Therefore, it is not economical.
Neither is it adequate, for we have already noted that
implicit or unnamed moieties can be nothing more than
structurally derivative products of relations of patri-
or matrifiliation among sections or subsections. So it
is highly improbable that the practice of "throwing
away the father" in cases of irregular marriage is
indicative of a basic, underlying matrilineal principle.
Moreover, this practice is a means of dealing with
exceptional cases - which are structurally exceptional
no matter how frequent they may be - and it is unclear
how this means can be taken as indicative of the principle or principles underlying "normal" practice.

If the practice of "throwing away the father" in
cases of structurally irregular marriage is not indicative of a basic, underlying matrilineal structure, how
otherwise may we account for it? First it should be
noted that this is only the "usual" practice in many

societies and not the universal practice in all societies with section or subsection systems. In some societies, Aranda for example, the standard practice in
cases of irregular marriage is "to throw away the
mother"; the offspring are assigned to the section or
subsection that stands in the 'man's child' relationship to that of their father. In yet other cases, it
is not clear that there is a general rule covering all
such cases. Piddington (1970: 332) reports that the
Nyamal and all other tribes in the Port Hedland region
of Western Australia, except for the Kariera, "throw
away the mother rather than the father," but among the
Karadjeri to the north "the usual practice is to throw
away the father." Piddington found, however, that in
some Karadjeri cases the mother was "thrown away."
According to Piddington, his informants had no explanation for this; "they merely accepted the fact that in
some cases a father might successfully insist that his
children should be placed in the other section of his
own patrimoiety."

Piddington notes also that a decision about section
affiliation may be left open until the time comes to
arrange the marriages of the parties concerned. In one
case he reports (1970: 333), two daughters of a couple
were eventually assigned to different sections, depending on the section affiliations of the men to whom they
were betrothed.[4] In this case an A man married a D
woman; their children would have been B through their

mother, or D through their father. Their section
affiliations remained ambiguous until they married.
One of the daughters was betrothed to an A man and then
regarded as B (through her mother); the other was
betrothed to a C man and then regarded as D (through
her father). In this way, one of the daughters ended
up in the same section as her mother, but both were
classified appropriately in relation to their husbands.
It is not improbable that arrangements like this are
more common than they would appear to be from the ethnographic literature and that, in general, the situation
when reckoning section or subsection affiliation is the
same as when reckoning kin-class status where more than
one possibility presents itself. There may well be a
rule-of-thumb that varies from one society to another,
but this rule is far from unexceptionable and the final
choice depends on evaluation of the perceived implications of one choice as opposed to another, and on
negotiation among the parties most immediately concerned. Piddington's report, that "in some cases a
father might successfully insist that his children
should be placed in the other section of his own patrimoiety," suggests that totemic organization is highly
relevant in this context. The normal arrangement is
for a man to belong to the section or subsection related
as 'father'-'man's child' to his father's section or
subsection, and this arrangement is ritually significant. Normally a man shares his father's cult-totem

and their sections or subsections together perform the
so-called increase rituals for that totem. This normal
arrangement may be preserved by assigning the offspring
of irregular marriages to the sections or subsections
paired as 'father'-'man's child' with their fathers'
sections or subsections (see also Meggitt 1962: 186;
Stanner 1936: 206; Maddock 1973: 91-4).

Of course, this normal arrangement would be preserved by a rule assigning all persons to the sections
or subsections paired as 'father'-'man's child' with
their fathers' sections or subsections, that is, by
consistently "throwing away the mother." But there
would be certain disadvantages in doing this. As Piddington (1970: 333-4) notes: "In traditional aboriginal
society it was the custom for males to marry at a more
advanced age than females. This means that husbands
would, in general, predecease their wives. This again
implies that women would generally experience more
serial marriages in the course of their reproductive
lives than men, their children going with them. Now if
one of these serial marriages were an irregular one,
endless complications would arise in regard to the terminology for half-siblings and other relatives acquired
through one or more regular marriages. Similar considerations would apply to section or subsection membership. If the father were thrown away, such complications would not arise so often." That is, women in
general have more husbands than men do wives, and many

of these marriages are irregular from the perspective
of section and subsection systems. Therefore, following the mother rather than the father in cases of irregular marriage will more often result in placing half-
siblings in the same category and, in general, in mini-
mizing the inconsistencies between the genealogical and
the section or subsection relations of individuals.

In short, section and subsection systems are based
on systems of kin classification and are integrated also
with the ritual-totemic system. Irregular marriages
result in dislocations among these several systems.
These dislocations may be minimized in the one case
(between kin classification and section or subsection
affiliation) by "throwing away the father," and in the
other (between the section or subsection system and the
ritual-totemic system) by "throwing away the mother."
In some societies, it seems, a choice has been made
between these two possible strategies, and that choice
has been made a rule-of-thumb for deciding section or
subsection affiliation in cases of irregular marriage.

Chapter 13

VARIATION IN SUBSECTION SYSTEMS

It seems probable that section and subsection systems
originated in, at most, one or two places and diffused
widely across the continent (Elkin 1970). They were
adopted by many peoples whose systems of kin classification already contained the superclasses that provided
the basis for the development of such systems. Presumably these peoples had little difficulty in understanding the systems and their articulation with systems of
kin classification. But these systems were adopted also
by many peoples whose systems of kin classification did
not contain the appropriate superclasses and who were
therefore not cognitively prepared to understand and use
these systems with ease. These peoples sometimes modified the section or subsection systems or their systems
of kin classification, or both, in efforts to make the
two systems more congruent with one another. As a consequence, in a number of Australian societies the relationships between the structures of the systems of kin
classification and the structures of the section or sub-

section systems are not the same as indicated in Chapter 12. Of course, these cases do not constitute evidence against the argument of Chapter 12. Indeed, the interpretation presented there permits us to see more clearly the nature of the problems that different peoples encountered (depending on the structures of their systems of kin classification) in making these adjustments, and to account for the alterations they made. To demonstrate this, we may consider the Murinbata, Gunwinggu, and Murngin, all of whom only recently adopted the subsection system.

MURINBATA

Prior to about 1930 the Murinbata had a simple Kariera-like system of kin classification, with neither the AGA nor the AGU rule. Any of a man's 'father's sisters' other than his own FZ and MBW could be assigned to him as a prospective WM. The children of 'father's sisters' who were potential WMs (all but FZ and MBW) were distinguished terminologically, as _purima_ and _nangqun_, from cross cousins (see Chapter 4). Not long before 1930 the Murinbata acquired the subsection system from the Djamindjung to the south (Stanner 1936). In 1935 the subsection system was still something of a "new fashion." Some men understood it well enough in the abstract and were able to recite the conventional matrifilial, patrifilial, and marital relations among the subsections, which they had learned by rote. But

the system was nowhere near fully incorporated into
Murinbata society. Indeed, some people did not understand the system at all and others were confused about
their subsection affiliations. People identified themselves and others by either of two subsection names
(related formally as MM and wDC), and for all practical
purposes the subsection system was treated as though it
were a section system.

It was recognized, by some people anyway, that the
formal relations among the kin classes and the formal
relations among the subsections corresponded in such a
way that a man of a specified subsection would find his
potential wives in two other subsections (in B1 and B2
for an A1 man). But the Murinbata refused to insist
that a man should surrender his rights to certain of his
potential wives just because these kinswomen happened
to belong to a subsection not formally paired with his
own by marriage. That is, it was recognized that an A1
man, for example, could be 'given' either a B1 or a B2
woman (but not just any B1 or B2 woman) as a wife.

In 1935 some adjustments had already been made in
the system of kin classification, such that it was possible to distinguish terminologically, in some cases,
between kinsmen in different subsections who had formerly belonged to the same terminological class. The most
notable adjustment was within the COUSIN class. Formerly the division was between MBC, FZC, and more distant
kintypes of the COUSIN class, but by 1935 the Murinbata

had adopted the Djamindjung term _pugali_ to designate MBC, FZC, and their parallel cousins (e.g., MFBSC, FFBDC, etc.), while retaining the terms _purima_ and _nanggun_ to designate MMBDC, FFZSC, etc. Thus an Al man would normally designate his COUSINS in B1 as _purima_ and _nanggun_ and his COUSINS in B2 as _pugali_. Despite this change, the marriage rule remained the same; a man's prospective WM could be any 'father's sister' other than his own FZ or MBW, so he could be given either a _purima_ or a _pugali_ to be his wife.

Corresponding to this terminological change, a further distinction was introduced between the daughters (e.g., MBDD, FZDD) of a man's female _pugali_ and the daughters of his _purima_ (MMBDDD, etc.). The former were designated by the special expression _wakal nginar_ 'man's daughter-potential WM' and the latter simply as _wakal_ 'man's daughter'. The husbands of the _wakal nginar_ came to be designated by the special expression _lambara_ 'sister's son-potential WF', while the husbands of the _wakal_ remained _mulok_ 'woman's (or sister's) son'. _Nginar_ was added also to _bipi_ 'father's sister' to distinguish the mother's (e.g., MMBD) of _purima_ from the mothers (e.g., FZ, FFBD) of _pugali_. The new term _ngaguluk_ 'potential WMB' was introduced to designate the brothers of _bipi nginar_ 'father's sister-potential WM'.

In 1935 there were no changes in the designations of cross grandparents and their siblings. But some people reasoned that those classificatory cross grand-parents who belonged to the same subsection as ego's purima, and who were of the appropriate age for ego to marry, were a 'little bit' like ego's purima. Thus, it was reasoned, a man should be able to marry a classifi-catory 'sister's son's daughter' tamoin, who was the daughter of his wakal nginar 'daughter-potential WM' and lambara 'sister's son-potential WF'. Stanner (1936: 203) suggests that acceptance of this marital possibil-ity facilitated acceptance of the subsection system itself and retention of "its conventional patterns in the Murinbata setting." That is, it may be that some men were motivated to accept the subsection system because, treated as a marriage-class system, it expanded the range of potential wives. However, as already noted, they were not prepared to let the structure of the subsection system interfere with and curtail already established marital possibilities.

Also in 1935 there was some disagreement about how subsection affiliation should be determined in instances of irregular marriage. Some people argued for "throwing away the father," others for "throwing away the mother," but the latter practice ("indirect patrilineal descent," as Stanner calls it) was gaining ground. This was because the Murinbata felt that subsections should not prevent a man's children from belonging to his moiety (Stanner 1936: 206).

In 1950 (see Falkenberg 1962: 205-37) the situation

was much the same. Although a few additional changes
had been made in the system of kin classification,
these were still only changes in subclassification and
neither the AGA nor the AGU rule had been introduced.
Men still had the right to claim either kind of COUSIN,
pugali or purima, as wives. So, as far as marriage was
concerned, the subsection system was still treated as
though it were a section system. Although all the
subsection names were in use, individuals were still
designated by either of two subsection names. The
practice in cases of irregular marriage was to follow
the father rather than the mother.

Falkenberg (1962: 224) summarizes the changes in
the system of kin classification by saying, "the tendency in the process of change has been to modify the
Kariera type structure in favour of an Aranda type,"
but he adds, "this goal has not been reached . . . ."
Perhaps the most notable additional change was the
introduction of a new term kawu to distinguish MM and
MMB and their reciprocals from FF and FFZ and their
reciprocals, who remained kanggul. Also, the expression
mangga was introduced to distinguish FM from MF, MFZ,
and FMB, who remained tamoin. That is, the parallel-
and cross-grandkin classes were subdivided more or less
along subsection lines by distinguishing between the
two sexes. Otherwise, the only notable change was that,
if a man had been promised a particular classificatory
'sister's sons's' daughter (e.g., a MBDDD) as a wife,

he designated her as <u>purima-nan</u> 'little <u>purima</u>' or 'little prospective wife'. The mother of such a woman was, in 1950, designated as <u>bipi</u> <u>nginar</u> 'father's sister-WM', rather than as <u>wakal</u> <u>nginar</u> 'daughter-WM' (as in 1935), and the father of such a woman was, in 1950, designated as <u>kaka</u> <u>kapi</u> 'mother's brother-WF' rather than as <u>lambara</u> 'sister's son-WF' (as in 1935). Otherwise, the daughters of a man's female <u>pugali</u> were designated simply as <u>wakal</u> 'man's child' and their husbands as 'woman's (or sister's) son".

In summary, the changes made by the Murinbata in their system of kin classification following the introduction of the subsection system were structurally quite minor. They were changes only in subclassification, and not changes in equivalence rules; those equivalence rules that typify Aranda-like systems, the AGA and AGU rules, remained alien to the Murinbata system.

GUNWINGGU

The Gunwinggu system of kin classification was not analyzed in the preceding chapters, but it would have been appropriate to discuss it in Chapter 5, along with the Nyulnyul and Mardudhunera systems. It has been described as a Kariera-type system (Warner 1933; Berndt and Berndt 1951) and as an Aranda-type system (Elkin, Berndt, and Berndt 1951; Berndt and Berndt 1970; R. Berndt 1971), depending on the ethnographer's construction of the associated "normal" or "preferred" type of interkin marriage. There is ample evidence that it is a relatively simple Kariera-like system; the designations of its principal classes are extended collaterally by the same rules of structural equivalence as are the corresponding terms of the systems discussed in Chapters 4 and 5. It differs from those systems only in how some of its specially designated subclasses are defined and distinguished from the residual subclasses of their respective superclasses. Because these differences are not relevant to the following discussion there is no need to specify them here. As already implied, there are no reasons to suppose that this system features either the AGA or the AGU rule. Appearances to the contrary in the published diagrams are attributable (again, see Chapter 5) to the possibility of intergeneration as well as intrageneration marriage.

The Gunwinggu adopted the subsection system sometime after 1912 from their neighbors to the south (Elkin, Berndt, and Berndt 1951: 260). It is not known what their system of kin classification was like prior to their adoption of the subsection system, but it is clear enough that it did not feature the AGA and AGU rules, and that the orthodox spouse was a 'cousin' (including MBD, FZD). It is not known in what form the subsection system was introduced to the Gunwinggu, but it is now somewhat different from the normal or standard system found in the Kimberleys and Central Australia (compare Figures 12.4 and 13.1). In the standard system, if a man belongs to, say A1, then so do his FF and SC; his MM, ZDC, etc., belong to A2. But in the Gunwinggu system if a man belongs to A1 his FF and SC as well as his MM and ZDC belong to A2. In the same way, in the standard system for an A1 man his FM, FMBSC, MMBDC, etc., belong to B1 and his MF, MBC, FZC, etc., belong to B2; but in the Gunwinggu system for an A1 man his cross cousins and second cross cousins are B1 and his FM and MF are in B2. That is, Gunwinggu sections are divided into subsections <u>by generation levels</u>. As a consequence, the usual two four-generation matrifilial cycles are complemented by two four-generation patrifilial cycles; although a man and his SS do not belong to the same subsection, a man and his SSSS do.

Clearly, as the Berndts point out, the Gunwinggu have adapted the subsection system to their system of kin classification and to their rule of marriage whereby a man may marry any 'cousin'. Unlike the Murinbata, they have avoided dividing COUSINS into two subclasses

Figure 13.1. Gunwinggu subsection system

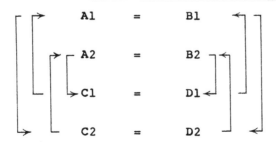

that would normally belong to different subsections.
This transformation of the subsection system probably
was not difficult to achieve, for it requires only that
the matrifilial relations among the subsections of <u>one</u>
of the two matrifilial cycles be reversed throughout
(again, compare Figures 12.4 and 13.1). In the standard
system the A-C matrifilial cycle is,

A1 → C1 → A2 → C2 → A1,

but in the Gunwinggu system it is,

A1 → C2 → A2 → C1 → A1.

That is, in the standard system A1 is 'mother' of C1
but in the Gunwinggu system A1 is the 'woman's child'
of C1, and so on throughout the A-C cycle. In the B-D
cycle both systems are the same.

Although the Gunwinggu subsection system and system of kin classification are, to a degree, correlated
with one another, the Gunwinggu subsection system is not
a logical development out of the Gunwinggu system of kin
classification, which did not and does not now contain
all of the requisite superclasses for the development
of a normal subsection system. Two of the sections of
the Gunwinggu subsection system correspond to the
FATHER-MAN'S CHILD and MOTHER-WOMAN'S CHILD superclasses
of the system of kin classification (e.g., D and C
respectively for a member of A), but the other two
sections (A and B) have no corresponding superclasses
in the system of kin classification (because the Gunwinggu PARALLEL-GRANDKIN class is <u>not</u> a subclass of the

SIBLING class, and the COUSIN class is not a subclass
of the CROSS-GRANDKIN class). Further, the kintype
compositions of the Gunwinggu FATHER-MAN'S CHILD and
MOTHER-WOMAN'S CHILD classes are not the same as the
kintype compositions of the corresponding classes in
Aranda-like systems (because they are determined in the
Gunwinggu system by the parallel-cross status-extension
rule, and in Aranda-like systems by the AGA and AGU
rules).

The Gunwinggu, like the Murinbata and many other
peoples, permit men to marry their classificatory
'sister's sons' daughters', as well as their 'cousins'.
Those 'cousins' ganjulg who are singled out as prospective wives are designated as gagali. Similarly, a
classificatory 'sister's son's daughter' maga who is
singled out as a prospective wife also is designated as
gagali. For a man of subsection A1, such a kinswoman
would normally belong to B2 so he may have gagali 'prospective wives' in both B1 and B2. To this extent
A1 = B2 is an "alternate" marital pair (which is not to
say that any A1 man may marry any B2 woman, for an A1
man's classificatory 'father's mothers' also are in B2
and he may not marry them). Extension of the designation 'prospective wife' to kinswomen of the second
descending generation does not rest on use of the AGA
or AGU rules, but instead on recognition of these women
as potential or prospective wives. If this is not
understood it is all too easy to misidentify the Gun-

winggu system of kin classification as an Aranda-like
system.

MURNGIN

In 1931 Warner described the subsection system of
several east Arnhem Land "tribes" as a conventional
Aranda-like system (see also Warner 1958: 116-23), but
he noted that a man is not restricted in his choice of
a wife to women of only one subsection; he may take a
wife from either of two other subsections. Thus an A1
may marry either a B1 or a B2 woman, and so may an A2
man. "This," he said, "reduces the subsections from
the point of view of marriage to the four sections of
the Kariera type," but "the descent rules follow the
usual Arunta method." That is, as in the normal subsection system the two matrifilial cycles are A1 → C1 →
A2 → C2 → A1, and B1 → D2 → B2 → D1 → B1. Warner diagrammed the system as a pair of concentric circles in
which A1-B1, A2-B2, and so on are the marital pairs.
These pairings seem arbitrary, because Warner, unlike
subsequent ethnographers among the so-called Murngin,
did not claim that the people themselves assert (somehow) that certain marital pairs are "normal" and others
"alternative." Indeed, his argument appears to be that
the Aranda system is based on the rule that A1 marries
only B1 while the Murngin system is based on the rule
that A1 marries either B1 or B2. But Warner did not
distinguish as clearly as he should have between his

informants' statements about formal relations among the
subsections themselves and their statements about how
persons of certain kin classes, who happen also to belong to certain subsections, might intermarry. If this
distinction is made, it is possible to suggest that
what Warner encountered was this. His informants (like
Stanner's Murinbata) knew the conventional arrangement -
thus his diagram of the system with A1 paired maritally
with B1, and so on - but they added that it was a matter
of indifference to them to which of two subsections a
man's rightful wives might belong; he still had a right
to them in either case. They knew that it is a feature
of the subsection system as a set of abstract categories
that, in principle, A1 marries B1, and so on, but for
them this was not a practical marriage rule. Presumably, if they said an A1 man may marry either a B1 or a
B2 woman, they did not intend to say that this was how
the subsections in the abstract are related to one another, but rather that, despite the formal A1-B1, etc.,
relations, an A1 man may marry a B2 woman if she happens
to be one of his potential wives by right of kin-class
status.

Shortly thereafter the same system was described
by T. Webb (1933), a local missionary. Webb reported
(or seemed to report) the same matrifilial cycles and
the same alternative marital pairs (A1 with B1 or B2,
etc.). However, he added that in each case one of the
pairs is the "normal" or "regular" one, whereas the

other is permissible "under certain circumstances,"
which he left unspecified. Elkin (1933a) then pointed
out that Webb's "normal" marital pairs (in Warner's
notation, A1-B2, A2-B1, but C1-D1, C2-D2) entail a
system that differs from the normal Aranda-like subsection system in that it has no patrifilial couples,
but instead has two four-generation patrifilial cycles,
as well as the usual two four-generation matrifilial
cycles. That is, the system entailed by Webb's "normal"
marital pairs, plus the two matrifilial cycles on which
he and Warner appear to agree, is a modified system of
the Gunwinggu type (see Figure 13.1). Elkin, arguing
on the assumption that elsewhere subsections are "part
of the machinery for preventing cross-cousin marriage"
(1933a: 415), suggested that this local modification
of the system is an adaptation to the practice of matrilateral cross-cousin marriage. In a normal subsection
system, he argued, a man's MBD is not in the subsection
formally paired with his own by marriage; an A1 man
could marry his MBD only if he were permitted to marry
a B2 woman (as well as a B1 woman). Or an A1 man's MBD
would be in B1 only if his MB in C2 were permitted to
marry a D1 woman (as well as a D2 woman) - which comes
to the same thing.

It is important to note, however, that Webb's
"normal" scheme would not result in every man's MBD
being in the subsection formally paired with his own by
marriage. Such a matching of MBDs with subsections

could be accomplished only through consistent alterna-
tion of Webb's "normal" and "alternative" marriages.
The feature of Webb's system that permits any man's
(but not every man's) MBD to be in the subsection
formally paired with his own by marriage is only the
possibility that A1 may marry B1 or B2, A2 may marry
B2 or B1, and so on. This feature is a part of Warner's
model also, and neither model has any structural advantage over the other in this respect. It may be worth
pointing out that, presumably, the "circumstance" under
which a man might marry a woman of the "alternative"
subsection is that she happens to be one of his galle
'potential wives' to whom no other man has a prior
rightful claim, or that if there is such a man he has
waived his claim to her.

Subsequently, the modified subsection system reported by Webb has been recorded also by Elkin (1950,
1953) and the Berndts (R. Berndt 1971) for other areas
in eastern Arnhem Land; and the "normal" system has
been recorded by Hiatt (1965) and Shapiro (1969) in the
western part of the so-called Murngin region. Apparently, some people in this area have adopted the subsection
system without modification but others have adapted it
slightly by reordering the formal relations among some
of the subsections. It is generally agreed, however,
that throughout the area the subsection system, whatever
its conventional or orthodox form is alleged to be, has
little or no practical effect on marriage. The critical

variable is the right to claim kinswomen classified as
galle (or by its cognate), and it is of no consequence
to which subsection such a kinswoman may belong (see
Warner 1958: 123; Shapiro 1969: 183). As far as the
practical business of arranging marriages is concerned,
the subsection system is not merely redundant, it is
largely irrelevant. By and large, however, insofar as
men marry kinswomen to whom they have rightful claims,
that is galle, they marry also in accord with the norms
of the subsection system.

Much subsequent discussion of the Murngin subsection system has rested on the assumption that the Murngin themselves have attempted to adapt the system to
their (alleged) rule of matrilateral cross-cousin marriage and to their system of kin classification which,
it is assumed, is predicated on this marriage rule. A
great deal of analytical ingenuity (for a summary of
which see Barnes 1967) has been exhibited in attempts
to develop a model of eight "marriage classes" linked
asymmetrically by the practice of MBD-FZS marriage. As
Barnes (1967) has shown, such a model can be constructed
but it is not, in any case, a model of the subsection
systems reported by Warner and Webb. It seems that all
attempts to find a systematic correspondence (no matter
how complex) between the interkin marriage rule, the
system of kin classification, and the subsection system
are doomed to failure, given the available ethnographic
data - which Barnes has shown are confused and confus-

ing. Barnes suggests that if the "Murngin problem" is to be resolved, further fieldwork is required. But perhaps not.

Barnes himself assumes there is "little ground for controversy on the marriage rule, as expressed in the terminological system" and "that the pattern of relations between kin among the Murngin is based on the assumption that orthodox marriage is between a man and one of his mother's brother's daughters" (1967: 5, 6). However, as we saw in Chapter 8, Warner's repeated assertions that the orthodox marriage is between MBD and FZS is seriously misleading and the system of kin classification is not predicated on such a rule. The principal marriage rule is that a man has a rightful claim to his MMBDD who is classified as _galle_, and the _galle_ class is a subclass of the CROSS-GRANDPARENT class. This subclass consists of ego's CROSS GRANDPARENTS who are not in G+2 but who are in G= or G-2, and it includes his MBS and MBD who are structurally equivalent to MF and MFZ under the AGA rule. The reciprocal of _galle_ is _due_, and the _due_ class is a subclass of the CROSS-GRANDCHILD class; it consists of ego's CROSS GRANDCHILDREN who are not in G-2 but who are in G= or G+2, and it includes his FZS and FZD who are structurally equivalent to mDC and wBDC under the AGA rule. Because a man's MBD is a member of the same subclass as his MMBDD, he may marry her instead of his MMBDD, but he has no right to his MBD as such.

The Murngin system of kin classification is not in any radical sense "asymmetrical." The appearance of asymmetry is attributable to the priority of the AGA rule (but not the AGU rule) over the parallel-cross status-extension rule, and to the general absence of self-reciprocity among the terms of the system. As a consequence, although the Murngin system of kin classification features the AGA and the AGU rules and the four major superclasses from which the normal subsection system was elsewhere derived, the Murngin four major superclasses do not have the same internal structure as the corresponding classes of, say, the Walbiri system. The kintype compositions of the FATHER-MAN'S CHILD superclasses are not the same as in the Walbiri system. The Murngin FATHER-MAN'S CHILD and MOTHER-WOMAN'S CHILD superclasses do not contain the special UTERINE SIBLING and AGNATIC SIBLING subclasses, and neither does the Murngin CROSS-GRANDKIN class, which includes the galle class. Because the four major superclasses of the Murngin system do not all subdivide in the same way, and therefore do not articulate with one another in the same way as the four major superclasses of Walbiri and other Aranda-like systems, these superclasses and their interrelations cannot be objectified as a normal subsection system. Inevitably, despite the pronounced but limited structural similarities between the Murngin and Walbiri systems of kin classification, the normal or standard subsection system must be "out of joint" with the Murngin system of kin classification.

The Murngin themselves did not develop the subsection system. They acquired it as an abstract set of
categories and intercategory relations from neighboring
peoples. The principal difficulty they must have encountered in understanding and using the normal system
is not that the normal system is designed (as some
anthropologists have speculated) to prevent cross-cousin
marriage whereas the Murngin "prescribe" such marriages
(they do not!). The real difficulty must have been that
the normal subsection system is predicated in part on a
division of the CROSS-GRANDKIN class that is substantially different from the way in which that class is
subdivided in the Murngin system. In Aranda-like systems of kin classification, on which the normal subsection system is based, MBD and MMBDD belong to different
subclasses of the CROSS-GRANDKIN class, but in the
Murngin system they belong to the same subclass. If the
normal subsection system were followed, a man of sub-
suction Al would have both galle (MBD, MMBDD) and due
(FZD, MFZDD) in subsections Bl and B2, and his own-
generation galle as well as his own-generation due would
be divided among the two subsections. Presumably, the
Murngin were not particularly disturbed by the fact that
a man's galle and due were in the same subsection(s),
so that both "prescribed" and "prohibited" kinswomen
were in the same category. But they may well have found
it anomalous that the normal subsection system divides

a man's own-generation <u>galle</u> between two subsections.

It would have required radical revision of the subsection system to create out of it a system of eight
classes with two four-generation matrifilial cycles and
in which MBC and FZC are in different classes. The
Murngin did not attempt such a revision of the subsection system, and it would be idle (in this context anyway) to speculate on the nature of the system they might
have created had they ever thought of trying. But it
was a relatively simple matter to revise the normal
subsection system in such a way that, in principle, all
of a man's own-generation <u>galle</u> would be in the same
subsection.

Beginning arbitrarily with some man and placing
him in some subsection - say A1 - his wife or rightful
wife (MMBDD) would then be placed in subsection B1,
following the normal subsection rules. It could then
be deduced that his MBD should be placed in subsection
B2. But this might be thought anomalous, because both
women are <u>galle</u> of ego's own generation, and he may
marry either of them. The anomaly may be resolved in
either of two ways. One way would be to give up the
idea that, in principle, men of one subsection marry
only women of one other subsection, and it might be
said, in effect, let an A1 man marry either a B1 or a
B2 woman, and so on, provided she is his <u>galle</u> and no
other man has a prior right to her. The rules of matrifiliation could be retained as received. Under these

rules, a person's subsection affiliation could be predicted, as it were, from knowledge of his mother's
subsection affiliation, but not from knowledge of his
father's, although in the normal system it could, again
in principle, be predicted from knowledge of either.
It may be that Warner's informants had worked out this
"compromise," or it may be that they were still living
with the anomaly.

The other possibility would be to revise certain
relations among some (but not all) of the subsections
in such a way that all of a man's own-generation galle
would, in principle, fall into the same subsection. As
in the Gunwinggu case, this could be done quite simply
by reversing the matrifilial relations among the subsections of one of the matrifilial cycles. One logical
consequence of this revision would be that, whereas
formerly Al was paired in marriage with Bl and A2 with
B2, now Al would be paired formally with B2 and A2 with
Bl. This is the arrangement reported by Webb (1933).

This assessment of the differences between the
systems reported by Warner and Webb may seem contra-
dicted by the appearance that they report the same two
matrifilial cycles. The contradiction is only apparent.
Figure 13.2 shows the conventional system reported by
Warner; Figure 13.3 shows the system constituted by
Webb's "normal" marital pairs. The only difference, in
these figures, is that in Webb's "normal" system Al
marries B2 rather than Bl, and A2 marries Bl rather

Figure 13.2. Murngin subsection system: Warner's version

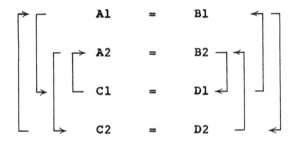

Figure 13.3. Murngin subsection system: Webb's version

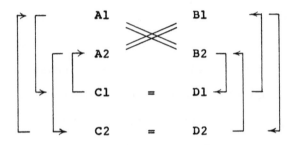

Figure 13.4. Murngin subsection system: resolution

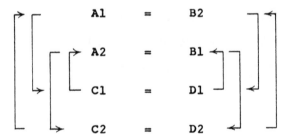

than B2. However, Webb's "normal" system may be re-
drawn, as in Figure 13.4, by placing B2 in the position
of B1 and vice versa. The difference then is that the
matrifilial relations among certain positions are reversed, although the relations among the names of the
positions are not. Thus Warner's and Webb's systems
only appear to have the same matrifilial cycles.

In short, by reversing the matrifilial relations
among the subsections of one matrifilial cycle, the
Murngin assure that, in principle if not in practice,
a man's own-generation galle will all be in one subsection. These are, after all, the kinswomen in respect
of whom he has marital rights, and it is perhaps only
expectable that the Murngin should not want the subsection system to divide this category. In insisting that
this jurally significant category should not in principle be divided by the subsection system, they are in
agreement with the Aranda and other peoples with normal
subsection systems - for in these other cases, too, the
kinswomen in respect of whom a man has marital rights
normally (or in principle if not in fact) belong to only
one subsection. Thus the rules of the subsection system
merely reiterate, as far as marriage is concerned, the
rules of interkin marriage.

Chapter 14

KINSHIP AND THE SOCIAL ORDER

Meyer Fortes (1969: 120), following Radcliffe-Brown,
has aptly characterized Australian society in general
and Walbiri society in particular as "an incontestable
limiting case of a social order wholly comprised within
a framework of kinship institutions." Meggitt (1972)
disagrees and argues that "an approach through kinship
analysis alone" - which he attributes to Radcliffe-
Brown - is inadequate to produce a coherent and intellectually satisfying account of Walbiri or any other
Australian society. As Meggitt sees it, by focusing
on kinship and attempting to generate "broader social
groupings" such as descent groups, moieties, sections,
and subsections "from the kinship system itself by
regarding them simply as aggregates of like kinship
relationships," we cannot bring together "materials on
totemism, kinship, and broader social groupings" except
"by means of various ad hoc functional explanations
whose logical or systematic connections with [the]
kinship analysis are at best strained or analogical."

A more satisfactory account, Meggitt argues, would show how these various institutions are articulated without artificially reducing one to another by relating them all to certain fundamental metaphysical postulates of Australian cultures.

The argument of this chapter is that the evidence presented in Meggitt's Desert People - in my estimation, the single most outstanding social anthropological account of an Australian society - may and must be interpreted as supporting Fortes' proposition. Such an interpretation is possible not only because of the quality of the ethnography but also because of Meggitt's exceptionally clear recognition (in 1962) of the ways in which the anthropological preoccupation with moieties, sections, and subsections has seriously distorted our understanding of Australian society. The following reanalysis of Meggitt's data on relationships between Walbiri kin classification and "broader social groupings" does not challenge his insistence that a satisfactory account of Australian society must include an account of Australian cosmological conceptions and their articulation with social categories. Indeed, this articulation is accomplished at least in part by the postulation of (metaphoric) kinship relations between men and Dreamtime beings.[1]

WALBIRI "MOIETIES"

According to Meggitt (1972) the "broader social

groupings" of Walbiri society include exogamous patrimoieties, exogamous matrimoieties, endogamous alternate-
generation moieties, and a set of eight subsections.
He states: "Those activities which are explicitly
concerned with the sacred or secret life that brings
man into conjunction with the dreamtime are organized
in the broadest sense in terms of patrimoiety affiliation. All dreamings or totems, like all Walbiri, belong
to one or another patrimoiety" (1972: 73). "Within each
patrimoiety," he further states (1966: 30; 1972: 69),
"is a number of patrilineages and the men of each patrilineage constitute a cult lodge, which is the custodian
of a dreaming or group of dreamings." Similarly, "each
matrimoiety contains about a score of small, unnamed,
genealogically unconnected and residentially dispersed
sets of people defined through matrilines of filiation."
The members of these sets "meet irregularly or contingently to deal with secular, public affairs such as
betrothals, marriages, mortuary duties, inheritance of
domestic equipment, and obligations of revenge" (1962:
69). Meggitt adds that such social events "are, through
their participants, organized in the widest sense in
terms of matrimoiety affiliation" (1972: 73). Finally,
he says that it is "an essential feature" of the subsection system that "it summarily marks a person's
simultaneous memberships of patrimoiety, matrimoiety,
and endogamous moiety of merged alternate generation
levels" (1972: 74).

All this may give the impression that the Walbiri
as a whole are divided into patrimoieties, matrimoieties, and alternate-generation moieties, that the patrimoieties are composed of numerous patrilineages, and
that the subsections are defined by the intersection of
the various kinds of moieties. Other anthropologists
have been quite explicit in describing other Australian
societies in this way, but as T. Strehlow (1965: 134-5)
has pointed out, referring in particular to the Aranda,
"the very terms used imply a system which - in the
absence of any recognizable tribal structure - could
not exist in actual fact, and which, if it could have
existed in theory, would not have worked in practice."

The cogency of Strehlow's observation is revealed
by the fact that the so-called moieties of Walbiri
society, like the Kariera and Aranda "moieties" noted
in Chapter 12, have no proper names. They are identified and distinguished only by reciprocal expressions,
some based on kinship terms. Like the corresponding
Kariera and Aranda categories, they cannot be anything
more than aggregates of kin classes, for no one can be
identified by either term of a pair except through
knowledge of how he or she is related, as a kinsman,
to the speaker or to some other person or set of persons
taken as the point of reference.

The expression _gira_, glossed by Meggitt as 'own
patrimoiety' (1972: 69), is also one of the expressions
for the father class and, of course, the FATHER class.

Meggitt (1962: 203) glosses it also as 'the fathers and
sons'. In ritual contexts it denotes the 'bosses',
masters, or custodians of the Dreaming whose ritual is
being performed, that is, the men of the "patrilodge"
or "patrilineage" that 'owns' that Dreaming. They may
invite their classificatory BROTHERS, including their
classificatory 'father's fathers' and 'mother's mother's
brothers' (AGNATIC and UTERINE SIBLINGS) from other
cult lodges, to participate in the ritual. All these
men are related to one another directly or indirectly
as one or another kind of FATHER-MAN'S CHILD, so it is
appropriate that they should refer to one another as
gira, 'the fathers and sons' (by neutralizing the opposition between 'father' and 'man's child' and permitting
the designation for the senior category to stand for the
oppositional category or the reciprocal set as a whole).

Gira, then, designates a set of SIBLINGS and their
FATHERS (including their AGNATIC and UTERINE SIBLINGS).
Again in ritual contexts, the opposed expression is
guru-ngulu or gurung-ulu, which Meggitt glosses as
'other patrimoiety' or 'those outside the ritual' (1966:
30). The men designated by this expression are the
'mother's fathers' and 'brothers-in-law' of the 'owners'
(Meggitt 1972: 69; 1966: 39), and their function is to
prepare and assist the 'owners' or gira. These 'workers' are, in the first instance, the close uterine kin
and in-laws of the 'owners' of the Dreaming whose ritual
is being performed. They, in turn, may be assisted by

their BROTHERS and FATHERS. Thus, the opposed set consists of the male MOTHERS and their FATHERS (including
their AGNATIC and UTERINE SIBLINGS) of the members of
the first set.

Coincidentally, the two opposed sets of 'owners'
and 'workers' include men of all eight subsections.
The 'owners' all belong to two subsections, say A1 and
D2 (see Figure 12.5), and their AGNATIC and UTERINE
SIBLINGS belong to two others, A2 and D1. Similarly,
the 'workers' all belong to the other four subsections.
It is possible, therefore, to describe the social organization of the ritual in terms of the subsection affiliations of the participants, but it would be quite misleading to describe the two sets of participants as
moieties or even as representatives of moieties. The
so-called moieties emerge contingently through the way
in which ritual activities are organized by kinship.
It is only as a consequence of the inclusion, by invi-
tation, of AGNATIC and UTERINE SIBLINGS of the 'owners'
of a Dreaming that men of subsections A2 and D1 participate as though they had a right to do so; and it is
only as a consequence of the inclusion, again by invi-
tation, of AGNATIC and UTERINE SIBLINGS of the 'workers'
that men of subsections C1 and B2 participate as though
they had a right to do so. The apparent existence of
the so-called moieties is, therefore, a function of the
way ritual is organized through kinship, and it is not
true that ritual is "organized in the broadest sense in

terms of patrimoiety affiliation" (Meggitt 1972: 73).

Neither is ritual activity organized (nor does
Meggitt claim that it is) through the subsection system.
The participants in a ritual may include men of all
eight subsections, divided into two complementary categories of four subsections each. But men do not participate, either as 'owners' or as 'helpers', by right of
subsection affiliation. Subsections are not jurally
significant units. Meggitt states: "Basically, [the
Walbiri] regard subsections as a summary expression of
social relationships that may sometimes be practically
useful but just as often is dangerously ambiguous.
When important questions arise concerning, for instance,
the disposal of a woman in marriage, the selection of a
circumciser, the avenging of a death or the organization
of a revalatory ceremony, specific genealogical connections and local community affiliations constitute the
approved frame of reference within which decisions are
made. Subsection membership as such is irrelevant, for
it does not make the fine discriminations among people
that are considered significant in such situations"
(1962: 169).

The so-called matrimoieties are likewise nothing
more than aggregates of kin classes. The expressions
wiarba and gundawangu, glossed by Meggitt as 'own matrimoiety', refer to a set of SIBLINGS and their MOTHERS
(including their AGNATIC and UTERINE SIBLINGS). The
opposed expressions guluwangu, gundwangga, magundawanu,

glossed by Meggitt as 'other matrimoiety', refer to the FATHERS and their MOTHERS (and their AGNATIC and UTERINE SIBLINGS) of the members of the first set. Again, because there is in principle a simple correlation between relative subsection and kin-class affiliations, these expressions may be used also to refer to the relations between sets of subsections. For a man in A1, the set to which he belongs is A1, A2, C1, and C2, and the other set is B1, B2, D1, and D2.

Meggitt (1972: 69) says, "each matrimoiety contains about a score of small, unnamed, genealogically unconnected and residentially dispersed sets of people defined through matrilines of filiation." The members of these sets "meet irregularly or contingently to deal with secular, public affairs such as betrothals, marriages, mortuary duties, inheritance of domestic equipment, and obligations of revenge" (1972: 69). Meggitt adds, such social events "are, through their participants, organized in the widest sense in terms of matrimoiety affiliation" (1972: 73). This phrasing lends a spurious grouplike quality to these sets, which is not dispelled by changing their designation from "matrilines of descent" to "matrilines of filiation" (Meggitt 1972: 84, note 7). The affairs in question focus on particular individuals, so the matrilines are not fixed sets but only the close uterine kin of those individuals, in particular their MBs, MZC, MMBs, and MMZDC (Meggitt 1962: 194). For example, in deciding on a betrothal

the principal actors are the MBs and MMBs of the prospective spouses (1962: 195); other so-called members
of the so-called matrimoieties have no say in the
matter. Meggitt (1962: 201) says, "whereas the casual
remarks of the people might suggest that matrimoiety
affiliation alone orients behavior in betrothal and
death, in fact the structural units primarily concerned
are the matrilines," that is, the close uterine kin of
the prospective spouses or the deceased. Meggitt does
not describe the "casual remarks" that might be misconstrued by outsiders as suggesting that in these contexts
individuals act as members of matrimoieties. We may
suppose, however, that in speaking about certain activities the Walbiri sometimes say that certain participants
are gundawangu or guluwangu in relation to the focal
parties or to one another. They say also that, for a
given individual, one set of four matrifilially related
subsections (the set to which that individual's subsection belongs) is gundawangu, and the other set is
guluwangu. If one accepts the conventional wisdom that
sections are divisions of moieties, and subsections are
divisions of sections, it is easy to misconstrue these
expressions as the designations of moieties, and on that
basis to misconstrue references to sets of uterine kin
as references to sets of persons acting as representatives of matrimoieties.

Finally, the so-called alternate-generation moieties consist of a given person's SIBLINGS (including

AGNATIC and UTERINE SIBLINGS) and their CROSS GRANDKIN
(so-called own moiety) in opposition to their FATHERS
and MOTHERS (and their AGNATIC and UTERINE SIBLINGS (so-
called other moiety). For a person in subsection A1,
his or her own set is A1, A2, B1, B2, and the other set
is C1, C2, D1, D2. Meggitt (1962: 188-90) reports that
these "moieties" are of no jural significance. Clearly,
the expressions serve only to indicate that the designated parties are of the same generation or alternate
generations with respect to one another, or that they
are of adjacent generations with respect to one another.

We have seen that certain "broader social categories" of Walbiri society - the so-called moieties and
matrilines - are structurally derived from the system
of kin classification, and that social relations between
persons as members of such categories are nothing more
or less than their social relations as members of kin
classes. We may now consider the so-called patrilines.

WALBIRI "PATRILINES"

Scattered throughout Walbiri country there are many
"spirit centers . . . in respect of which certain categories of people have particular rights and obligations"
(Meggitt 1972: 67). In his 1972 paper, Meggitt gives
the impression that these categories are the "patrimoieties," which are composed of numerous "patrilines,"
"patrilineages," or "patrilodges" (the initiated men of
a patriline). Dreamings (totems and their associated

rituals and ritual paraphernalia), he says, are 'owned'
by patrimoieties but the right and duty to execute the
rituals for a particular Dreaming are allocated to a
particular patriline "within" the 'owning' patrimoiety
(1972: 69, 73, 77; also 1966: 30-1). But in his 1962
account (pp. 68, 169, 186, 201-27) we are given to
understand that each Dreaming is 'owned' by a particular
patriline or patrilodge. The men of that lodge, and
only that lodge, have the right and duty to perform
certain rituals. They are assisted in this by men of
another lodge, usually one whose senior men are brothers-in-law of the senior men of the 'owning' lodge.
These assistants are 'workers'; they prepare the ritual
ground and paraphernalia and decorate the 'owner'-
performers. Men who belong to the same "patrimoiety"
as members of the 'owning' lodge - that is to say,
their classificatory 'father's fathers', etc., and their
classificatory 'mothers' mother's brothers', etc. - may
participate as though they, too, were 'owners'. But
they are not 'owners' in the strict sense. They are
not initiated into the lodge itself; they do not have
the right to perform its rituals; and, although they
may participate in these rituals, they do so only by
initiation (Peterson 1969: 30; Meggitt 1962: 224).
Similarly, men who belong to the same "patrimoiety" as
the 'worker' lodge - that is, their classificatory
'father's fathers', etc., and their classificatory
'mother's mother's brothers', etc. - may assist the

'workers' in carrying out their duties.

Meggitt's 1962 account is strongly confirmed by Peterson (1969) and Munn (1964, 1973: 21-7). Further, it should be clear by now that the so-called moieties exist, to the extent that they may be said to exist at all, only in ritual contexts. The appearance that they exist even there is attributable to the extension of the expression gira 'fathers and sons' (which designates the 'owners', and which Meggitt glosses as 'custodians' or 'bosses') to the participating classificatory 'father's fathers' and 'mother's mother's brothers', etc., of the 'owners'. Similarly the expression gurungulu (which designates the 'workers') is extended to their classificatory 'father's fathers', and so on. The Walbiri as such are not divided into patrimoieties; and no Walbiri is in any sense a "member" of a patri-moiety. It is merely that the participants in any ritual are divided into two categories: 'owners' and their classificatory 'father's fathers', etc., and 'workers' and their classificatory 'father's fathers', etc. The 'workers' are uterine kin and in-laws of the 'owners' (but the in-law classes here are subclasses of the uterine kin classes).

What then of the patrilines or patrilodges them-selves? Are they descent groups, or are they better understood as aggregates of men standing in particular kin-class relationships to one another? Fortes (1969: 113-31) is reluctant to describe the patriline as a

descent group but concedes that it has a "minimal corporate structure," is "visualized both by its members
and others as unilineally continuous," and "its members
are bound to one another as if they were a single personality in the ritual sense . . . ." Yet, he notes,
the patriline "is in fact recruited by exclusive patrifiliation in every generation, not by reference to a
patrilineal pedigree going back three or more generations - though never of course by other than kinship
criteria." Further, "if the notion of agnatic connections stretching back at least as far as the father's
father is implicit in patriline structure, it seems to
have only secondary relevance for the constitution and
activities of the group compared with the immediate
connection of father and son."

An important point not noted by Fortes is that the
patrilines also have no proper names. Such a group is
identifiable only by reference to the Dreaming it 'owns'
and which its members designate as 'father' or 'father's
father'; and the sole corporate function of the group
is to execute the rituals associated with that Dreaming.
Further, although Meggitt reports the expression banba,
used by men to designate their own "lodge-ceremony line"
(1962: 205; but cf. Munn 1973: 30, 183-210), it seems
that there is no generic Walbiri expression that may be
glossed as 'patriline' - unless it is, again, gira
'fathers and sons'. Meggitt (1962: 211-5) argues at
some length that the Walbiri themselves suppose that the

patrilines or patrilodges are perpetual corporations, but he goes on to note that the continuity of the lodge is dependent on the continuity of the Dreaming and that this is recognized by the Walbiri. They cannot imagine that a Dreaming could cease to exist, and they say that "at death [a] man's portion of the lodge's spiritual essence (<u>bilirba</u>) réturns to the dreaming site to merge again with the eternally existing pool of dreaming power and to await reincarnation in a future member of the lodge or in members of that totemic species" (1972: 76). Therefore they cannot imagine that a given patriline or patrilodge could cease to exist.[2] There is, however, no evident necessity to conceptualize these Walbiri ideas in terms of descent and descent groups.

The Walbiri assert that in the Dreamtime various beings emerged from, traveled about on, and reentered the earth at known points, which are now sacred sites.[3] The world as the Walbiri know it was given shape by their activities as they traveled about, in part by the performance for the first time of the rituals that Walbiri enact today. Enactment of these rituals by the Walbiri is said to be essential for the preservation of the world order established by the Dreamtime beings. It is possible for men to participate in this process of continuous creativity because they share in the power of the Dreamtime beings, as their offspring in a special sense. These beings deposited <u>guruwari</u>, particles of their powers of fertility, along their tracks or where they emerged from or reentered the earth. A pregnant woman passing a site where <u>guruwari</u> are lodged may be entered by them, so that her child when it is born also contains the <u>guruwari</u> of that Dreamtime being. Thus each Walbiri instantiates a particular Dreamtime being, and the being that an individual instantiates may or may not be the one instantiated by his or her human genitor or father. Neither men nor women procreatively transmit their <u>guruwari</u> to their offspring.[4]

The right and duty to conduct the rituals first performed by a particular Dreamtime being, which rituals now stimulate the activities of their <u>guruwari</u>, is allocated to a particular set of men. These men are the agnatic descendants of a man (the first man?)[5] who himself instantiated that being. They themselves may or may not instantiate that being by birth but they come to do so through the ritual process of initiation. A man who has himself been fully initiated into the secrets of that Dreamtime being has the right to have his sons initiated also, because they are of "one blood" (Munn 1973: 191-2). The initiation ritual includes circumcision, ritual killing, and rebirth, "as dreamtime power in the form of the lodge totem becomes part of [the initiate's] psyche or spirit" (Meggitt 1972: 76). This incorporation of the <u>guruwari</u> of the Dreamtime being in the initiate is accomplished through physical contact with certain ritual objects also called <u>guruwari</u> (Meggitt 1962: 207, 225; Munn 1973: 30). The spirit transferred in this way, and therefore shared by all men initiated into a totemic cult who are therefore "spiritual 'brothers'," is termed bilirba (Meggitt 1962: 207). It is said to be located in a man's stomach. At death it returns to the Dreaming site "to merge again with the eternally existing pool of dreaming power and to await reincarnation" (or, better, reinstantiation) in another human being or life form also associated with that Dreamtime being. Obviously, bilirba and guruwari are related: According to Meggitt (1962: 207), the bilirba of a particular Dreamtime being is "the concentrated expression" of all its guruwari existing in all its sites in the country of a community.[6]

Meggitt (1962: 207) says, "patrilineal descent is the main determinant of lodge membership," but it is doubtful that, in this context, it is necessary to speak of "patrilineality." This suggests that a man's entitlement to instantiate a Dreamtime being through the ritual process is dependent on something more than his presumptive paternity, his 'blood' relationship with his genitor. Yet Meggitt's, Munn's, and Peterson's accounts indicate that it is not. A man's entitlement to be initiated in this way, principally by his father with his mother's brother as the principal assistant (Meggitt 1962: 206), is predicated solely on the presumption that he is his father's son - and being initiated in this way makes him a (metaphoric) 'son' of the Dreamtime being who is then his (metaphoric) 'father', as well as his 'father's father'.

Of course, because each man's entitlement is dependent on his presumptive paternity, any man may say that not only he and his father, but also his father's father, have the same bilirba. Further, because no man may be ritually given that bilirba whose father did not also acquire it in the same way, it could be said that any man's acquisition of it is contingent on his father's father's, and his father's father's father's, and so on, ritual acquisition of it. And, finally, any man who has a particular bilirba must be related through his father, and his father's father, to all other men and their fathers and father's fathers, who also have that spirit. Yet all of this is a simple logical entailment of the genealogical basis of entitlement to ritual acquisition of the bilirba, which is patrifiliation, or being an acknowledged son of a man who holds the identity in question.

It is not hair splitting to say that the set of men who have the same patrispirit (bilirba) do not constitute a patrilineage or patrilineal descent group. The difficulty of deciding when a genealogically constituted group is or is not a descent group, and if not how then to describe it, is an old anthropological problem. It has become more acute in recent years as we have encountered cultural categories and social forms that are not readily describable in terms of analytic-descriptive concepts formulated to deal with other cases. But there need be no difficulty in making this distinction. We must distinguish (as per Sahlins 1965) between group constitution and group composition and allow for the possibility that groups (in different societies) may be similarly composed yet constitutionally quite different. Also, we must pay closer attention than we have in the past to the relevant cultural categories and their definitions. When we do, we find that it is essential to distinguish between genealogically defined categories that are egocentric (kinship categories) and those that are ancestor-oriented (descent categories), and among the latter between those that are patrilineally or matrilineally defined and those that are not (see also Schneider 1967; Scheffler 1973: 756-64, 774-80).

From this perspective Walbiri "patrilines" and "patrilodges" are not descent groups or patrilineages because they are not defined with reference to ancestors (much less specified human ancestors). Such a group is defined with reference to a specified and still-existent Dreamtime being and includes only those men who have been ritually reborn as his 'sons'. These men must perforce be related <u>inter</u> <u>se</u> as father and son, or as father's father and son's son, or in other ways, all of which are encompassed within the kin classes 'father's father', 'son's son', 'father', 'son', and 'brother'. The 'father's father' and 'son's son' classes, however, are included in the SIBLING class, so that at the super-

class level the men of these sets are only FATHERS,
SONS, and BROTHERS of one another. It is therefore
wholly appropriate that in ritual contexts these sets
are designated as gira 'fathers and sons', for the
father-son or patrifilial (not patrilineal) relationship is their singular genealogical structural basis.
Constitutionally, then, they are not descent groups but
are more like kindreds; the ego or propositus is the
still-existent Dreamtime being. Unlike most other
cultural categories described in the ethnographic
literature as "kindreds," these Walbiri categories are
mutually exclusive; by definition, no individual can be
a member of more than one. Because of this difference,
some anthropologists may want to avoid describing them
as kindreds or even as restricted kindreds. If a
general designation is necessary, "patrifilial kin
groups" seems adequate.

CONCLUSIONS

The preceding discussion of the so-called broader
social groupings of Walbiri society has demonstrated
that the categories in question are, indeed, "aggregates of like kinship relations," and they are based on
and structurally derived from the system of kin classification. The so-called moieties of Walbiri society
are nothing more than fairly complex kin classes. The
expressions glossed by Meggitt as 'own patrimoiety' and
'other patrimoiety' designate sets of men related among

themselves as FATHER-MAN'S CHILD and to members of the
other set as MOTHER-WOMAN'S CHILD. The expressions
glossed as 'own matrimoiety' and 'other matrimoiety'
designate sets of men related among themselves as
MOTHER-WOMAN'S CHILD and to members of the other set as
FATHER-MAN'S CHILD. The expressions glossed as 'own
alternate-generation moiety' and 'other alternate-
generation moiety' designate sets of men related among
themselves as SIBLING or CROSS GRANDKIN and to members
of the other set as FATHER-MAN'S CHILD or MOTHER-WOMAN'S
CHILD. Further, Walbiri "patrilines" and "patrilodges"
are not descent groups or patrilineages, but are small
groups of men who are the ritual 'sons' of a specified
Dreamtime being.

For a variety of reasons, Western anthropologists
have found it all too easy to understand (that is, to
misunderstand) these egocentric categories as lineally
defined or descent-ordered categories. This elementary
misunderstanding has given rise to numerous additional
false hypotheses and tangled theoretical constructs.
Although worse examples could easily be adduced, it was
precisely the central error in Radcliffe-Brown's thinking about Australian society that he failed to realize
that even the "patrilineal clan" is a kind of kin class.
This prevented him from realizing the full implications
of his brilliant "discovery" that Australian moieties
(so-called), sections, and subsections are structurally
derived from Australian systems of kin classification,

rather than, as it was then and is even now commonly
supposed, the other way around. Once it is realized,
however, that Radcliffe-Brown's "descent lines," far
from being the categories on which Australian systems
of kin classification are based, are themselves structurally derivative and dependent on rules of kin-class
definition and expansion, it must be acknowledged that
he was right to insist that in Australia social structure is equivalent to kinship.

There is no inconsistency between this proposition
and Meggitt's (1972: 66) insistence that a satisfactory
account of Australian society must include an account
of Australian cosmological conceptions and their articulation with Australian social categories. As we have
seen, Australian social categories are, virtually
exhaustively, kin categories. Their articulation with
cosmological conceptions is accomplished, at least in
part, through relations of "kinship" that Australian
cultures postulate between men and the beings who
populate the Dreamtime. The "kinship" postulated
between human and Dreamtime beings is, of course,
metaphoric. It is not at all the same thing as, or
even an attenuated form of, the kinship posited among
human beings. We should be clear about why it must be
described as metaphoric and about why it is important
to recognize that it is metaphoric.

Contrary to the fantasies of some Western observers, Australian concepts of kinship are rooted in con-

cepts of bisexual reproduction. As understood in Australian cultures, fertile sexual intercourse is necessary and sufficient to produce an animate human being.
The man and woman who produce such a being are known as
his or her 'father' and 'mother', and he or she is
known as their 'child'. In some areas these concepts
are supplemented by others that account not for how
human beings reproduce but for how each individual
acquires certain aspects of his or her social personality - through identification with a particular Dreamtime being, which relates him or her in a specific
socially significant way to the world of the Dreamtime.
The ways in which individuals are alleged to acquire
such identities vary from culture to culture, but
probably nowhere is it asserted that some Dreamtime
being (rather than some man) is the genitor of each
human being. What is asserted is that each human being
instantiates a Dreamtime being by incorporating a part
of its "essence." In some societies this incorporation
is said to occur intrauterine (not at the time of conception but at the time of the quickening), and in
others it is said to occur at the time of birth. In
either case it is accomplished without human interven-
tion, except for the activities required to recognize
its occurrence and the identity of the relevant Dreamtime being (which activities make it ultimately a
social process, even if not recognized as such by
Australians). Alternatively, incorporation of the

"essence" of a Dreamtime being may be accomplished
ritually through "initiation ceremonies" much later in
life. Where this incorporation is accomplished intrauterine, it is of course accomplished within the context of the human reproductive process. This has made
it easy for non-Australian observers to mistake this
theory for a theory of how humans reproduce, but it is
instead just a special case of a more widespread theory
about how humans come to instantiate Dreamtime beings,
and it is special only with regard to the putative time
of occurrence of incorporation of the "essence" of such
a being.

Drawing on their theories of how human beings
reproduce, some Australian cultures postulate an analogous process as part of their theories of the relationships between human and Dreamtime beings. In this
process the Dreamtime being whose "essence" has become
incorporated in an individual is likened to the individual's 'father' in the special sense that he is, as
it were, the individual's "spiritual genitor" - not the
genitor of his or her life and body, but the genitor of
certain other aspects of his or her being. Conversely,
the individual is said to be the 'child' of the Dreamtime being, in the special sense that he is, as it
were, the Dreamtime being's "spiritual offspring."
Thus, when used to express a relationship between a
Dreamtime being and one of his human representatives,
the expressions 'father' and 'child' are used in special

metaphorical senses. They refer not to genealogical relationships established through the human reproductive process but to similar relationships established through a similar process.

This metaphoric 'father-child' relationship confers on the individual certain rights and duties with respect to the Dreamtime being and the events (e.g., rituals), objects, and places associated with it. It confers also certain rights and duties with respect to other men and women, not only those who also instantiate that being and who are the individual's metaphoric 'brothers' and 'sisters' (as well as his genuine kin of these and other kinds). Kinship relations are posited among the Dreamtime beings themselves, and these are extended to their human 'children' who thereby stand in various other metaphoric kin-class statuses vis-à-vis other Dreamtime beings and one another. These metaphoric kin-class statuses in turn confer a variety of specific rights and duties largely confined to the sphere of ritual activity.

This is not to say that the ritual sphere is ordered wholly and solely by metaphoric kin-class statuses, and it should be emphasized that in Australia ritual and politics go hand in hand (cf. T. Strehlow 1947, 1971b; Pink 1936; R. Berndt 1965). In some Australian societies (Walbiri, for example) access to a socially significant status as an instantiator of a particular Dreamtime being is regulated by kinship proper. "Conception totems" (that is, totemic identities acquired intrauterine), although recognized, are discounted for purposes of access to valued knowledge. Only the sons of men who have been ritually 'reborn' as the 'sons' of Dreamtime beings are entitled to be so 'reborn' themselves, which further entitles them to access to the secrets of that being. This process of 'rebirth' does not sever but only reinforces the genealogical and social relationships between men and their fathers and other agnatic kin. Adding metaphoric kin-class statuses to the genuine kin-class statuses (often of the same kind) that already exist between the initiatee and established members of the cult group, it provides the latter with additional modes (both mystical and concrete) of control of the conduct of the initiatee. In other societies (Aranda, for example), access to sacred-secret knowledge - which is powerful in its own right - and, thereby, access to authority and power of more general social relevance, is at once both more and less restricted. Men may be initiated into the cults of their "conception totems" and into the cults of their own fathers' ritual 'fathers'. By getting themselves initiated into several cult groups, and by participating as much as possible in the ritual activities of others, men acquire various degrees of knowledge and thereby various degrees of informal power and influence. However, to be a 'proper man' (a rightful member) or an 'owner' of a body of knowledge,

objects, and places identified with a particular Dreamtime being, one must instantiate that being both through
"conception" and through initiation as the son of a man
who himself instantiated that being in both ways, and
so on. The status of genuine 'owner' is thereby available to only a few men. Because it confers the right
to control the initiation of other men, it confers
considerable power over many aspects of their lives;
and because it confers the right to determine when
rituals are performed, who participates in them, and
how, it confers considerable power over many related
aspects of the social life of the community as a whole.

There are, of course, other spheres of social life
in which human-Dreamtime being relationships do not
figure prominently, if at all - the "secular" spheres
(Meggitt 1972; Peterson 1969) of domestic life, arranging betrothals and marriages, meeting mortuary obligations, taking revenge, and so on. In these spheres,
too, kinship is the pervasive organizing principle. In
all these activities individuals participate and are
allocated various rights and duties vis-à-vis one
another as particular types or kinds of kin.

In a word, kin-class statuses are the elementary
structures of Australian social life. Originating in
nuclear families and in the genealogical relationships
among them, these statuses are extended to encompass
and to order virtually all social relations within
human communities of varying size. Beyond this they

are extended metaphorically to encompass and to order
social relations between human communities and the
community of the Dreamtime beings. In this way, Australian cultures establish a moral community that
embraces the cosmos.

In all this there is, of course, much that is
distinctive of and virtually unique to Australia, but
there is much also that is not unique. In postulating
a morally ordered universe and in attempting to comprehend it metaphorically, by analogy with the forms of
their own social life, Australians demonstrate not how
"primitive" they are but their intellectual kinship
with the rest of mankind. Mistaking a metaphor for a
theory of how humans reproduce, many Western observers
have failed not only to see this but have further misled themselves into attributing to Australians a mental-
ity they do not possess, concepts they do not know, and
social forms that would be ludicrously unworkable. The
misconception is so deeply rooted in anthropological
culture about Australian culture that its corollary
assumptions now stand virtually on their own. Dissent
from the opinion that "kinship" is a misnomer for any
aspect of Australian culture is likely now, no less than
in the past, to encounter charges of ethnocentrism, of
lack of imagination, and of imposing alien forms on
social and cultural categories that it is our first
responsibility as anthropologists to comprehend, as
the saying goes, "in their own terms." As it happens,

however, some of the concepts by which Australians order their social lives - indeed, some of the most fundamental concepts - are not at all alien to us or to the rest of mankind. Clothed in their own linguistic forms, elaborated upon in somewhat unusual ways, and incorporated metaphorically into Australian cosmology, concepts of kinship are as basic as they are pervasive in the organization of Australian social life.

NOTES

1:   Preliminary considerations

1. See Radcliffe-Brown (1930-31, 1951), Elkin (1964 [1938]), Fortes (1969: Chapter 7).

2. See Spencer and Gillen (1889, 1927), Fison and Howitt (1880), Lévi-Strauss (1969 [1949]), Maybury-Lewis (1967), and perhaps Maddock (1973).

3. For a highly sophisticated statement of this interpretation see Meggitt (1972).

4. For a brief but eloquent discussion of this Australian concept see Stanner (1956).

5. For an early attempt to justify expansion of the concept "kinship" to include such cases see Malinowski (1913: Chapter 6), but see also Scheffler (1973: 748-56).

6. Stanner (1965b) has exposed a similar bias in the Western treatment of Australian religion.

7. In this connection see also Evans-Pritchard (1933).

8. This suggestion is consistent with Meggitt's (1962: 272-3) observation that Walbiri men tend to volunteer information about the spiritual aspects of reproduction whereas women tend to volunteer information about the physical-sexual aspects - so that anthropologists and other outsiders, who have tended to rely on men as sources of information, are more likely to be told about the spiritual than about the physical-sexual aspects of the local theory.

9. For another case in point compare Stanner (1933)
and Falkenberg (1962) on the Murinbata and other
Daly River tribes. Stanner says: "It is clear
that two theories of sex exist side by side:
(a) a mystical theory of the type commonly found
in Australian cultures and (b) a barely understood,
confused version of the orthodox theory learnt
from the whites." Stanner neither describes this
latter theory nor produces evidence to show that
it was "learnt from the whites." It could be then
that this is just another physical-sexual folk
theory of reproduction similar enough to the European folk theory to be mistaken for it. Falkenberg shows that women know they are pregnant long
before they know anything about having been entered
by a spirit-child, but it is regarded as "improper"
for a woman's knowledge that she is pregnant to
become public knowledge before the identity of the
spirit of her child has been established and the
group has been informed of it by the old men. In
this way the Murinbata try to prevent a man other
than a woman's husband or one of his clan mates
from successfully claiming that he "found" the
spirit of her unborn child. It is clear enough
that nowadays the Murinbata theory is similar to
the Aranda theory, and it is not likely that this
similarity is a modern development. In the 1930s
the "mystical theory" was about what happened at
the time of the quickening and was therefore complementary rather than opposed to the physical-
sexual theory.

Elkin has written that the Karadjeri also are
ignorant of physical paternity, but in an unpublished field report (1928a) he notes that a Karadjeri woman, rather than her husband, sometimes
dreams that a spirit-child has come looking for
its mother and, when she answers its call, it
enters her body. She then tells her husband, who
can say that he "found" the child. One informant
is reported as saying that sometimes a man finds
the spirit-child but does not tell his wife about
his dream until she tells him that she has conceived (is pregnant). Plainly, the Karadjeri
theory about the role of spirit-children in the
process of reproduction is not a theory of conception per se.

10. For a description of these systems see Elkin
(1964: Chapter 4), also Chapter 12 herein.

11. Curr (1886) had already noted a number of apparent
exceptions to this proposition. Howitt (1891: 43)
also had noted instances of Australian societies
without "class systems" but with "systems of rela-

tionship" like those found among groups with "class systems," and he attributed this to the loss of "class systems" by the former groups. The historical evidence indicates, however, that "class systems" were spreading widely across the continent at the time of European contact, and their diffusion has continued in recent years (cf. Elkin 1970: 707-12). It is highly improbable that the "losses" posited by Howitt ever occurred.

12. Polysemy is the condition wherein a linguistic expression has two or more distinct significata (senses) that are related, by the sharing of some of their components, in such a way as to suggest derivation one from the other. The sense from which the others are derived may be described as the primary or structurally prior sense of the expression. See also Scheffler and Lounsbury (1971: 6-12; 59-66) and Waldron (1967).

13. Like narrowing or specialization, polysemy by generalization (also called widening, expansion, or extension) also is common in natural languages in domains other than kinship. Numerous examples are given in all the standard works on lexical semantics, for example, Stern (1965), Waldron (1967), Nida (1975), and Lyons (1977).

14. See also Maybury-Lewis's (1967: 489) interpretation of Kariera kinship terms.

15. This paucity of data is more probably attributable to the failure of anthropologists and others to appreciate the significance of certain linguistic facts than it is to the nonexistence of such facts.

16. In the early debates about the meanings of Australian (and other) "terms of relationship," it was sometimes argued that, if it is said that a term like Dieri apiri "means father," it is nonsense to say also that father's brother is "regarded as a father," or "as a kind of father," or "as a special kind of father" (cf. Thomas 1966 [1906]). The alleged nonsense is that such statements imply concepts of "group paternity." The alleged nonsense is allegedly even more obvious if such statements concern words alleged to mean "mother" and used to denote mother's sister, etc., as well as one's mother. But such statements imply no such thing as "group paternity" or "group maternity," if it is assumed that the words in question are polysemous, that is, that each designates at least two categories, only one of which (the structurally primary one) requires for membership that the deno-

tatum should be one's genitor or genetrix, as the case may be.

2: Types and varieties

1. The fullest expositions of Radcliffe-Brown's theory are in his 1930-31 and 1951 papers (the former subsequently issued as a monograph, now out of print). A. P. Elkin (1964, Chapter 3) offers a slightly different interpretation and is still, I think, the best introduction to the subject.

2. For recent discussions of Radcliffe-Brown's views see Hiatt (1962, 1966, 1968), Stanner (1965a), Birdsell (1970), and Tindale (1974).

3. For some of the relevant ethnographic data see Radcliffe-Brown (1913, 1918, 1923), Meggitt (1962), Hiatt (1965), Shapiro (1970), and Thomson (1972). Hiatt (1967) provides an important general discussion of rights of bestowal of women in marriage.

4. Occasionally the briefer gloss 'wife's mother' is substituted for 'prospective wife's mother', but without intending to imply that the relationship WM is the focus of the class whose designation is so glossed. In general, the prospective or potential in-law classes of Australian systems of kin classification are subclasses of kin or consanguineal classes and are genealogically (not "affinally") defined. See also Chapter 5, Nyulnyul. For comment on comparable cases see Scheffler and Lounsbury (1971: 92-8) and Scheffler (1976c: 273-4).

5. For additional discussion and application of the concept of superclass-subclass relationships, especially in the domain of kin classification, see Lounsbury (1964b: 364-6) and Scheffler and Lounsbury (1971: 90 ff.).

6. These observations on the Murawari, Wongaibon, and Nyulnyul systems are not intended to imply that the only differences between all so-called Aranda-type systems and Kariera-like systems in general are differences in how certain superclasses are divided into subclasses (see especially Chapters 5, 7, and 9).

3: Pitjantjara

1. This class is covert insofar as it is not, as far as we know, lexically realized by means of a simple expression that we could appropriately gloss as

'grandkin'. That is, it may be that in the Pitjantjara language "there is no word for this
concept." But of course the concept may be
expressible in a slightly more complex way, for
example, by the union of the male and female
grandkin terms, and it is, in any event, implied
by the reciprocal relations among the four terms.

2. I neglected to inquire whether this restriction is
entirely general or is confined only to situations
of direct address of MB and FZ.

3. For elaboration on this point in relation to other
systems of kin classification see Greenberg (1966)
and Scheffler and Lounsbury (1971: 74-6).

4. As defined in Scheffler and Lounsbury (1971: 116-
7), the analytic category "in-laws" consists, in
the first instance, of one's spouse's parents and
siblings and, conversely, of the spouses of one's
children and siblings. This analytic distinction
is descriptively relevant to many languages (see,
e.g., Scheffler 1972: 351). Its descriptive relevance is not dependent, however, on the presence
of a simple two-term lexical opposition between
"kin" and "in-laws." As Schneider (1968: 3) and
others have observed, not all significant cultural
categories and oppositions are verbally realized
in simple two-term oppositions, but may be expressed directly or indirectly in other ways.
Also, the relationship between the two categories
of "relative" is not everywhere the same. In
English, the "in-law" category is not a subclass
of "blood relatives" but a subclass of the category
"relative by extension" (cf. Scheffler 1976b).
Consistent with this, in most societies where
English is spoken people are not expected to marry
blood relatives; and so, in the vast majority of
instances, individuals are related to their in-laws
only by marriage. In contrast, where marriage
between persons who are genealogically related is
positively sanctioned, the category "in-laws" is
sometimes a subclass of the category "blood relatives." It seems likely that this is the arrangement in the Pitjantjara case (although the Pitjantjara concept of genealogical connection may
not be expressed in the idiom of "blood relationship"). Here, it seems, although specific in-law
categories such as wife's father are lexically
realized, the generic category itself is not. Of
course, it does not necessarily follow that if
wife's father, for example, is a subclass (in fact,
a subsubclass) of the category 'relative' (waltja),
it is also a subclass of some particular subclass

of 'relative' such as 'mother's brother'. There
is no norm that restricts a man's choice of a
father-in-law to members of the 'mother's brother'
class.

5. Note that in the case of ego's lineal kin seniority
   status necessarily coincides with the direction of
   generational removal (ascending-descending =
   older-younger). But in the case of ego's collateral kin this is not necessarily so. It is possible
   for ego's MB, for example, to be younger than ego.
   Thus it might be thought necessary to include
   direction, as well as degree of generational removal in dimension (3). Perhaps so, but in many Australian societies when situations like this arise
   the younger MB is designated as 'sister's child',
   that is, by the age-appropriate term, which is
   also the generationally appropriate term, if direction of generational removal is not a part of the
   definition of the term.

6. The appropriateness of describing these categories
   as "moieties" is questioned in Chapter 12 below.

7. The expression "spouse-equation rule of kin-class
   expansion" designates the general class of rules
   that specify structural equivalences between certain consanguineal relationships and certain spouse
   and in-law relationships. The relationships specified as structurally equivalent vary from case to
   case, and so does the directionality of the rule.
   In some systems the rule specifies that certain
   kintypes (e.g., MBD) are to be regarded as structurally equivalent to certain spouse or in-law
   types (e.g., W or WZ) when regarded as links to
   other kin and in-laws. Such rules not only assimilate in-law relationships to consanguineal relationships and categories but also govern the classifications of many more-distant kintypes. In
   other systems, however, the spouse-equation rule
   specifies only that ego's spouse is to be regarded
   as a member of a particular kin class. The corollaries of such a rule assimilate ego's in-laws to
   other specified kin classes, but they do not govern
   the classifications of any kintypes; that is, the
   rule operates only in the in-law domain and not in
   the consanguineal domain. It is shown in subsequent chapters that Australian spouse-equation
   rules are of this latter kind.

8. The following is a rather simplified treatment of
   Iroquois-like systems. There is much variation
   within this class, too, but it is not particularly
   relevant to the present discussion. For a slightly

different treatment of Iroquois-like systems and the structural relations between them and so-called Dravidian-type systems see Scheffler (1971b).

9. The Bardi system, recently described by Robinson (1973), may well be a singular exception, but because Robinson's report became available to me only when this volume was already in press the Bardi system cannot be discussed here.

4: Kariera-like systems

1. The relevant sources on the Kariera system are Radcliffe-Brown (1913), Bates (1913, Ms.) and Hale (Ms. 1).

2. Kay (1965, 1967) attempts to account for the full genealogical and in-law ranges of the terms of so-called Dravidian- (including Kariera-) and Iroquois-type systems by means of a "general definition" of "parallel vs. cross" and permutations thereon. The attempt is motivated by a false and, in this case largely covert, assumption that underlies much of the early work in "componential analysis" of kinship semantics, namely, that it is essential to specify one conjunctive definition for the total genealogical (and, if any, in-law) range of each term. If polysemy is recognized, it is assumed to be based on narrowing, and the possibility of polysemy by generalization is not taken into account. But of course it must be. When it is, it becomes fairly obvious that Dravidian- (including Kariera-) and Iroquois-like systems share the same "parallel vs. cross" opposition but differ in the rules whereby the categories defined (in part) by this opposition are widened. These rules do not permute the parallel-cross opposition itself.

3. Cf. Falkenberg's (1962: 218-20, 232-7) discussion of a number of relevant Murinbata cases. Beyond the range of one's own immediate in-laws, whether or not changes of designation are made depends in large part on the genealogical and social "closeness" of the parties concerned and on the perceived social advantages or disadvantages that would be incurred by a change of designation. Perhaps needless to say, the "politics" of acquiring and bestowing wives and prospective mothers-in-law are relevant in this context.

5: Nyulnyul and Mardudhunera

1. In this paragraph I have assumed that the special

subclasses are prospective in-law classes, and not more general potential in-law classes. It is not clear from the published and unpublished ethnographic accounts which they are.

7: Arabana

1. In addition to Radcliffe-Brown's published accounts (1923) of the Murawari and Wongaibon systems, I have had access to his field notes - now on deposit in the Department of Anthropology, University of Sydney. I am indebted to Peter Lawrence and Les Hiatt for copies of them.

2. Beckett (1959: 202) notes that his informant came from the far south of Wongaibon territory and Radcliffe-Brown's came from the far north, and he suggests there is "a slight but clear difference between the two groups of Wongaibon." Perhaps so, but the major difference between the two bodies of data appears to be that Radcliffe-Brown reports a number of prospective in-law terms and Beckett does not.

8: Yir Yoront and Murngin

1. Note that in Figure 8.1 the designation for mSC, ZSC is given as kar, rather than as marang as in Sharp's published diagram, where marang is presumably a misprint. Kar appears in this position in Sharp's unpublished dissertation (1937: 78-9) and was recorded also by Thomson (1972: 30) and Barry Alpher (personal communication) as the reciprocal of pa'a.

2. In January, 1972, I interviewed a Thayorre speaker, Bob Holroyd, in Brisbane. Thayorre speakers are northern neighbors of the Yir Yoront. In Thayorre, I was told, the arrangement is the same. Mayat (WM, MMBD) is a kind of 'father's sister' pinar; ta:man (WMB, MMBS) is a kind of 'father' nganip, and pa:ngun (wDH, FZDC) is a kind of 'man's child' nerngk. The special designations signify potential in-law statuses.

3. In his dissertation (1937: 88) Sharp is less equivocal on this point. He states virtually without qualification that line R-3 simply repeats line R-1 and, reciprocally, line L-3 simply repeats L-1.

4. If this is true of R-3 and L-3, it must be true also of R-2 and L-2, and R-1 and L-1; that is, any one of ego's close relatives, no less than his or her more distant relatives, may be married to a

woman or man of any one of several kin classes in
relation to ego. Presumably, within the range of
parents' cross cousins, however, the spouse-
equation rule and other extension rules operate to
assign a kin-class status via the marital relationship - so that within this range the system
(or any person's use of it) may appear to be neatly consistent with the alleged rule of marriage to
a matrilateral cross cousin.

5. My information on the Ngantjara system comes from
Peret Arkwookerum, with the assistance of J. R.
von Sturmer. The data were acquired in Brisbane
in January 1972.

6. Barnes (1967) provides a useful introduction to
the so-called Murngin controversy.

7. Shapiro (1969: 50 ff.) also reports the expression
marrkangga (Peterson's marganga 'partial'), which
he says is opposed in meaning to dangang ('full',
complete'). According to Shapiro these expressions may be preposed to kinship terms, for example
dangang waku and marrkangga waku. Shapiro inter-
prets the categories marked by dangang as the
focal, irreducible categories of the system of kin
classification, and the categories marked by
marrkangga as structurally derivative. According
to Shapiro, the distinction between the dangang
and marrkangga categories is based on "sib"
(lineage) affiliation. Thus, for example, the
dangang waku are the children of a person's
'sisters' of his or her own lineage, and the
marrkangga waku are all other waku. Although it
may be true that the opposition between the dangang
and marrkangga subclasses of the fully widened kin
classes is based on lineage affiliation (as well
as an egocentric genealogical connection), it is
highly improbable that the dangang classes are
irreducible and therefore focal.
    According to N. Peterson (personal communica-
tion, 1972), the Murngin distinguish not only
between dangang and marrkangga kin of each terminological category, but also between yuwalk 'true'
or miringkurr 'real' and all other kin of each
category. The 'true' or 'real' kin of each category are of course the genealogically closest kin
of that category, that is, wC in the case of waku,
F in the case of bapa 'father', M in the case of
arndi 'mother', and so on. Of course, the 'real'
or 'true' kin of each terminological class are
included in the dangang subclass of the expanded
class designated by that term, but this should not
be misconstrued as indicating that the dangang

class is structurally prior in relation to the
'true' or 'real' class. The 'true' or 'real'
class is not derived by narrowing or specialization
from the dangang class, but is the central class
on which the terminological class as a whole is
based; within this class a distinction may be made
between dangang and marrkangga subclasses - this
on the basis of lineage affiliations (for a detailed account of which again see Shapiro 1969:
50 ff.).

9: Walbiri and Dieri

1. For a very different analysis of the Aranda system
   see Hammel (1966). Hammel's analysis is based on
   the assumption that the system of kin classification is based on the subsection system. A number
   of analytical problems posed by Hammel are obviated
   by recognition that it is not and by recognition
   of the polysemy of the kinship terms.

10: Ngarinyin

1. Lucich's lists of kintypes and their designations
   (1968: 75-98) must be used with caution. At first
   glance it may appear that each and every kintype
   is designatable by more than one term, sometimes
   three or four. Careful study of the text reveals,
   however, that the listed terms often are not alternative designations for one kintype. Lucich
   accepts without question the received anthropological wisdom that rules of interkin marriage are
   basic structural principles of Australian systems
   of kin classification. He argues that, in 1963,
   there were two "correct" marriageable categories
   of kinswomen and, correspondingly, "two separate
   subsystems of affinal terminology." These false
   assumptions are incorporated in his "model" of the
   system and in his lists of kintypes and their designations. For each kintype he lists not only the
   consanguineally appropriate designation(s) but
   also the designations appropriate to the two marital relationships that may, he says, "correctly"
   exist between ego and persons of that kintype. So
   the several terms listed for a kintype are not
   always alternative designations for that kintype;
   more often they represent alternative designations
   for a person who happens to be related to ego in
   that way and perhaps by marriage in some other way,
   too.

12: Kin classification and section systems

1. The occurrence of this kind of interindividual

"inconsistency" in the designation of particular
relatives is the basis of Rose's proclamation,
"the method of kin classification among the Aborigines is not based primarily on genealogical considerations" (Rose 1976). Like so many other
observers of Australian society, Rose fails to
distinguish the classification of individuals from
the classification of relationships. He is not
unaware that two individuals may be related in
more than one way, partly because of the occurrence
of "irregular" or "exceptional" marriages, but he
insists that "terminological anomalies" are too
common to be accounted for in this way. He notes,
for example, that, in his Groote Eylandt research,
he found that "the actual mother of an individual
called 'wife' was in 39% of 175 cases called by a
term that could be translated as 'father's sister',
and in 36% of cases by a term meaning 'mother's
mother's brother's daughter'" (Rose 1976; also
1960: 60). But Rose (1960) and Turner (1974) provide ample evidence that the MMBD class is a special prospective WM subclass of the FZ superclass.
Rose's and Turner's ethnographies are full of unan-
alyzed evidence for this and other superclass-
subclass relationships which, when analyzed,
reveal that this is basically just another Kariera-
like system, although it features also a limited
Omaha-type skewing rule.

2. Elkin (1933b: 66) says the practice of such marriages "means that the system is failing to fulfill the purpose for which it was presumably
evolved." But he has also questioned that this -
the regulation of marriage - is the purpose for
which section and subsection systems were evolved
(1933b: 78, 79, 90).

3. For the best relevant ethnographic accounts see
Meggitt (1962) and Hiatt (1965). See also Hiatt
(1967).

4. Elkin (1950) reports similar cases from Arnhem
Land, albeit in areas where the subsection system
was still in the process of being adopted.

14: Kinship and the social order

1. Of course Walbiri cannot be regarded as the proto-
typical Australian society, any more than any other
Australian society may be so regarded. So whatever
may be demonstrated here for the Walbiri case may
or may not be true also of other Australian societies; its truth or falsity must be demonstrated
for each case individually. To my knowledge, how-

ever, there is very little in the ethnographic
accounts of other Australian societies to indicate
that the general features of the argument advanced
here are not applicable also in those cases - and
much to indicate that they are applicable. The
apparently most-contrary indications are that
moieties with proper names, genuine moieties that
is, and other apparently descent-ordered categories
such as "semimoieties," "phratries," and the like
have been reported for many Australian societies.
These, however, seem to be little more than schemes
for the classification of totems or Dreamings, and
thereby the persons related to them by putative
kinship ties. It is apparently just as true of
them as it is of the so-called moieties discussed
below, and of sections and subsections, that "membership" in them confers no rights of either an
individual or collective kind and entails no
duties, other than those arising from the specific
kinship statuses and totemic associations by which
their "members" are distinguished inter se (cf.
Fortes 1969: 113).

2. Meggitt (1962: 212, 215) was unable to obtain information on "group fission" or on procedures for
the reallocation of Dreamings in the event of ex-
tinction of the 'owning' patriline. We can only
suppose, however, that these processes are similar
to the Aranda arrangements described in detail by
Pink (1936) and T. Strehlow (1947, 1965, 1971b).

3. The following summary account of Walbiri "cult
totemism" is abstracted from the richly detailed
and eloquent accounts provided by Meggitt (1962,
1966, 1972), Munn (1964, 1971, 1973) and Peterson
(1969, 1972). I omit here any mention of the
sexual-procreative symbolism revealed in Munn's
penetrating and wholly convincing analyses.

4. Among the Walbiri a man's "conception totem" is
not ritually significant. A man whose "conception
totem" is not also his father's "cult totem" is
not initiated into the lodge of his own "conception
totem." Among the Aranda the arrangement is somewhat different. Aranda men are initiated into the
cults of their own "conception totems" and into the
cults of their fathers' ritual 'fathers', whether
or not these are the same cult groups. But a man
whose "conception totem" is not also his father's
ritual 'father' is disadvantaged. He is not regarded as a 'proper man' (a full member) of his
father's cult group. The man most rightfully
entitled to 'own' the knowledge and paraphernalia
of a particular Dreamtime being is one whose own

father and father's father instantiated that being
both through birth (that is, as a "conception
totem") and through ritual. See Pink (1936) and
T. Strehlow (1947: 84-176, "Tjurunga Ownership";
1965, 1971b).

5. The published accounts of Walbiri myth do not
comment on this question. According to T. Strehlow
(1964: 727, 731; 1971a: 619, n.623), in Aranda
myth human beings were not created in whole by the
Dreamtime beings but existed in "eternity" in a
largely undifferentiated mass, from which human
beings were "liberated" by actions of the Dreamtime beings. Subsequently, individuals came to
instantiate particular Dreamtime beings.

6. The expression _bilirba_ also designates the "matri-
spirits," which are transmitted procreatively,
but these "spirits" have no totemic associations
and they are said to dissipate after the posses-
sor's death has been avenged (Meggitt 1962: 192).

REFERENCES

BARNES, J. A. 1967. Inquest on the Murngin. Royal
    Anthropological Institute Occasional Paper no. 26.
    - 1973. Genetrix : genitor :: nature : culture. In
    Goody, J. (ed.), Contexts of kinship. Cambridge:
    Cambridge University Press.
BATES, D. M. 1913. Social organization of some western
    Australian tribes. Report of the Australasian
    Association for the Advancement of Science 14:
    387-400.
    - MSS. Collected papers. Australian National Li-
    brary, Canberra. MS-365 (file number).
BEAN, S. S. 1978. Symbolic and pragmatic semantics:
    a Kannada system of address. Chicago: University
    of Chicago Press.
BECKETT, J. 1959. Further notes on the social organ-
    ization of the Wongaibon of western New South
    Wales. Oceania 29: 200-7.
BERNDT, R. M. 1955. Murngin (Wulamba) social organi-
    zation. American Anthropologist 57: 84-106.
    - 1965. Law and order in aboriginal Australia. In
    Berndt, R. and C. (eds.), Aboriginal man in Aus-
    tralia. Sydney: Angus and Robertson.
    - 1971. Social relationships among two Australian
    aboriginal societies of Arnhem Land: Gunwinggu
    and 'Murngin'. In Hsu, F. (ed.), Kinship and
    culture. Chicago: Aldine.
BERNDT, R. M. and C. 1942-43. A preliminary report of
    field work in the Ooldea region, western South
    Australia. Oceania 13: 143-69.
    - 1951. Sexual behavior in western Arnhem Land.
    Viking Fund Publications in Anthropology no. 16.
    - 1964. The world of the first Australians. Sydney:
    Angus and Robertson.
    - 1970. Man, land and myth in North Australia: the
    Gunwinggu people. Sydney: Ure Smith.
BIRDSELL, J. R. 1970. Local group composition among

the Australian aborigines: a critique of the evidence from fieldwork conducted since 1930. Current Anthropology 11: 115-42.

COATE, H. H. J. and L. OATES. 1970. A grammar of Ngarinjin, Western Australia. Australian Aboriginal Studies no. 15. Australian Institute of Aboriginal Studies, Canberra.

CURR, E. M. (ed.). 1886. The Australian race. 4 vols. London: Trubner and Co.

DIXON, R. 1968. Correspondence: virgin birth. Man 3: 653-4.

- 1971a. A method of semantic description. In Steinberg, D. D. and L. A. Jakobovits (eds.), Semantics (a reader). Cambridge: Cambridge University Press.

- 1971b. The Dyirbal language of North Queensland. Cambridge: Cambridge University Press.

DUMONT, L. 1966. Descent or intermarriage? A relational view of Australian descent systems. Southwestern Journal of Anthropology 22: 231-50.

ELKIN, A. P. 1928a. Karadjeri social organization. Ms.

- 1928b. Nyulnyul social organization. Ms.
- 1928c. Bardi social organization. Ms.
- 1928d. Ngarinyin social organization. Ms.
- 1928e. Forrest River social organization. Ms.
- 1931a. The Dieri kinship system. Journal of the Royal Anthropological Institute 61: 493-8.
- 1931b. The social organization of South Australian tribes. Oceania 2: 44-73.
- 1932. Social organization in the Kimberley division, northwestern Australia. Oceania 2: 296-333.
- 1933a. Marriage and descent in east Arnhem Land. Oceania 3: 412-16.
- 1933b. Studies in Australian totemism. Oceania Monographs no. 2. Sydney.
- 1938a. Kinship in South Australia. Oceania 8: 419-52.
- 1938b. Kinship in South Australia. Oceania 9: 41-78.
- 1939. Kinship in South Australia. Oceania 10: 295-349.
- 1950. The complexity of social organization in Arnhem Land. Southwestern Journal of Anthropology 6: 1-20.
- 1953. Murngin kinship re-examined, and some remarks on some generalizations. American Anthropologist 55: 412-19.
- 1962. Aborigines: social organization. Australian Encyclopedia 1: 13-9.
- 1964. The Australian aborigines. New York: Doubleday. (Reprint of third edition, 1954; first edition, 1938.)
- 1970. The aborigines of Australia: "One in thought, word, and deed." In Wurm, S. A. and D. C.

Laycock (eds.), Pacific linguistic studies in honor
of Arthur Capell. Pacific Linguistics, series C,
no. 13.
ELKIN, A. P. and R. and C. BERNDT. 1951. Social organ-
ization in Arnhem Land. Oceania 21: 253-301.
EPLING, P. J. 1961. A note on Njamal kin-term usage.
Man 61: 152-9.
EVANS-PRITCHARD, E. E. 1933. The intellectualist
(English) interpretation of magic. Bulletin of
the Faculty of Arts, Egyptian University (Cairo)
1: 282-311.
FALKENBERG, J. 1962. Kin and totem. Oslo: Oslo
University Press.
FISON, L. and A. W. HOWITT. 1880. Kamilaroi and
Kurnai. Melbourne: George Robertson.
FORTES, M. 1969. Kinship and the social order.
Chicago: Aldine.
FRY, H. K. Ms. Fieldnotes on aborigines of South
Australia. On deposit South Australia Museum,
Adelaide.
GASON, S. 1874. The Dieyerie tribe. Adelaide. (Re-
printed in Woods, J. (ed.), The native tribes of
South Australia. Adelaide, 1879.)
- 1879. Dieri kinship terminology. In Taplin 1879.
GILLEN, F. Ms. Notes on some manners and customs of
the aborigines, McDonnell Ranges tribe of the
Arunta (subdivision known as the Arunta Eelpina
Ockneeyambanta group), 1894-96. 5 vols. On
deposit Barr Smith Library, University of Adelaide.
GLASS, A. and D. HACKETT. 1970. Pitjantjatjara
grammar. Australian Aboriginal Studies no. 34.
Australian Institute of Aboriginal Studies,
Canberra.
GOODALE, J. 1962. Marriage contracts among the Tiwi.
Ethnology 1: 452-66.
- 1971. Tiwi wives: a study of the women of Mel-
ville Island, north Australia. American Ethnological Society Monograph no. 51. Seattle: University of Washington Press.
GOODENOUGH, W. H. 1956. Componential analysis and the
study of meaning. Language 32: 195-216.
- 1964. Rethinking 'status' and 'role': toward a
general model of the cultural organization of
social relationships. In Banton, M. (ed.), The
relevance of models for social anthropology.
Association of Social Anthropologists Monograph 1.
London: Tavistock.
- 1970. Description and comparison in cultural
anthropology. Chicago: Aldine.
GREENBERG, J. H. 1966. Language universals. Current
Trends in Linguistics 3: 61-113.
HALE, K. L. 1966. Kinship reflections in syntax: some
Australian languages. Word 22: 318-24.
- 1971. A note on the Walbiri tradition of antonymy.

548

In Steinberg, D. D. and L. A. Jakobovits (eds.),
Semantics (a reader). Cambridge: Cambridge
University Press.
- Ms. 1. Language and kinship in Australia.
- Ms. 2. "cousin" and "fancyman."
HAMILTON, A. 1971. The equivalence of siblings.
Anthropological Forum 3: 13-20.
HAMMEL, E. A. 1966. A factor theory for Arunta kinship
terminology. Anthropological Records vol. 24.
Department of Anthropology, University of Califor-
nia, Berkeley.
HERCUS, L. and I. WHITE. 1973. Perception of kinship
structure reflected in the Adnjamathanha pronouns.
Papers in Australian Linguistics no. 6, Pacific
Linguistics, series A, no. 36.
HERNANDEZ, T. 1941. Social organization of the Drys-
dale River tribes. Oceania 11: 211-32.
HIATT, L. R. 1962. Local organization among the Australian aborigines. Oceania 32: 267-86.
- 1965. Kinship and conflict: a study of an aboriginal community in northern Arnhem Land.
Canberra: Australian National University Press.
- 1966. The lost horde. Oceania 37: 81-92.
- 1967. Authority and reciprocity in Australian
aboriginal marriage arrangements. Mankind 6:
468-75.
- 1968. Ownership and use of land among the Australian aborigines. In Lee, R. B. and I. DeVore
(eds.), Man the hunter. Chicago: Aldine.
HORTON, R. 1967. African traditional thought and
western science. Africa 37: 50-71, 155-87.
HOWITT, A. W. 1878. Notes on the system of consan-
guinity and kinship of the Brabrolong tribe, North
Gyppsland. In Smyth, R. B. (ed.), The aborigines
of Victoria. London: Trubner and Co.
- 1891. The Dieri and other kindred tribes of central Australia. Journal of the Royal Anthropological Institute 20: 30-104.
- 1904. The native tribes of southeast Australia.
London: Macmillan.
- Ms. The collected papers of A. W. Howitt. On
deposit Department of Anthropology, The Royal
Victoria Museum, Melbourne.
HUGHES, E. J. 1971. Nunggubuyu-English dictionary.
2 vols. Oceania Linguistic Monographs no. 14.
JOLLY, A. T. H. and F. G. G. ROSE. 1943-45. The place
of Australian aboriginal in the evolution of
society. Annals of Eugenics 12: 49-87.
- 1966. Field notes on the social organization of
some Kimberley tribes (1939-41). Ethnographisch-
Archäologische Zeitschrift 2:7: 97-110.
KAY, P. 1965. A generalization of the cross/parallel
distinction. American Anthropologist 67: 30-43.
- 1967. On the multiplicity of cross/parallel dis-

tinctions. American Anthropologist 69: 83-5.
KORN, F. 1971. Terminology and 'structure': the Dieri
case. Bijdragen tot dem Taal-Land-Volkenkunde
127: 39-81.
LANG, A. 1908. The origin of terms of relationship.
Proceedings of the British Academy 3: 139-58.
LAWRENCE, W. E. 1937. Alternating generations in
Australia. In Murdock, G. P. (ed.), Studies in
the science of society. New Haven: Yale Univer-
sity Press.
LAWRENCE, W. E. and G. P. MURDOCK. 1949. Murngin
social organization. American Anthropologist 51:
58-65.
LEACH, E. R. 1961. The structural implications of
matrilateral cross-cousin marriage. In Leach,
E. R., Rethinking anthropology. London: Athlone
Press. (First published in Journal of the Royal
Anthropological Institute 1951, 81: 23-55.)
- 1967. Virgin birth. Proceedings of the Royal
Anthropological Institute 1967: 39-50.
LÉVI-STRAUSS, C. 1969. The elementary structures of
kinship. Boston: Beacon Press. (Translation of
the 1967 French second edition. First French
edition 1947.)
LOUNSBURY, F. G. 1964a. The structural analysis of
kinship semantics. In Hunt, H. G. (ed.), Proceedings of the Ninth International Congress of Linguists. The Hague: Mouton.
- 1964b. A formal account of the Crow- and Omaha-
type kinship terminology. In Goodenough, W. (ed.),
Explorations in cultural anthropology. New York:
McGraw-Hill.
- 1965. Another view of the Trobriand kinship categories. American Anthropologist 67 (5, part 2):
142-85.
LOVE, J. R. B. 1941. Worora kinship gestures. Transactions of the Royal Society of South Australia
65: 108-9.
- 1950. Worora kinships. Transactions of the Royal
Society of South Australia 73: 280-1.
- Ms. Notes on the natives of Ernabella. South Australia Public Library, PRG 214.
LUCICH, P. 1968. The development of Omaha kinship
terminologies in three Australian aboriginal
tribes of the Kimberley Division, Western Australia. Australian Aboriginal Studies no. 15. Australian Institute of Aboriginal Studies, Canberra.
LYONS, J. 1977. Semantics. Vol. 1. Cambridge:
Cambridge University Press.
McCONNELL, U. 1934. Wikmunkan and allied tribes of
Cape York Peninsula. Oceania 4: 310-67.
- 1940. Social organization of the tribes of Cape
York Peninsula. Oceania 10: 434-55.
- 1950. Junior marriage systems. Oceania 21: 107-45.

McKNIGHT, D. 1971. Some problems concerning the Wikmungkan. In Needham, R. (ed.), Rethinking kinship
and marriage. London: Tavistock.
MADDOCK, K. 1970. Rethinking the Murngin problem: a
review article. Oceania 41: 77-89.
- 1973. The Australian aborigines: a portrait of
their society. London: Penguin.
MALINOWSKI, B. 1913. The family among the Australian
aborigines. London: Hodder. (Reprinted with an
introduction by J. A. Barnes. New York: Schocken
Books, 1963.)
MATHEW, J. 1910. Two representative tribes of Queensland. London: T. Fisher Unwin.
MATHEWS, R. H. 1905. Sociology of some Australian
tribes. Journal of the Royal Society of New South
Wales 39: 104-23.
MAYBURY-LEWIS, D. P. 1967. The Murngin moral. Transactions of the New York Academy of Sciences,
series II, 29: 482-94.
MEGGITT, M. 1954. Sign language among the Walbiri of
central Australia. Oceania 25: 1-16.
- 1962. Desert people. Sydney: Angus and Robertson.
- 1965. Marriage among the Walbiri of central Australia: a statistical examination. In Berndt, R.
and C. (eds.), Aboriginal man in Australia.
Sydney: Angus and Robertson.
- 1966. Gadjari among the Walbiri aborigines.
Oceania Monographs no. 14.
- 1972. Understanding Australian aboriginal society:
kinship systems or cultural categories? In Reining, P. (ed.), Kinship studies in the Morgan centennial year. Washington, D.C.: Anthropological
Society of Washington.
MONTAGU, M. F. ASHLEY. 1974. Coming into being among
the Australian aborigines. London: Routledge.
(Second revised edition; first edition 1937.)
MUNN, N. D. 1964. Totemic design and group continuity
in Walbiri cosmology. In Reay, M. (ed.), Aborigines now. Sydney: Angus and Robertson.
- 1965. A report on field research at Areyonga,
1964-65. Australian Institute of Aboriginal
Studies document 65/2. A.I.A.S. Library, Canberra.
- 1971. The transformation of subjects into objects
in Walbiri and Pitjantjatjara myth. In Berndt, R.
(ed.), Australian aboriginal anthropology.
Nedlands: University of Western Australia Press.
- 1973. Walbiri iconography. Ithaca, N.Y.:
Cornell University Press.
MURDOCK, G. P. (ed.). 1937. Studies in the science of
society. New Haven: Yale University Press.
NEEDHAM, R. 1963. A note on Wikmungkan marriage.
Man 63: 44-5.
- 1971. Introduction. In Needham, R. (ed.), Rethinking kinship and marriage. Association of Social

Anthropologists Monograph 11. London: Tavistock.
NIDA, E. 1975. Componential analysis of meaning: an introduction to semantic structures. The Hague: Mouton.
OATES, W. J. and L. F. OATES (eds.). 1970. A revised linguistic survey of Australia. Australian Aboriginal Studies no. 33. Australian Institute of Aboriginal Studies, Canberra.
PETERSON, N. 1969. Secular and ritual links. Mankind 7: 27-35.
- 1970. The importance of women in determining the composition of residential groups in aboriginal Australia. In Gale, F. (ed.), Women's role in aboriginal Australia. Australian Institute of Aboriginal Studies, Canberra.
- 1971. The structure of two aboriginal ecosystems. Ph.D. dissertation, University of Sydney.
- 1972. Totemism yesterday: sentiment and local organization among the Australian aborigines. Man 7: 12-32.
PIDDINGTON, R. 1932. Karadjeri initiation. Oceania 3: 46-87.
- 1937. Karadjeri kinship. Ms.
- 1950. Introduction to social anthropology. Edinburgh: Oliver and Boyd.
- 1970. Irregular marriages in Australia. Oceania 40: 329-43.
PINK, O. 1936. Land owners in the northern division of the Aranda. Oceania 6: 275-305.
RADCLIFFE-BROWN, A. R. 1913. Three tribes of Western Australia. Journal of the Royal Anthropological Institute 43: 143-94.
- 1914. The relationship system of the Dieri tribe. Man 14: 53-6.
- 1918. Notes on the social organization of Australian tribes. Journal of the Royal Anthropological Institute 48: 222-53.
- 1923. Notes on the social organization of Australian tribes. Journal of the Royal Anthropological Institute 53: 424-47.
- 1930-31. The social organization of the Australian tribes. Oceania Monographs no. 1. (Reprinted from Oceania 1: 34-63, 206-46, 322-41, 426-56.)
- 1951. Murngin social organization. American Anthropologist 53: 37-55.
- 1952. The study of kinship systems. In Radcliffe-Brown, A. R., Structure and function in primitive society. New York: Free Press. (First published in Journal of the Royal Anthropological Institute 1941, 71: 1-18.)
- 1956. On Australian local organization. American Anthropologist 58: 363-7.
REUTHER, J. R. Ms. Manuscripts and notes on the aborigines of South Australia. On deposit Department of

552

Anthropology, South Australia Museum, Adelaide.

RIVERS, W. H. R. 1907. On the origin of the classificatory system of relationships. In Thomas, N. W.,
et al. (eds.), Anthropological essays presented to
Edward Burnett Tylor. Oxford: Clarendon Press.

ROBINSON, M. V. 1973. Change and adjustment among the
Bardi of Sunday Island, northwest Australia. M.A.
thesis, University of Western Australia.

ROHEIM, G. 1938. The nescience of the Aranda. British
Journal of Medical Psychology 17: 343-60.

ROSE, F. G. G. 1960. Classification of kin, age structure and marriage amongst the Groote Eylandt aborigines: a study in method and theory of Australian kinship. New York: Pergamon Press; Berlin:
Academie-Verlag.

- 1976. Boundaries and kinship systems in aboriginal
Australia. In Peterson, N. (ed.), Tribes and
boundaries in Australia. Australian Institute of
Aboriginal Studies, Canberra.

ROTH, W. E. 1897. Ethnological studies among the
northwest central Queensland aborigines. Brisbane:
Government Printer.

- 1903. Superstition, magic, and medicine. North
Queensland Ethnography, Bulletin No. 5. Brisbane:
Government Printer.

SAHLINS, M. 1965. On the ideology and composition of
descent groups. Man 65: 104-7.

SCHEBECK, B. 1973. The Adnjamathanha personal pronoun
and the "Wailpi kinship system." Papers in Australian Linguistics no. 6, Pacific Linguistics,
series A, no. 36.

- 1974. Texts on the social system of the Adnjamathanha people. Pacific Linguistics, series D, no.
21.

SCHEFFLER, H. W. 1971a. Some aspects of Australian
kin classification: a correction. Mankind 8:
25-30.

- 1971b. Dravidian-Iroquois: the Melanesian evi-
dence. In Hiatt, L. R. and C. Jayawardena (eds.),
Anthropology in Oceania. Sydney: Angus and
Robertson.

- 1972a. Systems of kin classification: a struc-
tural semantic typology. In Reining, P. (ed.),
Kinship studies in the Morgan centennial year.
Washington, D.C.: Anthropological Society of
Washington.

- 1972b. Kinship semantics. Annual Review of
Anthropology 1: 309-28.

- 1972c. Baniata kin classification: the case for
extensions. Southwestern Journal of Anthropology
28: 350-81.

- 1973. Kinship, descent, and alliance. In Honig-
mann, J. (ed.), Handbook of social and cultural
anthropology. Chicago: Rand McNally.

- 1976a. Systems of kin classification. In Sturte-

vant, W. (ed.), Handbook of North American Indians.
Vol. 1. Washington, D.C.: Smithsonian Institu-
tion.
- 1976b. The "meaning" of kinship in American culture: another view. In Basso, K. (ed.), Meaning
in anthropology. Albuquerque: University of New
Mexico Press.
- 1976c. On Wordick's review of A study in structural semantics. International Journal of American Linguistics 42: 273-7.
- 1976d. On Krikáti and Siriono: a reply to Lave.
American Anthropologist 78: 338-41.
- 1977. Kinship and alliance in South India and
Australia. American Anthropologist 79:
SCHEFFLER, H. W. and F. G. LOUNSBURY. 1971. A study
in structural semantics: the Siriono kinship
system. Englewood Cliffs, N.J.: Prentice-Hall.
SCHNEIDER, D. M. 1965. Kinship and biology. In Coale,
A. J. (ed.), Aspects of the analysis of family
structure. Princeton, N.J.: Princeton University
Press.
- 1967. Descent and filiation as cultural constructs. Southwestern Journal of Anthropology
23: 65-73.
- 1968. American kinship: a cultural account.
Englewood Cliffs, N.J.: Prentice-Hall.
SCHNEIDER, D. M. and R. T. SMITH. 1973. Class differences and sex roles in American kinship and family
structure. Englewood Cliffs, N.J.: Prentice-Hall.
SCHULZE, Rev. L. 1891. The aborigines of the upper
and middle Finke River. Transactions of the Royal
Society of South Australia 14: 210-46.
SHAPIRO, W. 1967. Preliminary report on fieldwork in
northeast Arnhem Land. American Anthropologist
69: 353-5.
- 1968. The exchange of sister's daughter's daugh-
ters in northeast Arnhem Land. Southwestern
Journal of Anthropology 24: 346-53.
- 1969. Miwuyt marriage: social structural aspects
of the bestowal of females in northeast Arnhem
Land. Ph.D. dissertation, Australian National
University.
- 1970. Local exogamy and the wife's mother in
aboriginal Australia. In Berndt, R. (ed.),
Australian aboriginal anthropology. Nedlands:
University of Western Australia Press.
SHARP, L. S. 1934. The social organization of the Yir
Yoront tribe: kinship and the family. Oceania
4: 404-31.
- 1937. The social anthropology of a totemic system
in North Queensland, Australia. Ph.D. dissertation, Harvard University.
SILVERSTEIN, M. 1976. Shifters, linguistic categories,
and cultural description. In Basso, K. and H.

Selby, (eds.), Meaning in anthropology.
Albuquerque: University of New Mexico Press.
SMYTHE, W. E. 1948. Elementary grammar of the
Gumbainggar language. Oceania Monographs, no. 8.
SORAVIA, G. 1969. Tentative Pitjantjatjara-English
dictionary (Wharburton Range dialect). Ms.
Australian Institute of Aboriginal Studies Library,
Canberra.
SPENCER, B. and F. J. GILLEN. 1899. The native tribes
of central Australia. London: Macmillan.
- 1927. The Arunta: a Stone Age people. 2 vols.
London: Macmillan.
SPIRO, M. 1968. Virgin birth, parthenogenesis, and
physiological paternity: an essay in cultural
interpretation. Man 3: 242-61.
STANNER, W. E. H. 1933. The Daly River tribes: a
report of field work in North Australia. Oceania
3: 377-405.
- 1936. Murinbata kinship and totemism. Oceania
7: 186-216.
- 1956. The Dreaming. In Hungerford, T. (ed.),
Australian signpost. Melbourne: F. W. Cheshire.
- 1965a. Aboriginal territorial organization:
estate, range, domain, and regime. Oceania 36:
1-26.
- 1965b. Religion, totemism, and symbolism. In
Berndt, R. and C. (eds.), Aboriginal man in
Australia. Sydney: Angus and Robertson.
STREHLOW, C. 1913. Die Aranda- und Loritja-Stamme in
Zentral-Australian. Vol. 1. Das Soziale Leben.
Frankfurt am Main: Stadtischen Volker-Museum.
STREHLOW, T. G. H. 1947. Aranda traditions.
Melbourne: Melbourne University Press.
- 1964. Personal monototemism in a polytotemic
community. In Haberland, E., M. Schuster, and
H. Straube (eds.), Festschrift für Ad. E. Jensen.
Vol. 2. Munich: Klaus Renner Verlag.
- 1965. Culture, social structure, and environment.
In Berndt, R. and C. (eds.), Aboriginal man in
Australia. Sydney: Angus and Robertson.
- 1971a. Songs of central Australia. Sydney:
Angus and Robertson.
- 1971b. Geography and the totemic landscape in
central Australia: a functional study. In Berndt,
R. (ed.), Australian aboriginal anthropology.
Nedlands: University of Western Australia Press.
STERN, G. 1965. Meaning and change of meaning.
Bloomington: University of Indiana Press.
(First published in 1931.)
TAPLIN, G. (ed.). 1879. Folklore, manners, customs,
and languages of the South Australian aborigines.
Adelaide.
TERWEIL-POWELL, F. 1975. Developments in the kinship
system of the Hope Vale aborigines. Ph.D. disser-

tation, University of Queensland.

THOMAS, N. W. 1966. Kinship organizations and group marriage in Australia. Cambridge: Cambridge University Press. (First published 1906.)

THOMSON, D. F. 1935. The joking relationship and organized obscenity in North Queensland. American Anthropologist 37: 460-90.

- 1936. Fatherhood in the Wik Monkan tribe. American Anthropologist 38: 374-93.

- 1946. Names and naming in the Wik Monkan tribe. Journal of the Royal Anthropological Institute 76: 57-67.

- 1955. Two devices for the avoidance of cross-cousin marriage among the Australian aborigines. Man 55: 39-40.

- 1972. Kinship and behaviour in North Queensland: a preliminary account of kinship and social organization on Cape York peninsula. Australian Aboriginal Studies no. 51. Australian Institute of Aboriginal Studies, Canberra.

TINDALE, N. 1974. Aboriginal tribes of Australia. Berkeley: University of California Press.

TURNER, D. H. 1974. Tradition and transformation: a study of the aborigines in the Groote Eylandt area, northern Australia. Australian Aboriginal Studies no. 53. Australian Institute of Aboriginal Studies, Canberra.

WALDRON, R. 1967. Sense and sense development. London: A. Deutsch.

WARNER, W. L. 1930. Morphology and functions of the Australian Murngin type of kinship, I. American Anthropologist 32: 207-56.

- 1931. Morphology and functions of the Australian Murngin type of kinship, II. American Anthropologist 33: 172-98.

- 1933. Kinship morphology of forty-one Australian tribes. American Anthropologist 35: 63-86.

- 1958. A black civilization. New York: Harper and Row. (First edition 1937.)

WEBB, T. T. 1933. Tribal organization in eastern Arnhem Land. Oceania 3: 406-11.

WHITE, H. C. 1963. An anatomy of kinship: mathematical models for structures of cumulated roles. Englewood Cliffs, N.J.: Prentice-Hall.

local organization, 44n
Lounsbury, F. G., 3, 105, 231, 387, 396, 397
Lucich, P., 388, 389, 407, 408, 415

McConnell, U., 151, 153, 155, 160
Maddock, K., 318, 433, 475
marked terms, 17, 18, 59, 78, 94, 154, 229
marriage
    between offspring of female cousins, 43, 54, 122,
        188, 193
    between ZSD and FMB, 146, 179, 188-91, 205-6,
        232, 234
    intergenerational, 173-4, 179-80, 205, 233, 248
    intergroup, 14-5
    irregular, 279, 280, 319, 475-80, 485
    orthodox, 142, 146, 152, 205, 209, 278, 280
    rightful claims in, 51, 52-3, 147-9, 190-1, 195,
        203, 284, 298, 314
marriage classes, see sections and subsections
marriage rules
    Aranda type, 42, 54, 55, 81
    and descent lines, 41-2, 52-6, 65, 70, 81-2
    interkin, 51-7
    Kariera type, 41, 54, 81
    and rules of kin-class expansion, 51-2, 64-5,
        75-6, 135-7, 142, 148, 169-70
    and section and subsection systems, 4, 15, 469-74
    and subclassification, 57-64, 76-7, 171, 431
    (see also Tribal Name Index)
    and terminological equivalence, 46, 51-2
matricouples, 434
matridetermination, 318-25
matrifiliation, 434, 436, 440, 449-50, 459
matrimoieties, 433, 438, 506, 510-12, 523
Meggitt, M., 2, 470, 479, 504
    on theories of procreation 7, 9n
    on Walbiri, 328, 329, 332ff, 505ff
metaphor, 32, 33, 35
metaphors of kinship, 505, 519, 524, 526-30
moieties
    derivation of, 47-8
    implicit (unnamed), 433-6, 440, 476
    and kin classification, 1, 14-5, 30, 47-9, 443-4,
        507, 522-3
    and section and subsection systems, 433-40, 454,
        475, 507-13
monosemy, 27
Montagu, M. F. Ashley, 5, 6, 7, 9, 12
mother-in-law
    avoidance, 59, 154, 185, 195, 216
    bestowal, 59, 152, 194-5, 302
mourning terms, 161, 461
Munn, N., 88, 89, 97, 100, 106-7, 515, 516, 518

neutralization, 101, 103, 115, 125

Omaha-type skewing rule, 381, 395-8, 416
owners (of rituals), 508, 509, 514, 528

parallel-cross status, 105, 134n
parallel-cross neutralization rule, 116, 132, 145
parallel-cross status-extension rule, 138-9, 145
patricouples, 434
patrifiliation, 434, 436, 440, 449-50, 459, 520
patrilines, see cult lodge, descent lines, patrilodge
patrilodge, 508, 513-22
patrimoieties, 433, 435-6, 438, 443, 454, 506, 507-10,
        522
Peterson, N., 7, 287, 293, 296, 308, 316, 358, 514,
        515, 529
physical paternity
        ignorance of, 5, 6, 8-10
Piddington, R., 210, 214, 216, 217, 219, 220, 232, 477
polysemy, 18n, 62
        by narrowing, 17-20
        by widening, 21
potential in-laws, see in-laws and Tribal Name Index
procreation
        sexual, 7-8, 10, 525
        spiritual, 7, 8, 12
        theories of, 5-13, 524-6
pronominal systems, 462-9
prospective in-laws, see in-laws and Tribal Name Index

quickening, 7, 525 (see also procreation, theories of)

Radcliffe-Brown, A. R.
        on Aranda-type systems, 42, 53-4, 77
        on derivation of section and subsection systems,
                47-8, 432, 437-45
        on descent lines and kin classification, 43-51,
                70-2, 81, 86
        on Dieri, 45, 49, 367, 368, 372
        on functions of section and subsection systems, 4
        on Karadjeri-type systems, 42, 68-70, 208
        on Kariera, 162-3, 435
        on Kariera-type systems, 41
        on Kukata, 79-81
        on Kumbaingeri-type systems, 41, 57-68, 148
        on Mardudhunera, 191, 194-7, 200
        on marriage rules and kin classification, 51-7,
                79, 86
        on method in the study of systems of kin classi-
                fication, 82-5
        on Murawari, 259
        on Murngin, 70-4, 291, 295, 324
        on Ngarinyin, 386, 387, 406-7
        on superclasses, 72-4, 86

on relations between kin classification and section and subsection systems, 2, 20-1
his theory of Australian kin classification, 1,
40ff, 86
on Wongaibon, 259
on Yaralde-type systems, 42-3, 415, 416
reciprocal relations
as evidence for superclasses, 95-6, 124-7, 213,
215, 391-2
relationship terms, 2, 14
relative age, 96, 100n, 151, 156, 162, 247-8, 308
ritual
organization of, 508-10, 513-20

same-sex sibling-merging rule, 115
Scheffler, H. W., 3, 11, 29, 33, 50, 68, 100, 101, 440,
521
Scheffler, H. W., and Lounsbury, F. G., 3, 18, 19, 32,
64, 85, 225, 402
section systems
structure of, 443-52
section and subsection systems
derivation of, 49, 432, 437-43
functions of, 4, 432, 472, 474
and implicit categories, 22-4
and implicit moieties, 433-6, 476
and moieties, 432-4
origins of, 481
and regulation of marriage, 469-74
and systems of kin classification, 1-2, 4, 14,
16-26, 30, 48-9, 437ff, 481-2
in Gunwinggu, 487-92
in Murinbata, 482-7
in Murngin, 492-503
in Walbiri, 507-13
theories of the structure of, 432-43
senses
basic, 18
derivative, 24, 26
expanded, 26, 29, 60
metaphoric, 32-3, 35-6
narrowed, 17-20, 24
primary, 17, 20, 21, 24, 25, 26, 63, 64
restricted, 17, 63
specialized, 17-21
Shapiro, W., 287ff, 495-6
Sharp, L., 70, 71, 264, 269, 273, 276, 283
sibling-merging rule, 103
sister exchange, 123
prohibition of, 42, 208, 219, 406
sister's-daughter exchange, 180, 190, 207, 220, 276,
317, 381
social categories, 1, 29-35, 47, 505, 513, 524

TRIBAL NAME INDEX

568

OTHER TITLES IN THIS SERIES

*Also published in paperback